Optical Wireless Communication Theory and Technology

Series Editor

Xizheng Ke, School of Automation and Information Engineering, Xi'an University of Technology, Xi'an, Shaanxi, China

The book series Optical Wireless Communication Theory and Technology aims to introduce the key technologies and applications adopted in optical wireless communication to researchers of communication engineering, optical engineering and other related majors. The individual book volumes in the series are thematic. The goal of each volume is to give readers a comprehensive overview of how the theory and technology in a certain optical wireless communication area can be known. As a collection, the series provides valuable resources to a wide audience in academia, the communication engineering research community and anyone else who are looking to expand their knowledge of optical communication.

Xizheng Ke

Coding Theory in Optical Wireless Communication Systems

Volume I

Science Press
Beijing

Springer

Xizheng Ke
School of Automation and Information
Engineering
Xi'an University of Technology
Xi'an, Shaanxi, China

ISSN 2731-5967 ISSN 2731-5975 (electronic)
Optical Wireless Communication Theory and Technology
ISBN 978-981-19-9836-2 ISBN 978-981-19-9837-9 (eBook)
https://doi.org/10.1007/978-981-19-9837-9

Jointly published with Science Press
The print edition is not for sale in China mainland. Customers from China mainland please order the print book from: Science Press.

This Springer imprint is published by the registered company Springer Nature Singapore Pte Ltd.
The registered company address is: 152 Beach Road, #21-01/04 Gateway East, Singapore 189721, Singapore

Contents

Chapter 1
Response Characteristics of Semiconductor Lasers and Photodetectors

Lasers and detectors are the main devices of transmitters and receivers in optical wireless communication systems, respectively, and represent important parameters of the transmitters and receivers. This chapter establishes equivalent-circuit models based on the rate equation and analyzes the response characteristics of general double-heterojunction semiconductor lasers and PIN photodetectors from three aspects: DC characteristics, transient characteristics, and AC characteristics. It then focuses on the influence of lasers on DC, AC, and pulse signals, and discusses the factors affecting the performance of optoelectronic devices.

1.1 Basic Theory

Semiconductor lasers and photodetectors have attracted extensive attention as key devices in optical wireless communication systems because of their influence on common signals, particularly pulse signals.

1.1.1 Operating Principles

(1) Quantum jumps

The energy band of a semiconductor is formed by a series of discontinuous energy levels. Particles are located in different energy bands according to the different amounts of energy they carry. The lower energy position is called the valence band, E_v (ground state), the higher energy position is called the conduction band, E_c (excited state), and the gap between them is called the energy-band gap (forbidden band). The transitions of particles between the conduction band and valence band are called quantum transitions, which are mainly divided into the following three types.

X. Ke, *Coding Theory in Optical Wireless Communication Systems*, Optical Wireless Communication Theory and Technology, https://doi.org/10.1007/978-981-19-9837-9_1

Fig. 1.1
Spontaneous-emission jump

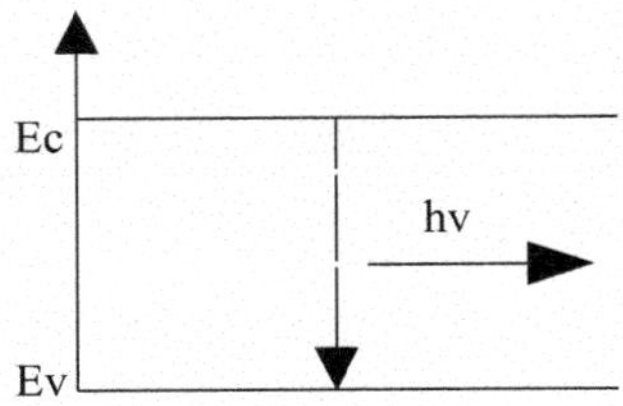

(1) Spontaneous emission: An example is shown in Fig. 1.1. Because the electron energy in the conduction band is high and in an unstable state, an electron spontaneously transitions to the valence band, recombines with a hole, and releases energy by radiating incoherent light in the form of a photon (hv, in the figure). This phenomenon is called a spontaneous-emission transition. A light-emitting diode (LED) is based on the principle of spontaneous emission.
(2) Stimulated radiation: As shown in Fig. 1.2, when the particle-number inversion distribution is realized in the active region—i.e., excess carrier particles are present—the electrons in the conduction band transition to the valence band and recombine with holes under the action of an incident photon; they radiate a coherent photon with the same direction, phase, and polarization as the incident photon. This phenomenon is called a stimulated-radiation transition. Semiconductor lasers operate on the principle of stimulated radiation.
(3) Stimulated absorption: As shown in Fig. 1.3, when the energy of an incident photon causes the valence-band electrons to transition to the conduction band, the active region will absorb the light. The valence-band electrons absorb the photon energy and transition to the conduction band to produce electron–hole pairs. This phenomenon is referred to as stimulated-absorption transition. Photodetectors are based on the principle of stimulated absorption.

Fig. 1.2
Stimulated-radiation jump

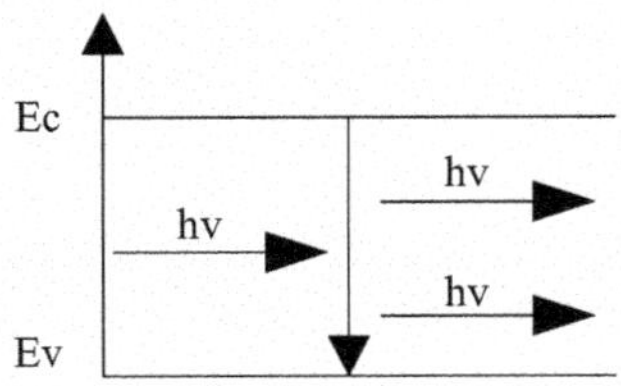

Fig. 1.3
Stimulated-absorption jump

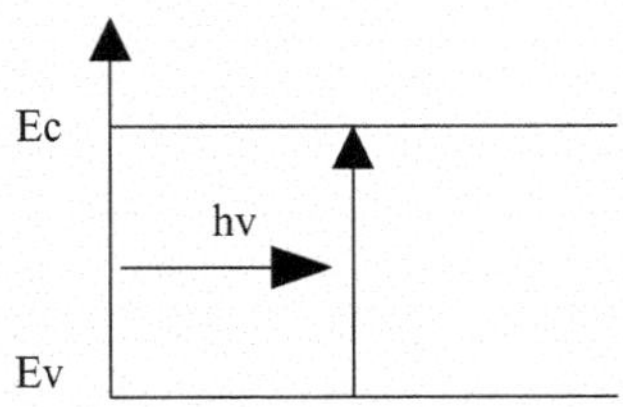

1.1.1.1 Luminescence Conditions of Semiconductor Lasers

A semiconductor laser uses the stimulated-radiation light-amplification principle of the PN junction to generate a coherent-radiation light source. Its normal operation must satisfy the following three conditions.

(1) An inversion distribution of the number of particles must be present in the active region. In the thermal-equilibrium state, the number of valence-band particles in the active region is greater than that in the conduction band, and a stimulated-radiation effect cannot be produced at this time. By injecting a large number of carriers into the active region, the low-energy electrons in the valence band are excited into the unstable conduction band. At this time, the particle-number distribution is opposite to that of its counterpart in thermal equilibrium. This phenomenon is called an inversion distribution of the number of particles in the active region. Under these conditions, the conduction-band electrons can combine with the valence-band holes to produce coherent photons under the action of incident photons.

(2) Laser oscillations must be formed in the cavity. To obtain an outgoing laser beam, it is not sufficient to achieve a particle-number inversion distribution in the active region; the laser must also oscillate. The working material causing the particle-number inversion distribution must be placed in the optical resonator, which can continuously amplify and output photons of a specific frequency and direction.

(3) The threshold current must be reached. Owing to the photon-absorption loss and laser-output loss on the cavity surface in the resonant cavity, the laser must have sufficient gain to enable a stable laser-oscillation output; this requires the injection of external energy. By injecting a sufficiently strong current, the active region of the laser can always realize a particle-number inversion distribution during operation. The higher the degree of the particle-number inversion, the greater the gain in the active layer. However, the input current of the laser must reach the threshold current before it can operate normally.

1.1.2 Basic Structure of Semiconductor Lasers

Depending on the structure, semiconductor lasers can be divided into vertical-cavity surface-emitting (VCSE) lasers, Fabry–Perot lasers, distributed-feedback (DFB) lasers, and quantum-cascade lasers (QCLs). The light emitted by traditional semiconductor lasers is parallel to the substrate, whereas the light emitted by VCSE lasers is perpendicular to the substrate. This difference gives VCSE lasers many advantages; for example, the cavity length enables dynamic single-longitudinal-mode operation, a small laser-divergence angle, and good coupling efficiency.

A distributed-feedback laser uses holographic lithography to convert the surface of a semiconductor crystal into a periodic ripple shape, which is called a Bragg

Fig. 1.4 Structure of a semiconductor laser

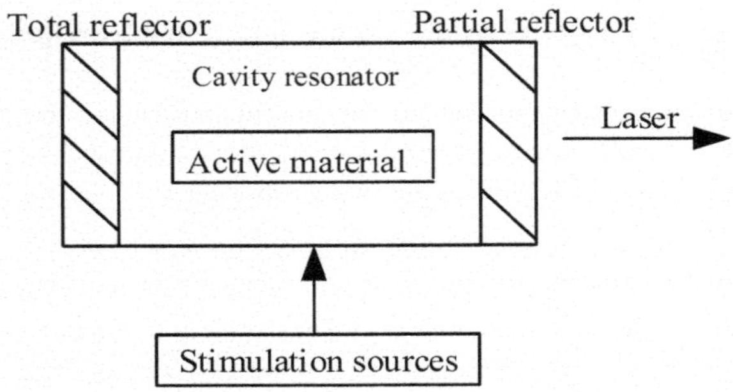

grating. The Bragg grating can select the frequency, amplify light waves, and realize single-longitudinal-mode operation in the cavity.

Unlike the semiconductor lasers mentioned above, the electrons of quantum-cascade lasers do not transition between energy bands, but between sub-bands in the conduction band. The photon energy is determined by the thickness of the quantum wells, rather than the width of the band gap. By changing the quantum-well thickness, the laser wavelength can be controlled, such that the semiconductor laser can cover the mid- and far-infrared bands.

The basic structure of a semiconductor laser is shown in Fig. 1.4. It is generally composed of a working material, an excitation source, and a resonator.

(1) The working material provides the material basis for the particle-number inversion distribution and the stimulated-radiation effect in the laser; it is also known as the gain medium. The working material should be selected to maximize the degree of the particle-number inversion distribution in the active area so that it can continue during the laser-emission process. At present, dozens of semiconductor materials, such as gallium arsenide (GaAs), indium arsenide (InAs), aluminum gallium arsenide (A1GaAs), and cadmium sulfide (CdS), can be made into laser-active materials.
(2) The resonant cavity is generally composed of two mirrors placed in parallel, one of which is a total mirror and the other is a partial mirror. Some mirrors partially reflect the light so that a small part of the light is transmitted through them. The distance between the two mirrors is called the laser-cavity length. The conduction-band electrons and valence-band holes in the working material combine to produce coherent light under the action of the incident photons. The coherent photons oscillate back and forth along the axis in the resonant cavity. The light is amplified by inducing new stimulated radiation through the working material multiple times. When the light increases to a certain extent, photons are emitted through the mirror. The incident photons generally originate from a spontaneous-emission effect in the active region.
(3) When the excitation source is in thermal equilibrium, an inversion distribution of the particle number cannot occur in the laser. Therefore, it is necessary to provide energy through an external device to pump the low-energy particles to a high-energy level. This device is referred to as an excitation source. The

excitation modes of semiconductor lasers mainly include optical-pump, electric-injection, and high-energy electron beam–excitation types; among these, the most important excitation mode is the electric-injection type.

1.1.3 Characteristic Parameters of Semiconductor Lasers

Compared with other types of lasers, semiconductor lasers have a special composition structure and working principle; therefore, they have characteristic parameters that are different from those of other lasers. The inherent characteristic parameters of several semiconductor lasers are described briefly below.

- Threshold current. The threshold current is defined as the current when the stimulated radiation of the semiconductor laser produces a stable laser beam. In general, the smaller the threshold current of a laser, the better it is.
- Conversion efficiency. The main function of a semiconductor laser is to convert electrical signals into optical signals. Therefore, under a certain incident electric power, the conversion efficiency determines the laser performance. The external-differential quantum efficiency, $\eta 1$, is used to express the size of the optical power from the micro level. It is defined as the probability of converting a pair of electron–hole pairs into photons. The power efficiency, η, is defined as the ratio of the output optical power to the input electrical power.
- Longitudinal-mode characteristics of semiconductor lasers. Light waves with multiple frequencies oscillate back and forth along the axis of the resonator. The oscillation mode of different stable light waves is called the longitudinal mode. Multiple longitudinal modes adversely affect the monochromaticity and coherence of an outgoing laser. When designing a laser, the device structure should be optimized to suppress multiple longitudinal modes in the cavity.
- Spontaneous-emission coefficient, β. In addition to stimulated radiation, the spontaneous-emission effect also produces photons in the resonant cavity. However, because the photons produced by spontaneous emission have different directions, phases, and polarization states, only a small number of them are coupled into the lasing mode. β represents the proportion of photons generated by spontaneous emission coupled into and out of the laser.
- Gain compression factor, ε. The gain-compression factor, also known as the gain-saturation factor, is used to describe the variation of the mode gain with the light intensity in a laser.
- Differential gain coefficient, a_g. This coefficient is used to describe the gain-recovery time of the carriers in a laser.

1.1.4 Operating Principles of Photodetectors

As an optical-receiving device in an optical wireless communication system, the main function of a photodetector is to restore an optical signal carrying information into an electrical signal to allow the subsequent circuit to process the information. The photodetector is the main component of the receiver. The following section introduces the working principles of common PIN photodetectors and avalanche photodetectors.

(1) Operating principle of a PIN photodetector

A PIN photodetector is composed of doped semiconductors (P-type and N-type) and an intrinsic semiconductor (I-type). The P-type and N-type semiconductors are separated by the I-type semiconductor. P-type semiconductors have more holes than free electrons, which means that positive charges can flow freely; N-type semiconductors have more free electrons than holes, which means that negative charges can flow freely; intrinsic semiconductors have no impurities, which means that they have neither positive nor negative charges.

Figure 1.5 shows the working principle of the PIN photodetector. V_1 is the reverse-bias voltage, W_i is the space-charge region width (I-region width), and hv is the energy supplied by the incident photons. The operating process of a PIN photodetector is as follows.

Because there is an electron-concentration difference between regions N and I, and a hole-concentration difference between regions P and I, electrons in the N region and holes in the P region diffuse to the I region. Positively charged holes and negatively charged electrons meet and compound in the I region. At this time, a space-charge region from the N region to the P region is generated, which is also known as the built-in electric field. When the built-in electric-field strength reaches a certain degree, owing to diffusion, it produces an electric-field force opposite its direction for the electrons and holes. Electrons drift to the N region and holes drift to the P region. Diffusion and drift form a stable built-in electric field during this dynamic-equilibrium process.

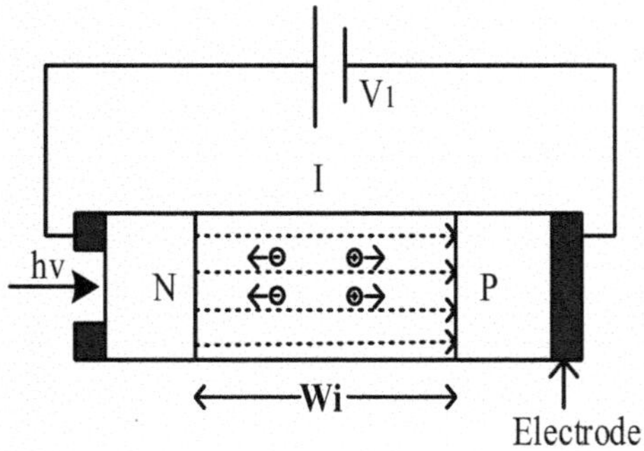

Fig. 1.5 Operating principle of a PIN photodetector

Because region I is an intrinsic semiconductor with high-impedance characteristics, the reverse-bias voltage V1, added to the periphery of the PIN photodetector, falls into this region. The reverse-bias voltage V1 is consistent with the direction of region I, which increases the field strength and width of region I. When the incident-photon energy hv is greater than the semiconductor band gap, most photons are absorbed into the space-charge region to produce electron–hole pairs. Under the action of an electric field, the electrons and holes drift to the N and P regions, respectively.

The non-space-charge region will also have an absorption–radiation effect that produces electron–hole pairs. However, because no external electric field is present, the photogenerated carriers in the P and N regions can only undergo diffusion. The minority carrier holes in the N region first diffuse to the boundary of the I region, then drift to the P region, due to the electric-field force in the space-charge region, and continue to move in the P region; the movement of minority electrons from the P region to the N region follows the same process.

The photogenerated carriers in region I move directionally under the force of an electric field to form a drift current. The time of the drift-current movement is called the transit time. The photogenerated carriers in regions P and N move directionally under the concentration difference to form a diffusion current. The time of the diffusion-current movement is called the diffusion time. The speed of the drift current movement is much faster than that of the diffusion current.

The generated diffusion and drift currents interact with the external circuit to form an output signal to complete the conversion of the photoelectric signal.

(2) Working principle of an avalanche photodetector

The structure of an avalanche photodetector (APD) is similar to that of a PIN photodetector, except that a lightly doped P-type semiconductor is added between region N and region I as region M. When the APD is applied with reverse-bias voltage V1, the semiconductors in each region bear different voltage intensities, owing to different doping concentrations, and region M becomes a high electric-field region, also known as the avalanche-gain region.

When the incident photons are absorbed and radiated in region I to generate carriers, the free electrons drift to region M under the action of the built-in electric field. The electrons enter the high electric field and accelerate continuously under the action of the strong electric-field force. These electrons that obtain ultra-high energy continuously collide with the lattice atoms in the movement process, prompting them to ionize continuously to produce electron–hole pairs. The newly generated electron–hole pairs are also accelerated by the electric-field force to again participate in the collision ionization, and the number of carriers is avalanche amplified; hence, it is called the avalanche-multiplication effect.

Figure 1.6 shows a photodetector with separate absorption and multiplication regions (SAM-APD). Electrons participate in the avalanche effect, which is also called electron-multiplication APD. Similarly, if the M region is placed between the I and P regions and holes participate in the avalanche gain, it is called a hole-multiplication APD. Regardless of the type of avalanche photodetector, the output

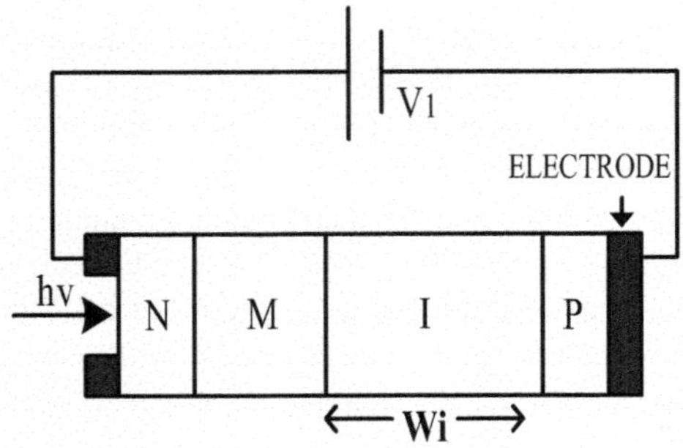

Fig. 1.6 Working principle of an avalanche photodetector

photocurrent increases geometrically at a given input optical power, compared to the PIN photodetector.

1.1.5 Performance Parameters of Photodetectors

The performance of a photodetector is mainly characterized by the following parameters.

(1) Responsivity and quantum efficiency

The main function of a photodetector is to convert a photoelectric signal. The conversion degree is measured by the responsiveness. Therefore, the responsiveness is defined as the ratio of the output current I_{out} to the input optical power P_{in}, using the following formula [1]:

$$R = \frac{I_{out}}{P_{in}}. \tag{1.1}$$

Quantum efficiency describes the photoelectric-conversion efficiency from a micro perspective; it is defined as the ratio of the number of photons involved in destroying the electron–hole pair to the total number of incident photons, as shown in the following formula [1]:

$$\eta = \frac{I_{out}/q}{P_{in}/hv} \tag{1.2}$$

where h is Planck's constant and v is the incident photon frequency. q is charge. The quantum efficiency and responsivity are collectively referred to as the sensitivity of a photodetector.

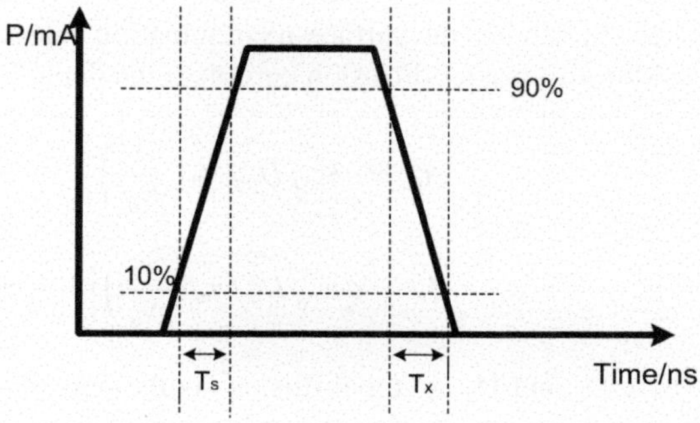

Fig. 1.7 Definition of impulse response time

(2) Response speed

The response of a photodetector to a high-speed modulated optical signal is characterized by the response speed. The faster the response speed, the better the performance of the photodetector [2]. When the incident-light signal is a pulse, the response speed is characterized by the pulse response time, which includes the rise and fall times. The rise time is defined as the time T_s required for the output electric pulse signal to increase from 10 to 90% of the steady-state value, and the fall time is defined as the time T_X required to decrease from 90 to 10% of the steady-state value. When the pulse rise time is the same as the fall time, the response time is $\tau = 2.2Ts$ (Fig. 1.7).

When the incident-light signal is sinusoidal, the response speed is characterized by a 3-dB frequency bandwidth. The response speed of the photodetector is mainly determined by three aspects: the carrier transit time in the space-charge region, carrier diffusion time in the non-space-charge region, and RC time constant of the photodetector.

(3) Dark current

When a reverse-bias voltage is applied to the photodetector, a small amount of current flows out of the output port without the entry of incident light. This weak current is called the dark current. The generation of the dark current is related to the thermal movement of electrons in the photodetector, which is mainly composed of the following three parts.

Generation–recombination current [3]:

$$I_{gr} = \frac{qnAw}{\tau_{gr}}\left[1 - \exp\left(\frac{-qV}{kT}\right)\right], \tag{1.3}$$

where A is the junction area, τ_{gr} is the effective carrier lifetime, n_i is the intrinsic carrier concentration, k is the Boltzmann constant, T is the temperature, V is the reverse-bias voltage, and w is the depletion-region volume.

Surface-leakage current [3]:

$$I_{surf} = \frac{Aqn_i S_0}{2}, \tag{1.4}$$

where S_0 denotes the surface-recombination velocity.

Minority-carrier diffusion current at the depletion-region boundary [3]:

$$I_n = qn_i^2\sqrt{D_p/\tau_p}\left(\frac{A}{N_d}\right)\left[1 - \exp\left(\frac{-qV}{kT}\right)\right] \quad (1.5)$$

$$I_P = qn_i^2\sqrt{D_n/\tau_n}\left(\frac{A}{N_a}\right)\left[1 - \exp\left(\frac{-qV}{kT}\right)\right], \quad (1.6)$$

where D_p and D_n are the diffusion coefficients of holes and electrons, respectively; N_d and N_a are the doping concentrations of holes and electrons, respectively, and τ_d and τ_a are the lifetimes of holes and electrons, respectively.

(4) Multiplication factor

The multiplication factor is used to characterize the extent to which the avalanche photodetector uses the avalanche-gain effect to increase the output photocurrent; it is defined as the ratio of the photocurrent I_f after multiplication to the photocurrent I_b before multiplication: $M = I_f/I_b$.

1.2 Response Characteristics of Semiconductor Lasers

Based on the rate equation, the equivalent circuit model of a double-heterojunction semiconductor laser diode (DH-LD) was established, using the circuit-simulation method to analyze its response.

1.2.1 Equivalent Circuit Model of a DH-LD

Semiconductor lasers convert photoelectric signals based on the quantum-transition theory. The rate equation describes the interaction between the photons and carriers in the laser. This is the most commonly used method for discussing the dynamic characteristics of semiconductor lasers. To simplify the analysis process, the following assumptions were made [4]:

(1) Single-mode oscillation in the resonator;
(2) The photons and electrons in the laser are in an ideal uniform-distribution state, and the changes in the carrier and photon densities are only time-dependent;
(3) The lateral diffusion of carriers along the resonator is ignored.

The rate equation of a DH-LD is simplified as follows [5]:

$$\frac{dn}{dt} = \frac{I_j}{qV} - R_n(n) - R_r(n) - Gs \quad (1.7)$$

$$\frac{ds}{dt} = \Gamma Gs + \Gamma\beta R_r(n) - \frac{s}{\tau_p}, \tag{1.8}$$

where n is the carrier density, I_j is the current injected into the laser, q is the electron charge, V is the volume of the active region, G is the optical gain, R_n is the nonradiative recombination term in the laser, and R_r is the radiative recombination term. s is the photon concentration, Γ is the light-limiting factor, which represents the proportion of photons in the active region, β is the spontaneous-emission coefficient, and τ_p is the photon lifetime.

Equation (1.7) indicates that the carrier density in the laser will increase with an increase in the injection current; however, the carrier density will simultaneously decrease, owing to the occurrence of spontaneous emission, stimulated emission, and non-radiation effects in the active region. Equation (1.8) shows that the photons generated by stimulated radiation in the active region and the photons coupled into the lasing mode by spontaneous radiation will increase the photon concentration in the active region; however, various losses in the resonator (light-output loss and photon-absorption loss from the light-emitting surface) will simultaneously decrease the photon concentration.

To construct the equivalent-circuit model, the classical Shockley formula was introduced to convert an equal number of carriers into circuit variables [6]:

$$n = n_e\left[\exp\left(\frac{V_j}{\eta V_T}\right) - 1\right], \tag{1.9}$$

where n_e is the equilibrium minority carrier density, V_j is the junction voltage, η is the ideal factor, $V_T = KT/q$, K is Boltzmann's constant, and T is the temperature. We apply Eqs. (1.9) to (1.7) and simplify it to obtain

$$\frac{dV_j}{dt}\frac{qVn_e e^{\frac{V_j}{\eta V_T}}}{\eta V_T} = I_j - qVR_n(n) - qVR_r(n) - qVGs. \tag{1.10}$$

If we set $\frac{qVn_e e^{\frac{V_j}{\eta V_T}}}{\eta V_T} = C_d$, $qVR_n(n) = I_n$, $qVR_r(n) = I_r$, and $qVGs = I_s$, the carrier rate equation can be converted into a circuit loop:

$$I_j = C_d\frac{dV_j}{dt} + I_n + I_r + I_s. \tag{1.11}$$

Similarly, we multiply both sides of Eq. (1.8) by qV to obtain

$$qV\frac{ds}{dt} = \Gamma qVGs + \Gamma\beta qVR_r(n) - \frac{qVs}{\tau_P}. \tag{1.12}$$

If we set $s = V_p/qV$, $C_p = 1$, and $R_p = \tau_p$, the photon rate equation can be converted into an optical circuit loop:

$$\Gamma I_s + \Gamma\beta I_r = C_p\frac{dV_p}{dt} + \frac{V_p}{R_p}, \tag{1.13}$$

where voltage V_p, capacitance C_p, and resistance R_p are the variables introduced in the equivalent-transformation process.

V_p is a microscopic quantity that cannot be measured directly. When the reflectivity of the two mirrors constituting the resonator in the laser is the same, the relationship between V_p and the output optical power P of the laser is expressed as [1]

$$P = \delta V_p = \frac{V\eta_0 hv}{2\Gamma\tau_p}V_p, \tag{1.14}$$

where η_0 is the differential quantum efficiency, v is the laser frequency, and h is Planck's constant. When the reflectivities of the internal mirrors of the laser are different, the relationship between V_p and the output optical power P_L and P_R of the left and right ports of the laser is [7]

$$P_L = \frac{cE_p}{2LV_T}\cdot In\left(\frac{1}{R_rR_l}\right)\cdot\frac{\sqrt{R_r}(1-R_l)}{\sqrt{R_r}(1-R_l)+\sqrt{R_l}(1-R_r)}V_p \tag{1.15}$$

$$P_R = \frac{cE_P}{2LV_T}\cdot In\left(\frac{1}{R_lR_r}\right)\cdot\frac{\sqrt{R_l}(1-R_r)}{\sqrt{R_l}(1-R_r)+\sqrt{R_r}(1-R_l)}V_p, \tag{1.16}$$

where R_l is the reflectivity of the left end face, R_r is the reflectivity of the right end face, E_p is the photon energy, c is the speed of light, and L is the laser-cavity length.

Equations (1.1), (1.12) and (1.13) can be used to construct the equivalent-circuit model of the double-heterojunction semiconductor laser. Equation (1.11) is converted from the carrier rate equation to characterize the variation law of the number of carriers in the laser. Therefore, it is equivalent to an electrical-circuit loop, in which $C_d\, dV_i/dt$ has the dimension of capacitance and is equivalent to capacitance C_d. I_n, I_r, and I_s are equivalent to voltage-controlled current sources.

Equation (1.13) is converted from the photon rate equation to characterize the variation law of the number of photons in the laser, so it is equivalent to an optical circuit loop, in which $C_p\, dV_p/dt$ has the dimension of capacitance and is equivalent to capacitance C_p. V_p/R_p has the dimension of resistance and is equivalent to resistance R_p. ΓI_s and $\Gamma\beta I_n$ are equivalent to a voltage-controlled current sources. Equation (1.14) describes the output optical power of the laser, which is equivalent to an under-controlled voltage source.

The chip and package parasitic parameters in the laser must also be considered to ensure that the equivalent-circuit model better reflects the response characteristics of an actual semiconductor laser. The chip parasitic parameters include the following:

(a) Parasitic capacitance, $C_t = C_d + C_s + C_e$, where C_d is the junction-diffusion capacitance, C_s is the space-charge capacitance, and C_e is the parallel capacitance introduced by the contact between the chip and the electrode;

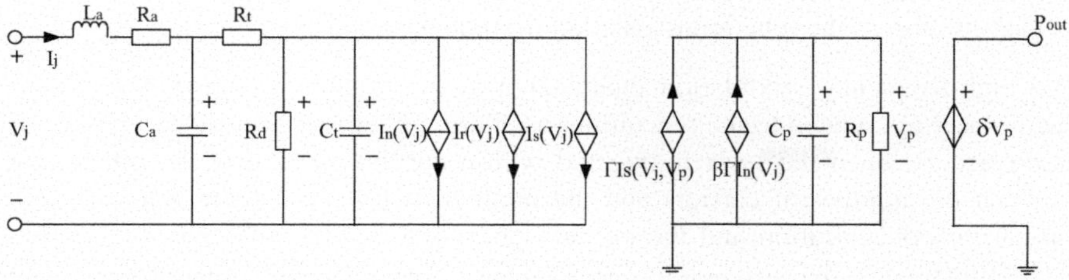

Fig. 1.8 Equivalent circuit model of a DH-LD

(b) Series resistance R_t. A European model contact resistance between the chip and the electrode; and
(c) Parasitic leakage-current equivalent resistance R_d.

The package parasitic parameters mainly come from the carrier and gold wire introduced in the chip-packaging process. When the chip pad is in contact with the carrier, a parallel capacitor C_a is introduced. A gold lead wire is used to connect the external circuit and chip. It presents inductance and resistance, which are represented by the series inductance L_a and series resistance R_a, respectively [8].

From Eqs. (1.12), (1.13), (1.14), and the above analysis, the equivalent-circuit model of the DH-LD is shown in Fig. 1.8. The equivalent-circuit model is a three-port model, where the two left ports correspond to the positive and negative poles of the actual laser and the right port is used to simulate the output optical power of the laser.

The equivalent-circuit model is derived from the rate equation, which describes the variation law of the carriers and photons in the laser. Each component in the model is a mathematical expression of various characteristics of the equivalent laser, and the dispersion parameters of the actual circuit are treated as lumped parameters.

1.2.2 Modulation Characteristics of a DH-LD

The internal modulation of a semiconductor laser directly modulates the signal by changing the laser-injection current. The quality of the modulation characteristics directly determines the transmission rate of the optical-communication system. Therefore, it is very important to study the modulation characteristics of lasers. Modulation characteristics, also known as small-signal frequency–response characteristics, can directly reflect the characteristics of semiconductor lasers in the frequency domain. The modulation characteristics of the laser are related to the relaxation–oscillation phenomenon of the device and the parasitic parameters introduced by the package.

(1) Solution of the relaxation–oscillation frequency

The high-frequency modulation characteristics of an electric-injection semiconductor laser are related to the relaxation–oscillation phenomenon. Relaxation oscillation occurs when the laser is injected with a current. Owing to the interaction between the number of carriers and the number of photons in the active region, the carrier concentration and the output optical power must undergo a damping-oscillation process before reaching a steady state. The damping-oscillation process of the output optical power is called relaxation oscillation and the oscillation frequency is called the relaxation–oscillation frequency.

The intrinsic frequency–response bandwidth of semiconductor lasers is limited mainly by the relaxation–oscillation frequency, f_O. It can be obtained from the rate equation that describes the dynamic characteristics of the laser [9]:

$$\frac{dN}{dt} = \frac{I}{qV} - A(N - N_{om})S - \frac{N}{\tau_N} \tag{1.17}$$

$$\frac{dS}{dt} = \Gamma\beta\frac{N}{\tau_N} - \frac{S}{\tau_p} + \Gamma A(N - N_{om})S, \tag{1.18}$$

where N is the carrier concentration, S is the photon density, I is the current injected into the active region, N_{om} is the transparent carrier concentration, and q is the electron charge. The volume of the active region $V = WLD$, where W, L, and D are the width, cavity length, and thickness of the active region, respectively. A is the gain factor, τ_p and τ_n are the photon and carrier lifetimes, respectively; Γ is the light-limiting factor and β is the spontaneous-emission coefficient.

Under small-signal modulation, the variables in Eqs. (1.17) and (1.18) can be expressed in the following form [10]:

$$\begin{cases} I(t) = I_0 + I_1 \exp(i\omega t) \\ N(t) = N_0 + N_1 \exp(i\omega t) \\ S(t) = S_0 + S_1 \exp(i\omega t), \end{cases} \tag{1.19}$$

where I_O, N_O, and S_O are the current intensity, carrier concentration, and photon density at steady state, respectively, and $I_1\exp(i\omega t)$, $N_1\exp(i\omega t)$, and $S_1\exp(i\omega t)$ correspond to AC components in the transient state.

Substituting Eq. (1.19) into Eqs. (1.17) and (1.18), we obtain

$$\begin{aligned} i\omega N_1 e^{i\omega t} = {} & \frac{I_0}{qV} + \frac{I_1 e^{i\omega t}}{qV} - \frac{N_0}{\tau_n} - \frac{N_1 e^{i\omega t}}{\tau_n} - \big[AS_0(N_0 - N_{om}) + AS_1(N_0 - N_{om})e^{i\omega t} \\ & + AN_1 S_0 e^{i\omega t} + AS_1 N_1 \left(e^{i\omega t}\right)^2\big] \end{aligned} \tag{1.20}$$

$$\begin{aligned} i\omega S_1 e^{i\omega t} = {} & \Gamma\beta\frac{N_0}{\tau_n} + \Gamma\beta\frac{N_1 e^{i\omega t}}{\tau_n} - \frac{S_0}{\tau_P} - \frac{S_1 e^{i\omega t}}{\tau_p} + \Gamma[AS_0(N_0 - N_{om}) + AS_1(N_0 - N_{om})e^{i\omega t} \\ & + AN_1 S_0 e^{i\omega t} + AS_1 N_1 \left(e^{i\omega t}\right)^2\big]. \end{aligned} \tag{1.21}$$

When the laser is steady, Eqs. (1.17) and (1.18) can be rewritten as follows [11]:

$$0 = \frac{I_0}{qV} - \frac{N}{\tau_n} - A(N_0 - N_{om})S_0 \tag{1.22}$$

$$0 = \Gamma A(N_0 - N_{om})S_0 - \frac{S_0}{\tau_p} + \Gamma\beta\frac{N_0}{\tau_n}. \tag{1.23}$$

We substitute the above formula into Eqs. (1.20) and (1.21), and set $N_{om} \approx 0$ and $\Gamma \approx 1$ for simplicity:

$$i\omega N_1 = \frac{I_1}{qV} - \frac{N_1}{\tau_n} - AS_1N_0 - AN_1S_0 - AS_1N_1\left(e^{i\omega t}\right)^2 \tag{1.24}$$

$$i\omega S_1 = AS_1N_0 + AS_0N_1 + AS_1N_1\left(e^{i\omega t}\right)^2 - \frac{S_1}{\tau_p} + \beta\frac{N_1e^{i\omega t}}{\tau_n}. \tag{1.25}$$

When $\beta \approx 0$, ignoring the second-order small quantity, the simplification is as follows:

$$i\omega N_1 = \frac{I_1}{qV} - (\frac{1}{\tau_n} + AS_0)N_1 - AS_1N_0 \tag{1.26}$$

$$i\omega S_1 = (AN_0 - \frac{1}{\tau_p})S_1 + AS_0N_1. \tag{1.27}$$

Let us introduce the gain-saturation term [12], $A = a_g/(1 + \varepsilon S_0)$, and solve Eqs. (1.26) and (1.27), where the small-signal modulation response of the laser can be obtained as follows:

$$S_1(\omega) = \frac{(I_1S_0a_g)/qV(1 + \varepsilon S_0)}{a_gS_0/\tau_p(1 + \varepsilon S_0) - \omega^2 - i\omega\left[1/\tau_n + a_gS_0/(1 + \varepsilon S_0)\right]}. \tag{1.28}$$

Taking the minimum value of the denominator of Eq. (1.28), the relaxation–oscillation frequency can be obtained as follows:

$$f_0 = \omega = \left(\frac{a_gS_0}{\tau_p(1 + \varepsilon S_0)}\right)^{1/2}. \tag{1.29}$$

It can be seen that the relaxation–oscillation frequency f_0 is directly proportional to the differential-gain coefficient a_g and steady-state photon density S_0, and is inversely proportional to the photon lifetime τ_p and gain-saturation factor ε.

(2) Simulation

The factors limiting the modulation rate of double-heterojunction semiconductor lasers come from two aspects. One is the intrinsic frequency–response bandwidth

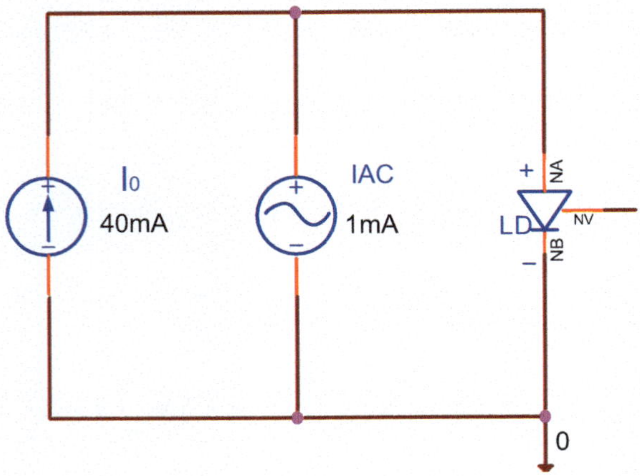

Fig. 1.9 DH-LD AC small-signal simulation circuit

determined by the parameters of the laser chip, which is mainly limited by the relaxation–oscillation frequency [13]. The other is the influence of the parasitic parameters [14] (parasitic resistance, capacitance, and inductance) introduced by the laser packaging on the frequency–response bandwidth.

However, the rate equation cannot analyze the influence of external factors on the frequency–response characteristics of the laser. The circuit in Fig. 1.9 was used to simulate the modulation-response characteristics of a double-heterojunction semiconductor laser. The peripheral circuit is as follows: I_0 is a 40-mA DC bias current source; IAC is a 1-mA AC modulation current source.

Table 1.1 shows the main characteristic parameters of a DH-LD, which can be simulated by incorporating them into the equivalent-circuit model shown in Fig. 1.9. By modifying the parameters in Table 1.1, the circuit can be used to simulate different devices.

The frequency–response characteristics (modulation-response characteristics) of the DH-LD can be obtained using the simulation circuit in Fig. 1.9. In Fig. 1.10, the horizontal axis represents the modulation frequency ω, the longitudinal axis represents the frequency response (output optical power P), and f_0 represents the relaxation–oscillation frequency.

As can be seen in Fig. 1.10, when the modulation frequency ω is in the low-frequency band, the curve is very flat and can transmit information well, when ω is equal to the relaxation–oscillation frequency f_0. A higher resonant peak of the curve produces a modulation distortion of ω in the high-frequency band, and the curve response decreases rapidly. Because the modulation frequency ω is generally located

Table 1.1 DH-LD characteristic parameters [15]

L	β	ε	Γ	N_{om}	W	λ	L_a	R_a	C_a
250 μm	1×10^{-3}	1×10^{-25}	0.2	1.2×10^{24}	1 μm	1.3 μm	0.5 nH	10 Ω	0.15 pF

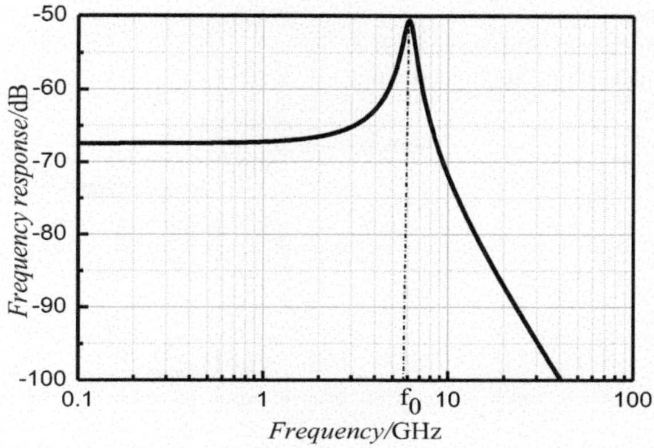

Fig. 1.10 Frequency response characteristic curve of a DH-LD

in the flat section of the response curve, it is necessary to improve the modulation frequency ω of the laser. The resonant peak of the curve must be shifted to the right, and the adjustable low-frequency band can be extended by increasing the relaxation–oscillation frequency f_0.

The circuit in Fig. 1.9 can also be analyzed by calculating the 3-dB frequency–response bandwidth (the bandwidth corresponding to twice the root sign of the maximum value of the frequency–response characteristic curve). The same conclusion can be obtained: The modulation bandwidth of the laser can be improved by increasing the relaxation–oscillation frequency f_0.

Figure 1.10 shows the basic characteristics of a DH-LD. It can be seen that the relaxation–oscillation frequency f_0 is related to the modulation frequency and modulation bandwidth. The influence of the relaxation–oscillation parameters (S_0, a_g, ε, and τ_p) on the modulation-response characteristics of the laser verifies the correctness of the relaxation–oscillation frequency theory.

Because the steady-state photon density S_0 is a function expression, it cannot be directly scanned using the PSpice software; however, the bias current I_0 can be scanned. Figure 1.11 shows the influence of the laser on its modulation-response characteristics under different bias currents, I_0. The initial conditions are as follows: $\varepsilon = 1 \times 10^{-25}$, $R = 0.3$, $a_g = 1.4 \times 10^{-12}$, $T = 300$ K, $L_a = 0.5$ nH, $R_a = 10\ \Omega$, $C_a = 0.15$ pF, and $I_0 = 40$, 50, 60, and 70 mA.

As can be seen from Fig. 1.11, as the bias current I_0 changes from 40 to 70 mA, the relaxation–oscillation resonance peak shifts to the right, and the relaxation–oscillation frequency f_0 increases from 6.1 to 12.5 GHz. Hence, the relaxation–oscillation frequency f_0 can be increased by increasing the bias current I_0 (steady-state photon density S_0).

The essence of the relaxation oscillation is the interaction between the carriers and photons in the laser. When the bias current I_0 injected into the laser increases, the number of carriers in the cavity per unit time increases (the growth rate increases), leading to an increase in the carrier damping-oscillation frequency. The relaxation–oscillation frequency f_0 also increases because of the interaction between the carriers

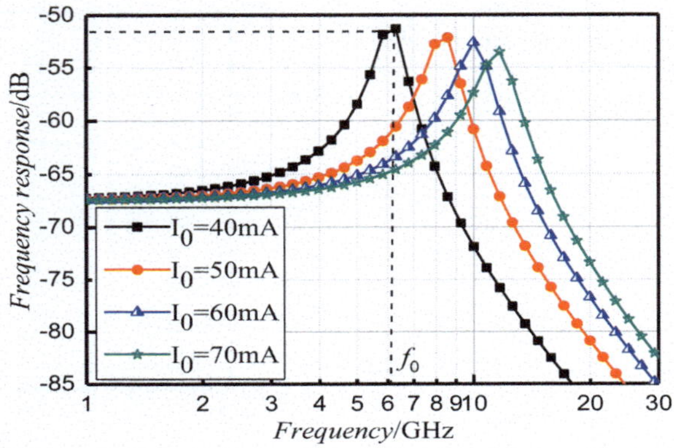

Fig. 1.11 Frequency–response curves under different bias currents I_0

and photons. Therefore, to improve the modulation frequency of the laser, the steady-state photon density S_0 can be increased by increasing the bias current I_0 and the relaxation–oscillation frequency f_0.

Similarly, it can be seen from Fig. 1.11 that as the bias current I_0 changes from 40 to 70 mA, the relaxation–oscillation frequency f_0 increases from 6.1 to 12.5 GHz, and the 3-dB frequency–response bandwidth increases from 3.4 to 8.5 GHz. The same conclusions can be drawn.

Figure 1.12 shows the influence of the laser on its modulation-response characteristics under different differential-gain coefficients, a_g. The initial conditions are as follows: $I_0 = 40$ mA, $\varepsilon = 1 \times 10^{-25}$, $R = 0.3$, $T = 300$ K, $L_a = 0.5$ nH, $R_a = 10$ Ω, $C_a = 0.15$ pF, and $a_g = 1.4 \times 10^{-12}$, 2.4×10^{-12}, 3.4×10^{-12}, and 4.4×10^{-12}.

As can be seen from Fig. 1.12, when the differential-gain coefficient a_g varies from 1.4×10^{-12} to 4.4×10^{-12}, the relaxation–oscillation resonance peak shifts to the right, and the relaxation–oscillation frequency f_0 increases from 8.5 to 18 GHz. Thus, f_0 can be increased by increasing the differential-gain coefficient a_g. This is because

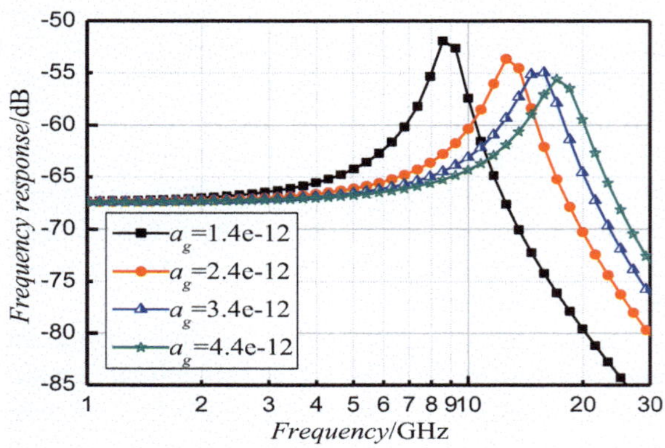

Fig. 1.12 Frequency–response curves under different differential-gain coefficients a_g

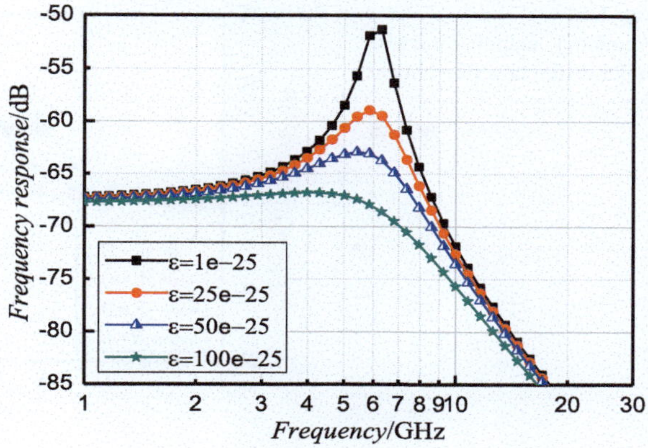

Fig. 1.13 Frequency–response curves under different gain-saturation factors ε

increasing a_g effectively reduces the carrier gain-recovery time. When the gain-recovery time is reduced, the number of carriers in the laser per unit time increases and the damping-oscillation frequency increases. f_0 also increases because of the interaction between the carriers and photons. Therefore, the modulation frequency of the laser can be improved by increasing the differential-gain coefficient a_g and increasing f_0.

Figure 1.13 shows the laser at different gain-saturation factors, ε. The initial conditions are as follows: $I_0 = 40$ mA, $a_g = 1.4 \times 10^{-12}$, $R = 0.3$, $T = 300$ K, $L_a = 0.5$ nH, $R_a = 10\ \Omega$, $C_a = 0.15$ pF, and $\varepsilon = 1 \times 10^{-25}$, 25×10^{-25}, 50×10^{-25}, and 100×10^{-25}.

From Fig. 1.13, as the gain-saturation factor ε changes from 1×10^{-25} to 100×10^{-25}, the relaxation–oscillation resonance peak shifts to the left, and the relaxation–oscillation frequency f_0 decreases from 6.55 to 5 GHz. It can be observed that f_0 increases as the gain-saturation factor ε decreases. ε describes the degree of mode-gain saturation in the laser. When the laser gain ε increases, the gain in the laser decreases and the growth rate of carriers and photons decreases; thus, the damping oscillation of the carriers and the relaxation–oscillation frequency f_0 of the photons also decrease.

The reasons for the gain-saturation effect mainly include the role of parasitic parameters in the high-frequency band, gain reduction caused by heating in the active region, and the influence of various nonlinear effects in the laser [10]. Therefore, to improve the modulation frequency of the laser, the gain-saturation factor ε can be reduced to increase f_0.

By reducing the photon lifetime τ_p, the modulation waveform of the laser can be optimized and the pulse response and distortion can be suppressed. From Eq. (1.30), τ_p can be reduced by appropriately shortening the cavity length L of the laser or reducing the reflectivity R of the resonator mirror:

$$\tau_p = \frac{n_g}{c\left[a_0 - \frac{In(R_1 R_2)}{2L}\right]}. \tag{1.30}$$

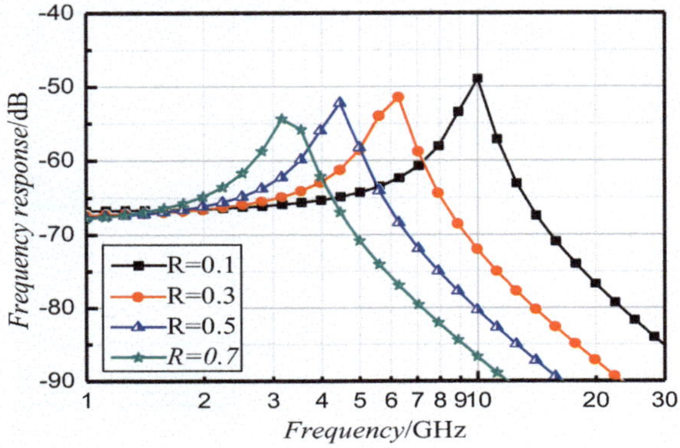

Fig. 1.14 Frequency–response curves under different reflectivities R

Because τ_p is a function expression, the photon lifetime cannot be calculated directly using PSpice software. It can be determined by scanning the reflectivity R of the resonator-plane mirror. Figure 1.14 shows the influence of the laser on its modulation-response characteristics under different reflectivities, R. The initial conditions are as follows: $I_0 = 40$ mA, $a_g = 1.4 \times 10^{-12}$, $\varepsilon = 1 \times 10^{-25}$, $T = 300$ K, $L_a = 0.5$ nH, $R_a = 10\ \Omega$, $C_a = 0.15$ pF, and $R = 0.1, 0.3, 0.5$, and 0.7.

From Fig. 1.14, as R changes from 0.1 to 0.7, the relaxation–oscillation resonance peak shifts to the left, and the relaxation–oscillation frequency f_0 decreases from 10 to 3.1 GHz. It can be seen that f_0 decreases with an increase in the reflectivity R. This is because the loss of the laser decreases with the increase in R, which reduces the growth rate of the output photons per unit time, as well as f_0. Therefore, to improve the modulation frequency f_0 of the laser, the photon lifetime τ_p can be reduced by reducing the reflectivity R of the plane mirror of the laser cavity or by shortening the length L of the laser cavity. Furthermore, the relaxation–oscillation frequency f_0 can be increased.

The influence of package parasitic parameters (L_a, C_a, and R_a) on the modulation-response characteristics of double-heterojunction semiconductor lasers was analyzed. Figure 1.15 shows the influence of the laser on its modulation-response characteristics under different parasitic inductances, L_a. The initial conditions are as follows: $a_g = 1.4 \times 10^{-12}$, $\varepsilon = 1 \times 10^{-25}$, $I_0 = 40$ mA, $R = 0.3$, $T = 300$ K, $R_a = 10\ \Omega$, $C_a = 0.15$ pF, and $L_a = 0.5$ nH, 1.7 nH, 2.4 nH, 3.5 nH, and 4.6 nH.

From Fig. 1.15, when the parasitic inductance $L_a = 0.5$n H, the 3-dB frequency–response bandwidth is 5.5 GHz; when $L_a = 1.7$ nH, the response bandwidth is 8.3 GHz; when $L_a = 2.4$ nH, the response bandwidth is 6.1 GHz; when the parasitic inductance is greater than 3.5 nH, the maximum response bandwidth is only 3.4 GHz.

It can be seen that with an increase in the parasitic inductance L_a, the response bandwidth first increases and then decreases. This is because the package parasitic parameters (L_a, C_a, and R_a) constitute an RLC resonant circuit, which changes the resonant peak of the frequency–response curve through parasitic oscillation, resulting in a change in the frequency–response bandwidth of the laser. Therefore,

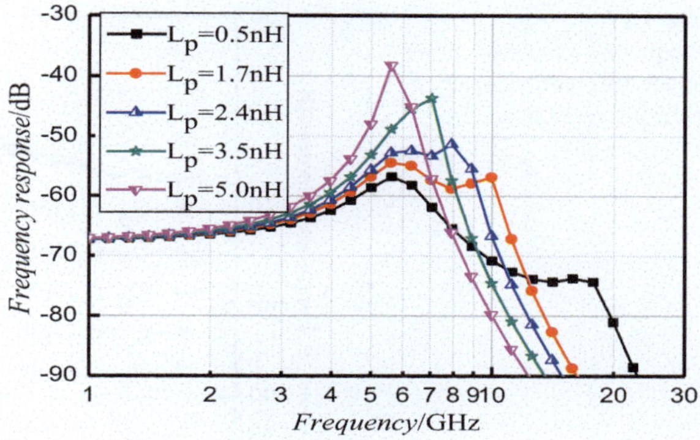

Fig. 1.15 Frequency–response curves under different package parasitic inductances L_a

by changing the packaging mode of the laser chip and reasonably controlling the value of the parasitic inductance L_a, the frequency–response bandwidth of the laser can be compensated.

The influence of the parasitic capacitance on the initial response of the laser under different packaging conditions is shown in Fig. 1.16. The initial conditions are as follows: $a_g = 1.4 \times 10^{-12}$, $\varepsilon = 1 \times 10^{-25}$, $I_0 = 40$ mA, $R = 0.3$, $T = 300$ K, $R_a = 10\ \Omega$, $L_a = 0.5$ nH, and $C_a = 0.15$ pF, 0.23 pF, 0.30 pF, 0.47 pF, and 0.60 pF.

From Fig. 1.16, when the parasitic capacitance C_a is 0.15 pF and 0.23 pF, the 3-dB frequency–response bandwidth is 9.1 GHz and 12 GHz, respectively; when the capacitance C_a is equal to 0.30 pF, the response bandwidth is 10.2 GHz; and when the capacitance C_a is 0.47 pF and 0.60 pF, the response bandwidth is 7.8 GHz and 6.1 GHz, respectively. It can be seen that as the parasitic capacitance C_a increases, the response bandwidth first increases and then decreases. Therefore, there is an optimal parasitic capacitance C_a that can maximize the 3-dB frequency–response bandwidth.

This phenomenon is also caused by the *RLC* resonant circuit, which is composed of the package parasitic parameters (L_a, C_a, and R_a). Therefore, the modulation

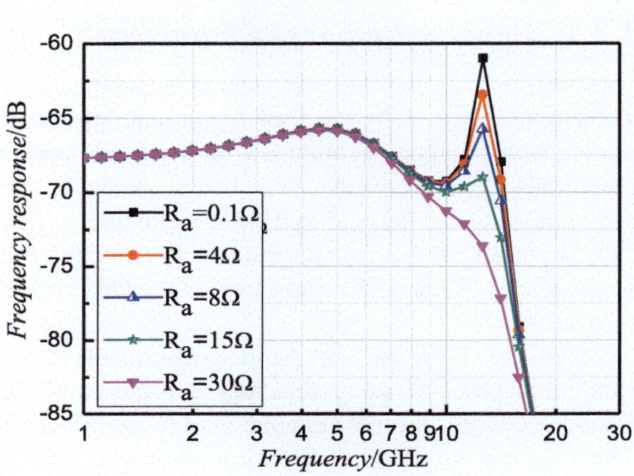

Fig. 1.16 Frequency–response curves under different package parasitic capacitances C_a

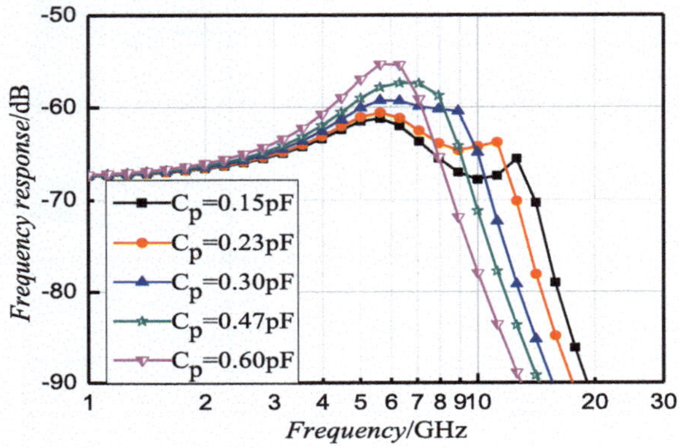

Fig. 1.17 Frequency–response curves under different package parasitic resistances R_a

bandwidth of the laser can be improved by changing the packaging mode of the laser chip and controlling the value of the parasitic capacitance C_a.

Figure 1.17 shows the influence of the laser on its modulation-response characteristics under different package parasitic resistances, R_a. The initial conditions are as follows: $a_g = 1.4 \times 10^{-12}$, $\varepsilon = 1 \times 10^{-25}$, $I_0 = 40$ mA, $R = 0.3$, $T = 300$ K, $C_a = 0.15$ pF, $L_a = 0.5$ nH, and $R_a = 0.1\ \Omega$, $4.0\ \Omega$, $8.0\ \Omega$, $15\ \Omega$, and $30\ \Omega$.

As can be seen from Fig. 1.17, when the parasitic resistance R_a is less than 4 Ω, the maximum 3-dB frequency–response bandwidth is 4.8 GHz; when the parasitic resistance R_a is 15 Ω and 30 Ω, the response bandwidth is 14.5 GHz and 10.4 GHz, respectively; finally, when the parasitic resistance R_a is equal to 15 Ω, the maximum response bandwidth is 15.6 GHz. It can be seen that with the increase of parasitic resistance R_a, the response bandwidth first increases and then decreases. Therefore, to improve the modulation bandwidth of the laser, a certain value of parasitic resistance R_a can be selected.

1.2.3 Impulse-Response Characteristics of a DH-LD

The circuit in Fig. 1.18 was used to simulate the pulse-response characteristics of a double-heterojunction semiconductor laser. The peripheral circuit has the following conditions: I_0 is a 30-mA DC bias current source; IPulse is the pulse current source; I1 and I2 indicate that the peak and initial currents are 10 mA and − 10 mA, respectively; TD, TR, and TF indicate that the delay, rise, and fall times are 1 ns, 0.1 ns, and 0.1 ns, respectively; PW and PER indicate that the pulse width and period are 1 ns and 3 ns, respectively.

From Fig. 1.19, the impulse response of the DH-LD goes through a transient-oscillation process before reaching a steady state. A series of spike pulses with frequency f_0 and gradual attenuation of the amplitude were output at the rising edge. This transient-oscillation attenuation process is known as the relaxation–oscillation phenomenon of a laser. This phenomenon produces serious waveform distortion

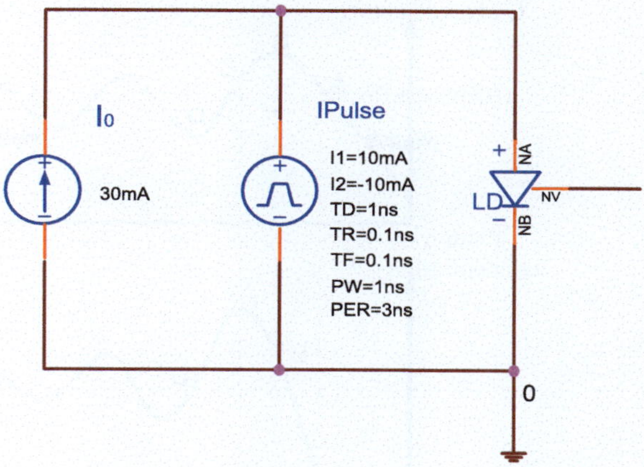

Fig. 1.18 Transient-characteristic simulation circuit of a DH-LD

when a laser is modulated at a high frequency. Therefore, to realize the high-speed transmission of information, the pulse-waveform distortion must be suppressed.

Relaxation oscillation limits the modulation frequency of the laser in the frequency domain. The relaxation–oscillation frequency parameters (S_0, a_g, ε, and τ_p) influence the pulse-response characteristics of the laser.

As shown in Fig. 1.20, when a current is injected into the laser, the carrier concentration begins to increase. When threshold point *A* is reached, the number of photons in the cavity begins to rise from A′. When the number of photons increases to B′, the number of carriers consumed by the stimulated radiation is equal to the number of injected carriers. The carrier concentration reaches a maximum at point B; however, it is still greater than the threshold, so the photon concentration continues to rise, and the number of carriers begins to decline.

When the carrier concentration falls to the threshold, the photon concentration stops increasing and reaches the maximum C′. At this time, the number of carriers consumed by the stimulated radiation is still greater than the number of injected

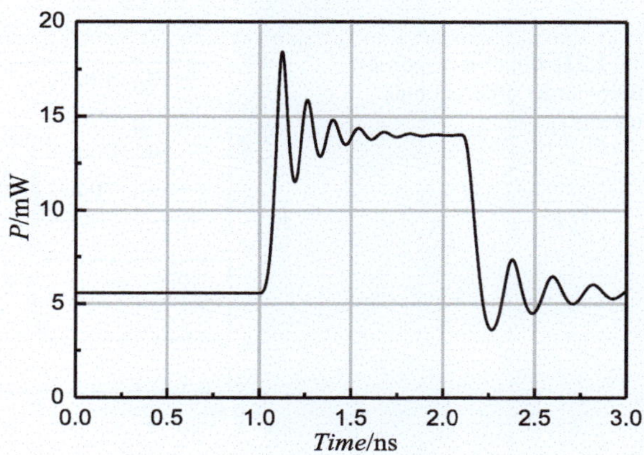

Fig. 1.19 Impulse-response characteristics of a DH-LD

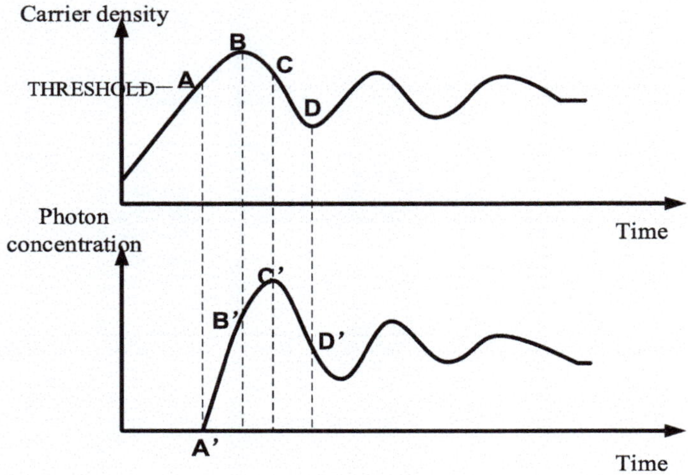

Fig. 1.20 Relaxation–oscillation formation diagram

carriers, and the number of carriers continues to decline. When the carrier concentration falls below the threshold, the stimulated radiation stops, and the photon concentration decreases until D′. Then, the number of carriers injected begins to be greater than the number of carriers lost by the stimulated radiation, and the number of carriers begins to rise, after falling to the lowest point D, to reach the steady state repeatedly. It can be observed that the relaxation–oscillation phenomenon is caused by the photoelectric interaction in the laser.

Figure 1.21 shows the influence of the laser on its impulse-response characteristics under different bias currents, I_0. The initial conditions are as follows: $\varepsilon = 1 \times 10^{-25}$, $R = 0.3$, $a_g = 1.4 \times 10^{-12}$, $L_a = 0.5$ nH, $R_a = 10\ \Omega$, $C_a = 0.15$ pF, $T = 300$ K, and $I_0 = 40$ mA, 50 mA, 60 mA, and 70 mA.

As shown in Fig. 1.22, as the bias current I_0 changes from 40 to 70 mA, the output optical power P (steady-state photon density S_0) increases from 13.5 to 27

Fig. 1.21 Impulse-response curves under different bias currents I_0

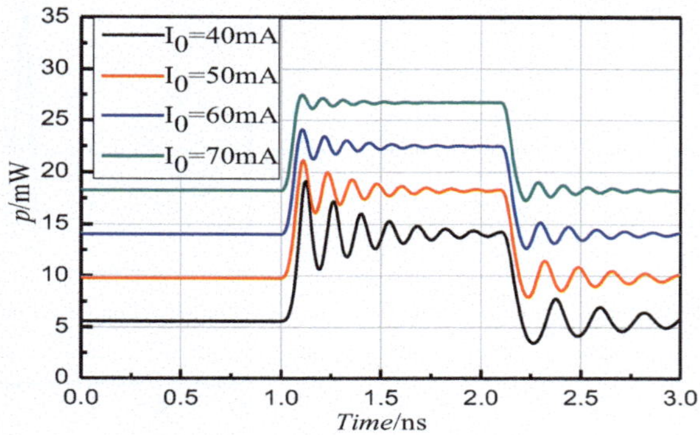

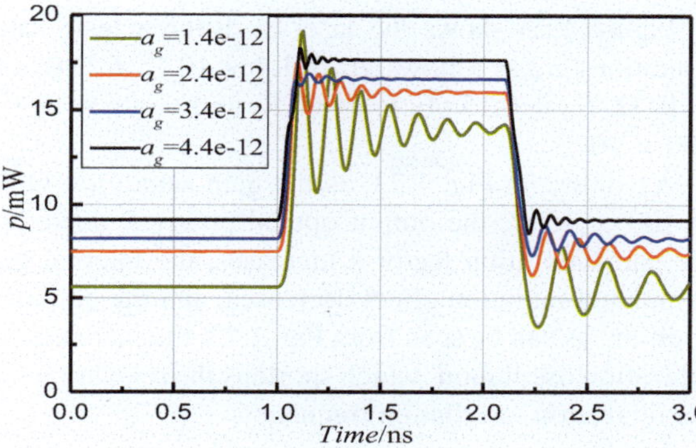

Fig. 1.22 Impulse-response curves under different differential-gain coefficients a_g

mW. This is because the recombination rate of the electron–hole pair stimulated radiation increases with the increase in the number of carriers in the laser. In addition, it can be seen from Fig. 1.21 that increasing the bias current I_0 can increase the relaxation–oscillation frequency, reduce the relaxation–oscillation amplitude, shorten the relaxation–oscillation process, and significantly improve the waveform distortion.

The increase in the relaxation–oscillation frequency is consistent with the previous frequency-domain simulation results, which will not be explained in detail here. The decrease in the relaxation–oscillation amplitude is caused by the gain-saturation effect of the laser. When the output optical power P increases, the gain in the laser decreases. The decrease in the gain suppresses the jitter of the output optical power. The more serious the decrease, the stronger the inhibition, and the relaxation–oscillation amplitude decreases. Therefore, the pulse-waveform distortion can be suppressed by increasing the steady-state photon density S_0 (bias current I_0).

Figure 1.22 shows the influence of the laser on its impulse-response characteristics under different differential-gain coefficients, a_g. The initial conditions are as follows: $\varepsilon = 1 \times 10^{-25}$, $R = 0.3$, $I_0 = 40$ mA, $L_a = 0.5$ nH, $R_a = 10\ \Omega$, $C_a = 0.15$ pF, $T = 300$ K, and $a_g = 1.4 \times 10^{-12}$, 2.4×10^{-12}, 3.4×10^{-12}, and 4.4×10^{-12}.

As shown in Fig. 1.22, when the differential-gain coefficient a_g is changing from 1.4×10^{-12} to 4.4×10^{-12}, the output optical power P increases from 14 to 17.5 mW. The increase in the differential-gain coefficient a_g can effectively reduce the gain-recovery time of the carriers, which will increase the number of carriers participating in the stimulated radiation per unit time and increase the output optical power P. In addition, it can be seen from Fig. 1.22 that increasing the differential-gain coefficient a_g can increase the relaxation–oscillation frequency, reduce the relaxation–oscillation amplitude, shorten the relaxation–oscillation process, and eliminate the waveform distortion.

An increase in the output optical power P causes a decrease in the gain of the laser. The decrease in the gain suppresses the jitter of the output optical power, and the relaxation–oscillation amplitude is reduced. Therefore, the pulse-waveform distortion can be suppressed by increasing the differential-gain coefficient a_g.

Figure 1.23 shows the laser at different gain-saturation factors, ε. The initial conditions are as follows: $a_g = 1.4 \times 10^{-12}$, $R = 0.3$, $I_0 = 40$ mA, $L_a = 0.5$ nH, $R_a = 10\ \Omega$, $C_a = 0.15$ pF, $T = 300$ K, and $\varepsilon = 1 \times 10^{-25}$, 25×10^{-25}, 50×10^{-25}, and 100×10^{-25}.

As shown in Fig. 1.23, as the gain-saturation factor ε changes from 1×10^{-25} to 100×10^{-25}, the output optical power P decreases from 14 to 13 mW. When the gain-saturation factor ε increases, the internal gain of the laser decreases, the stimulated-radiation effect decreases, and the output optical power P decreases. In addition, as can be seen from Fig. 1.23, the increase in ε has a damping effect on the relaxation oscillation, which shortens the relaxation–oscillation process and reduces the relaxation–oscillation frequency.

The photon lifetime is determined by scanning the reflectivity R of the resonator-plane mirror τ_p for analysis. Figure 1.24 shows the influence of the laser on its impulse-response characteristics under different reflectivities, R. The initial conditions are as follows: $a_g = 1.4 \times 10^{-12}$, $\varepsilon = 1 \times 10^{-25}$, $I_0 = 40$ mA, $L_a = 0.5$ nH, $R_a = 10\ \Omega$, $C_a = 0.15$ pF, $T = 300$ K, and $R = 0.1$, 0.3, 0.5, and 0.7.

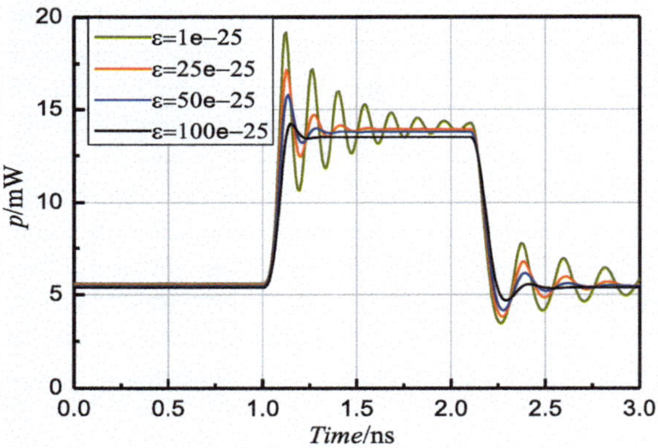

Fig. 1.23 Impulse-response curves under different gain-saturation factors ε

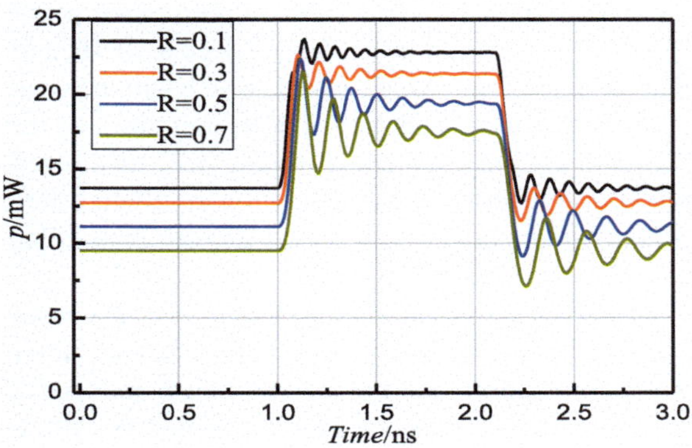

Fig. 1.24 Impulse-response curves under different reflectivities R

As can be seen from Fig. 1.24, as the reflectivity R changes from 0.1 to 0.7, the output optical power P decreases from 22.5 to 17.5 mW. This is because the output loss of the laser end-face decreases with an increase in the reflectivity R of the plane mirror. In addition, with an increase in reflectivity R, the relaxation–oscillation frequency decreases, the relaxation–oscillation amplitude increases, and the waveform distortion becomes increasingly serious. The increase in the relaxation–oscillation amplitude is caused by the increase in the gain of the laser caused by the decrease in the output power. Therefore, to suppress the waveform distortion, we can reduce the photon lifetime τ_p (plane-mirror reflectivity R).

1.2.4 DC Characteristics of a DH-LD

The circuit in Fig. 1.25 was used to simulate the P-I characteristics of a semiconductor laser: I_{DC} is a 30-mA DC current source. The working state of each component in the circuit under the excitation of a DC current source is analyzed.

The P-I characteristic curve of a semiconductor laser can be obtained using the simulation circuit in Fig. 1.25, as shown in Fig. 1.26. The horizontal axis represents the DC current I_{DC} and the vertical axis represents the output optical power P.

The simulation of the P-I characteristics of the laser is mainly used to discuss the external differential quantum efficiency and threshold characteristics. The external differential quantum efficiency is defined as the slope of the linear region of the P-I curve. As can be seen from Fig. 1.26, the DH-LD is a threshold device with a threshold current of 21 mA. When it is lower than the threshold current, only a spontaneous-emission effect can be generated in the active area, and the photon number is almost zero. When it is greater than the threshold current, the particle-number inversion distribution in the active layer begins to be stimulated, radiates, and generates coherent photons, and the output power gradually increases.

In addition, it can be seen from Fig. 1.26 that the external differential quantum efficiency is 30%. At present, the threshold current of the LD must be as small as

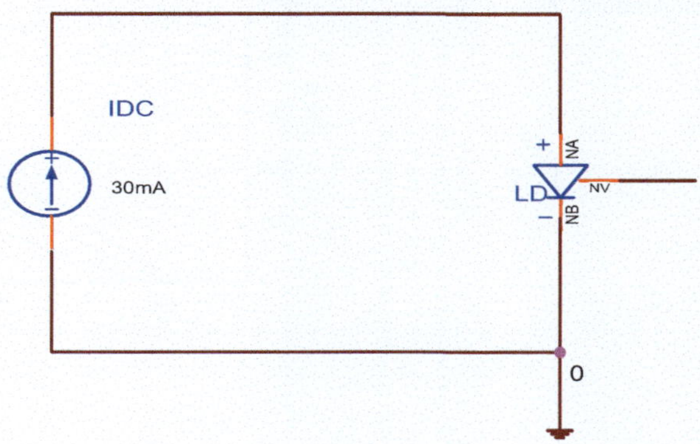

Fig. 1.25 DC-characteristic simulation circuit of a DH-LD

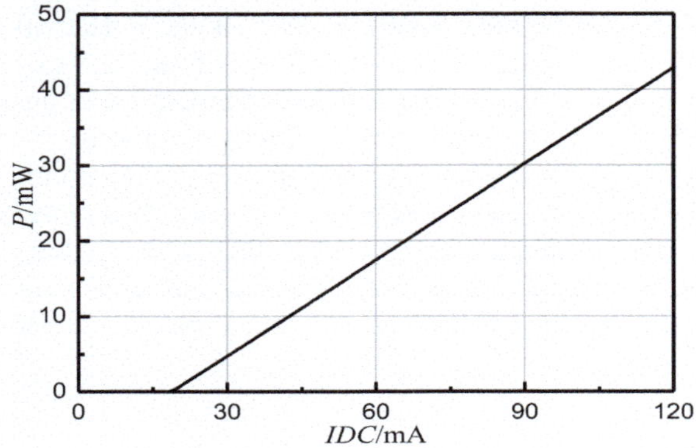

Fig. 1.26 P-I characteristics of a DH-LD

possible and the external differential quantum efficiency must be as large as possible; that is, a very high output power can be obtained through a very small injection current.

The P-I characteristics of semiconductor lasers are significantly affected by the temperature. Figure 1.27 shows the influence of the laser on its P-I characteristics at different temperatures, T. The initial conditions are as follows: $\varepsilon = 1 \times 10^{-25}$, $R = 0.3$, $a_g = 1.4 \times 10^{-12}$, $I_0 = 40$ mA, $L_a = 0.5$ nH, $R_a = 10\ \Omega$, $C_a = 0.15$ pF, and $T = 280$ K, 330 K, 380 K, and 430 K.

As can be seen from Fig. 1.27, as T changes from 280 to 430 K, the threshold current increases from 16 to 26 mA, and the external differential quantum efficiency decreases from 45 to 25%. Therefore, increasing the temperature increases the LD threshold current and reduces the external differential quantum efficiency. This is because an increase in temperature T leads to a large absorption loss and high leakage current, which have an adverse impact on the threshold current and external differential quantum efficiency.

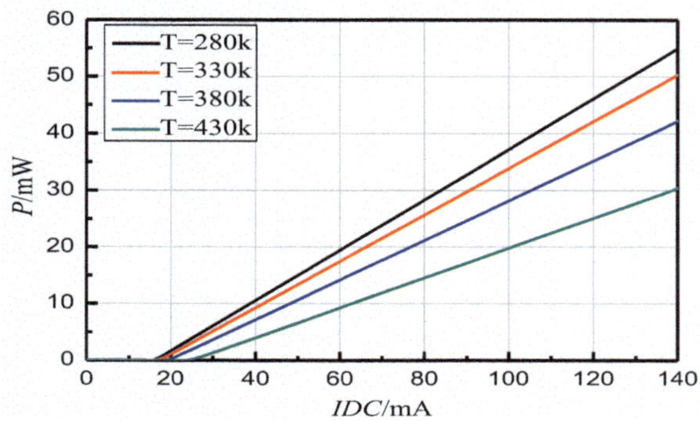

Fig. 1.27 P-I characteristic curves under different temperatures T

1.3 Response Characteristics of a PIN Photodetector

Based on the carrier rate equation, an equivalent-circuit model of a PIN photodetector was established, and the response characteristics of the PIN photodetector to incident light were studied. The results provide theoretical guidance for optimizing the dynamic characteristics of a PIN photodetector. The response of the photodetector is the primary index of the receiver. Figure 1.28 shows the equivalent circuit model of a PIN photodetector.

1.3.1 Equivalent-Circuit Model of a PIN Photodetector

Considering the principle of electrical neutrality in region I, it is assumed that

(1) the expansion of the depletion layer in the N and P regions is negligible, relative to the width of region I and
(2) the electric field in region I is uniform, and the electric fields in regions N and P are zero.

Light is incident on region N in Fig. 1.28. The following rate equation was used to establish the equivalent-circuit model of the reverse-biased PIN photodetector [1]:

N:

$$\frac{dP_n}{dt} = \eta_n P_{in} - \frac{P_n}{\tau_p} - \frac{I_p}{q} \tag{1.31}$$

P:

$$\frac{dN_p}{dt} = \eta_p P_{in} - \frac{N_p}{\tau_n} - \frac{I_n}{q} \tag{1.32}$$

I:

$$\frac{dN_i}{dt} = \eta_i P_{in} - \frac{N_i}{\tau_{nr}} - \frac{N_i}{\tau_{nt}} + \frac{I_n}{q}, \tag{1.33}$$

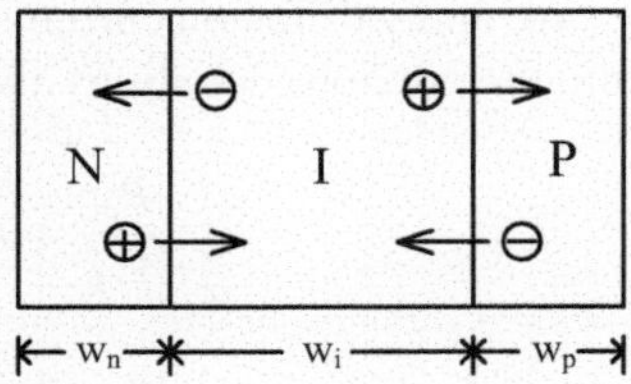

Fig. 1.28 One-dimensional structure of a PIN photodetector

where P_n and N_p are the total number of excess holes and electrons in the N and P regions, respectively; τ_n and τ_p are the lifetimes of electrons and holes in the P and N regions, respectively; I_n and I_p are the electron and hole currents of minority carriers in the N region, respectively, and q is the electron charge. P_{in} is the input optical power; N_i is the total number of electrons in the I region, τ_{nr} is the electronic recombination life of the I region, and τ_{nt} is the electron drift velocity in the I region.

The generation rates of electron–hole pairs in each region are [2]

$$\eta_n = (1-r)\left[1-\exp(-\alpha_n W_n)\right]/h\nu \tag{1.34}$$

$$\eta_p = \frac{(1-r)\left[1-\exp\left(-\alpha_p W_p\right)\right]}{\exp(\alpha_n W_n + \alpha_i W_i)\cdot h\nu} \tag{1.35}$$

$$\eta_i = \frac{(1-r)\left[1-\exp(-\alpha_i W_i)\right]}{\exp(\alpha_n W_n)\cdot h\nu}, \tag{1.36}$$

where r is the reflectivity, α_n, α_p, and α_i are the light-absorption coefficients of the N, P, and I regions, respectively, $h\nu$ is the photon energy, and W_n, W_p, and W_i are the widths of the N, P, and I regions, respectively. To convert parameters such as hole electrons into circuit variables, the constant C_{nc} was introduced, as follows [7]:

$$P_n = \frac{V_p}{q}C_{nc} \quad N_p = \frac{V_n}{q}C_{nc} \quad N_i = \frac{V_i}{q}C_{nc}. \tag{1.37}$$

We apply Eqs. (1.37), (1.36), (1.35), and (1.34) to Eqs. (1.31), (1.32), and (1.33) to obtain

$$\frac{dV_p}{qdt}C_{nc} = \frac{P_{in}(1-r)}{h\nu}\left[1-\exp(-\alpha_n W_n)\right] - \frac{V_p}{q\tau_p}C_{nc} - \frac{I_p}{q} \tag{1.38}$$

$$\frac{dV_n}{qdt}C_{nc} = \frac{P_{in}(1-r)\left[1-\exp\left(-\alpha_p W_p\right)\right]}{h\nu\exp(\alpha_n W_n + \alpha_i W_i)} - \frac{V_n}{q\tau_n}C_{nc} - \frac{I_n}{q} \tag{1.39}$$

$$\frac{dV_i}{qdt}C_{nc} = \frac{P_{in}(1-r)\left[1-\exp(-\alpha_i W_i)\right]}{h\nu\exp(\alpha_n W_n)} - \frac{V_i}{q\tau_{nr}}C_{nc} - \frac{V_i}{q\tau_{nt}}C_{nc} + \frac{I_n}{q}. \tag{1.40}$$

We apply diffusion currents [16] $I_n = \beta_n P_{in} + V_n/R_{nd}$ and $I_p = \beta_p P_{in} + V_p/R_{pd}$ into the previous equations and conclude

$$\frac{P_{in}}{V_{op}} = C_{nc}\frac{dV_p}{dt} + \frac{V_p}{R_P} + \beta_p P_{in} + \frac{V_p}{R_{pd}}, \tag{1.41}$$

where $V_{op} = \frac{h\nu}{q(1-r)\left[1-\exp(-\alpha_n W_n)\right]}$, $R_p = \frac{\tau_p}{C_{nc}}$, $I_{op} = \frac{P_{in}}{V_{op}}$, $R_{pd} = R_p\left[\mathrm{ch}\left(W_n/L_p\right)-1\right]$, and

$$\frac{P_{in}}{V_{on}} = C_{nc}\frac{dV_n}{dt} + \frac{V_n}{R_n} + \beta_n P_{in} + \frac{V_n}{R_{nd}}, \tag{1.42}$$

where $V_{on} = \frac{h\nu \exp(\alpha_n W_n + \alpha_i W_i)}{q(1-r)\left[1-\exp(-\alpha_p W_p)\right]}$, $R_n = \frac{\tau_n}{C_{nc}}$, $I_{on} = \frac{P_{in}}{V_{on}}$, $R_{nd} = R_n\left[\mathrm{ch}\left(W_p/L_n\right) - 1\right]$, and

$$\frac{P_{in}}{V_{oi}} + \beta_n P_{in} + \frac{V_n}{R_{nd}} = C_{nc}\frac{dV_i}{dt} + \frac{V_i}{R_{nr}} + \frac{V_i}{R_{nt}}, \tag{1.43}$$

where $V_{oi} = \frac{h\nu \exp(\alpha_n W_n)}{q(1-r)\left[1-\exp(-\alpha_i W_i)\right]}$, $R_{nr} = \frac{\tau_{nr}}{C_{nc}}$, $R_{nt} = \frac{\tau_{nt}}{C_{nc}}$, $I_{oi} = \frac{P_{in}}{V_{oi}}$, and $I_i = \frac{V_i}{R_{nt}}$.

β_n and β_p are the relationship coefficients between the incident light power and diffusion current, and L_n and L_p are the diffusion lengths of the electrons and holes, respectively.

Taking the N-I interface as an example, the output current is $I_j = I_i + I_p$. In addition, the chip parasitic parameters and package parasitic parameters in the photodetector need to be considered. The chip parasitic parameters can be expressed by C_c, R_c, and R_d. $C_{pn} = \varepsilon_0\, \varepsilon_r\, A/W_i$ is the junction capacitance, and parallel capacitance C_s is caused by chip and electrode welding; R_c is the series resistance of the chip, including the European-mode contact resistance and body resistance between the chip and the electrode. The parallel resistance R_d is used to characterize the tunneling current and leakage current in a PIN-PD.

Package parasitic parameters originate mainly from the carrier and lead wires introduced in the packaging process. When the chip pad is in contact with the carrier, a parallel capacitor C_e is introduced. A lead wire is used to connect the external circuit and chip. It presents inductance and resistance, which are expressed in series L_e and R_e, respectively [18], [19].

The equivalent-circuit model of a PIN photodetector can be obtained from Eqs. (1.41), (1.42), and (1.43) and the above analysis. As shown in Fig. 1.29, the left branch is the virtual end, which is used to simulate the input optical signal, the middle branch is the variation of the electron holes, and the right branch is the actual output current. $\mathrm{R_{op}}$, $\mathrm{R_{oi}}$, and $\mathrm{R_{on}}$ are the virtual resistors in the P, I, and N regions, respectively.

1.3.2 Impulse-Response Characteristics of PIN-PDs

The response of a PIN photodetector to a high-speed incident-light signal is characterized by the response speed. The faster the response speed, the greater the incident-light frequency that the photodetector can respond to. The response speed is expressed by the impulse-response time and the 3-dB frequency–response bandwidth in the time and frequency domains, respectively. Therefore, this section begins with impulse-response characteristics, and then simulates and analyzes the factors affecting the impulse-response time.

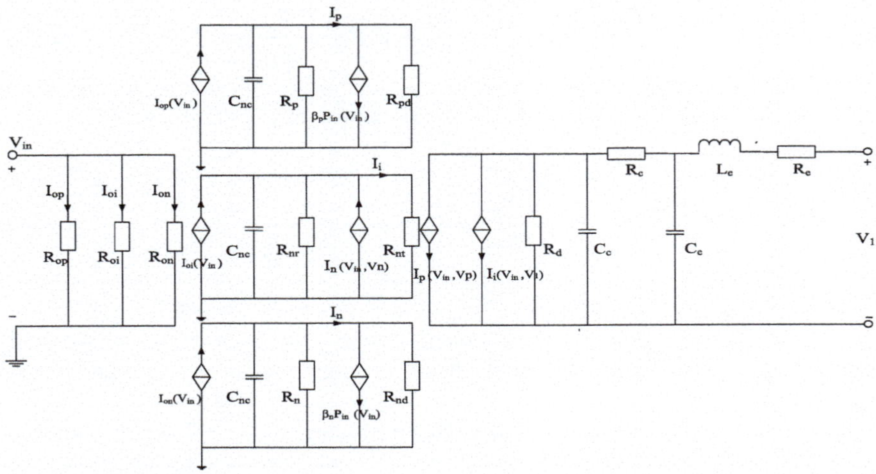

Fig. 1.29 Equivalent-circuit model of a PIN photodetector

The circuit in Fig. 1.30 was used to simulate the impulse-response characteristics of a PIN photodetector. The peripheral circuit is as follows: V_1 is a 5-V reverse-bias voltage; R is a 50-Ω sampling resistance; V-pulse is a pulse voltage source; V1 and V2 indicate that the initial voltage and pulse voltage are 0 and 1 V, respectively; the rest is consistent with Fig. 1.18.

Table 1.2 shows the main characteristic parameters of a PIN-PD, which can be simulated by incorporating them into the equivalent-circuit model in Fig. 1.29. By modifying the parameters in Table 1.2, they can be used to simulate different devices.

The impulse-response characteristic curve of a PIN photodetector can be obtained using the simulation circuit in Fig. 1.30, as shown in Fig. 1.31. The horizontal axis represents time T and the vertical axis represents the output photocurrent I.

Fig. 1.30 Transient-characteristic simulation circuit of a PIN photodetector

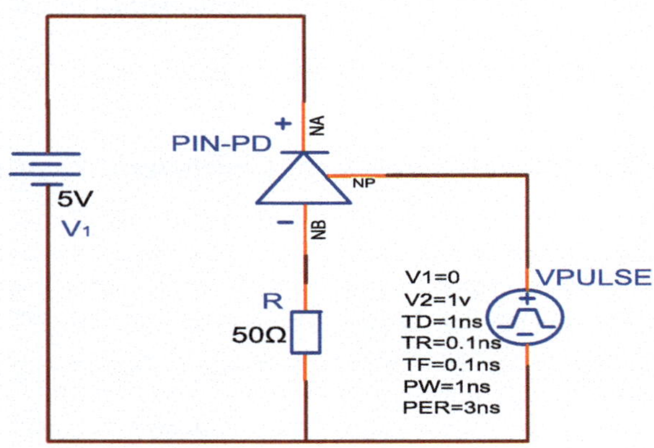

Table 1.2 Main characteristic parameters of a PIN-PD [2, 7]

A	W_i	α_i	r	τ_n	τ_p	τ_{nr}	L_e	C_e	R_e
12×10^{-9}	4.2 μm	3000	0.1	0.5 μs	0.5 μs	0.63 μs	1 nH	0.15 pF	10 Ω

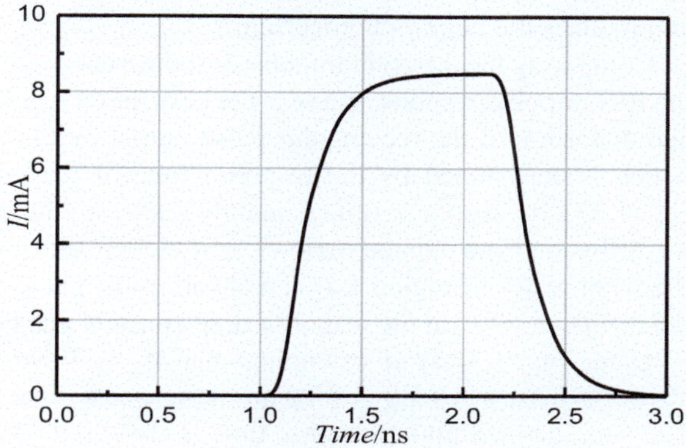

Fig. 1.31 Impulse-response characteristic curve of a PIN photodetector

As seen in Fig. 1.31, the impulse-response curve of the PIN photodetector exhibits a slight distortion, which is determined by the impulse-response time of the photodetector. The longer the response time, the more serious the pulse-waveform distortion. The response time is mainly related to the carrier transit time and RC time constant in region I [17], which are analyzed in the next section.

Figure 1.32 shows the influence of the PIN photodetector on its impulse-response characteristics under different widths, W_i, of region I. The initial conditions are as follows: $V_I = 5$ V, $A = 12 \times 10^{-9}$, $R_c = 5\ \Omega$, $C_c = 1.0$ pF, $R_e = 10\ \Omega$, $C_e = 0.15$ pF, $L_e = 1$ nH, $V_{input} = 0.4$ V, and $W_i = 3$ μm, 10 μm, 50 μm, and 100 μm.

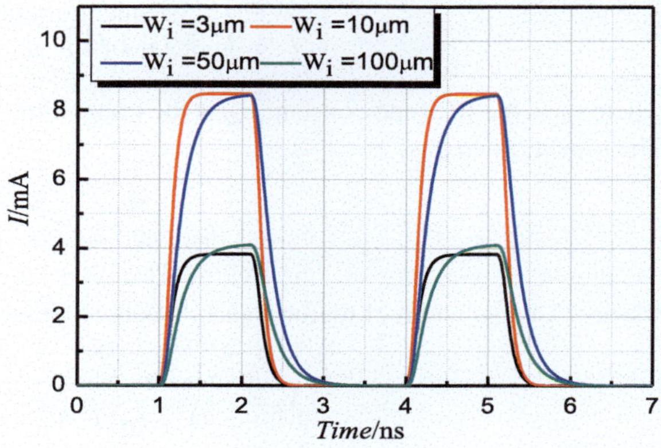

Fig. 1.32 Impulse-response curve under different widths W_i

As can be seen from Fig. 1.32, as the width W_i of region I varies from 3 μm to 100 μm, the output photocurrent first increases from 3.8 to 8.4 mA and then decreases to 4 mA. This is because, with the increase in the width W_i of region I, more photons are stimulated and absorbed to produce electron–hole pairs. In addition, when the reverse-bias voltage is constant, the field strength of region I decreases with an increase in W_i, which reduces the carrier accumulation rate in the photodetector and reduces the output photocurrent.

Comparing the different impulse-response curves in Fig. 1.32, when W_i is 10 μm, the PIN impulse-response curve is the best, and when W_i is 100 μm, the deformation and distortion of the rectangular pulse output by the detector are the most serious, which is determined by the response time of the pulse waveform. As shown in Eq. (1.37), the response time is mainly related to the carrier transit time in region I, the diffusion time outside region I, and the RC time constant. However, because of the wide width of region I, it can absorb most photons; therefore, the influence of the diffusion time on the response time constant can be almost ignored.

Using Fig. 1.37 as an example, when $W_i = 3$ μm, the impulse-response time is about 0.51 ns; when $W_i = 50$ μm, the response time is about 0.84 ns; when $W_i = 100$ μm, the maximum response time is about 0.99 ns; and when $W_i = 10$ μm, the minimum response time is about 0.33 ns. The reason for this phenomenon can be seen in Eqs. (1.45), (1.47), and (1.48).

When the width of region I increases, the decrease in the junction capacitance leads to a decrease in the RC time constant and a decrease in the response time. However, with an increase in the W_i value, the increase in the carrier transit time in region I becomes more severe, so the impulse-response time first decreases and then increases. From the above analysis, it can be seen that by selecting an appropriate W_i, the impulse-response time is minimized, and the waveform distortion can be effectively suppressed.

In addition, the influence of W_i on the output photocurrent of the photodetector should also be considered.

The photodetector response time [18] is

$$t = \left(t_{rc}^2 + t_{di}^2 + t_{dr}^2\right)^{1/2}, \tag{1.44}$$

where t_{rc} is the RC time constant, t_{di} is the transit time, and t_{dr} is the diffusion time. The transit time t_{di} is

$$t_{di} = \frac{W_i}{V_n}, \tag{1.45}$$

where W_i is the width of region I and V_n is the carrier-drift velocity of region I [7]:

$$V_n = \frac{\mu_n F}{\sqrt{1 + (\mu_n F / V_s)}}, \tag{1.46}$$

where V_s is the saturation drift velocity. Electric-field strength $F = (V_1 + V_{BI})/W_i$, where V_{BI} is the built-in potential, μ_n is the electric-field mobility, and the RC response time constant is [17]

$$t_{rc} = R_c(C_s + C_{pn}), \tag{1.47}$$

where C_s is the parasitic capacitance, C_{pn} is the junction capacitance, and R_c is the parasitic resistance. The junction capacitance C_{pn} is

$$C_{pn} = \frac{\varepsilon_r \varepsilon_0 A}{W_i}, \tag{1.48}$$

where A is the photosensitive surface, ε_0 is the vacuum dielectric constant, and ε_r is the dielectric constant of the material.

Figure 1.34 shows the influence of the PIN photodetector on its impulse-response characteristics under different reverse-bias voltages, V_1. The initial conditions are as follows: $W_i = 4.2$ μm, $A = 12 \times 10^{-9}$, $R_c = 5\ \Omega$, $C_c = 1.0$ pF, $R_e = 10\ \Omega$, $C_e = 0.15$ pF, $L_e = 1$ nH, $V_{input} = 0.4$ V, and $V_1 = 2$ V, 4 V, 6 V, and 8 V.

As shown in Fig. 1.33, as the reverse-bias voltage V_1 changes from 2 to 8 V, the output photocurrent increases from 9 to 22.5 mA. With an increase in the reverse-bias voltage, the carrier accumulation rate in the photodetector increases, which increases the output photocurrent. For the impulse-response curve with a reverse-bias voltage of 2 V, it can be seen that the rise time is 0.21 ns and the fall time is 0.41 ns. At this time, the output pulse waveform is slightly deformed. For the impulse-response curve with a reverse-bias voltage of 8 V, the response rise time is 0.19 ns and the fall time is 0.18 ns, and the pulse waveform is the best.

It can be seen from Eqs. (1.21) and (1.22) that the greater the reverse-bias voltage, the stronger the electric field in region I, the faster the carrier drift speed in region I by the electric-field force, and the shorter the transit time. It can be seen from the above analysis that by increasing the reverse-bias voltage, the impulse-response time can

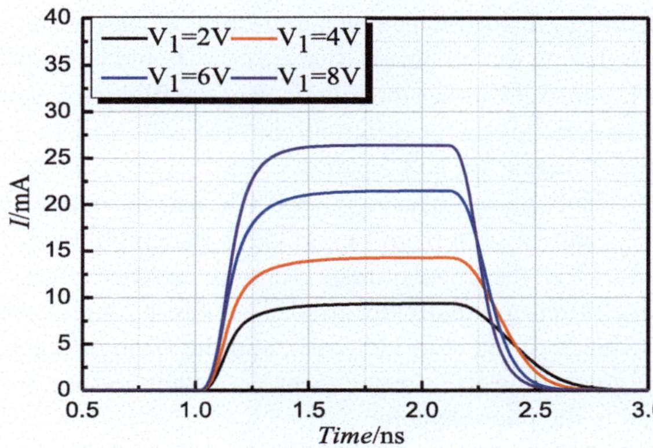

Fig. 1.33 Impulse-response curves under different reverse-bias voltages V_1

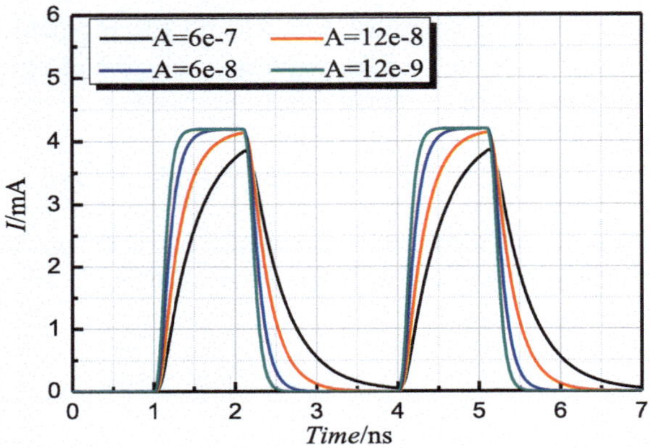

Fig. 1.34 Impulse-response curves under different photosensitive surfaces *A*

be reduced and the pulse waveform of the detector can be optimized. This method can only optimize the pulse waveform to a certain extent because the field strength in region I tends to saturate with an increase in the reverse-bias voltage.

Figure 1.34 shows the influence of the PIN photodetector on its impulse-response characteristics under different photosensitive surface areas, *A*. The initial conditions are as follows: $W_i = 4.2\ \mu\text{m}$, $V_I = 5$ V, $R_c = 5\ \Omega$, $C_c = 1.0$ pF, $R_e = 10\ \Omega$, $C_e = 0.15$ pF, $L_e = 1$ nH, $V_{input} = 0.4$ V, and $A = 6 \times 10^{-7}$, 12×10^{-8}, 6×10^{-8}, and 12×10^{-9}.

As can be seen from Fig. 1.34, as the photosensitive surface area *A* increases from 12×10^{-9} to 6×10^{-7}, the impulse-response time increases from 0.33 ns to 1.65 ns, the response time of the photodetector becomes longer, and the more serious the pulse-waveform distortion is. According to Eqs. (1.47) and (1.48), the junction capacitance in the photodetector increases with an increase in the photosensitive surface area *A*. The stronger the junction-capacitance effect, the slower the charge–discharge process, the longer the carrier transmission time, and the greater the response delay.

From the above analysis, it can be seen that by reducing the photosensitive surface area *A* of the photodetector, the impulse-response time can be reduced and the pulse-waveform distortion can be suppressed. However, the optical-fiber coupling caused by reducing the photosensitive surface should be considered.

Figure 1.35 shows the influence of the PIN photodetector on its impulse-response characteristics under different chip parasitic resistances, R_c. The initial conditions are as follows: $W_i = 4.2\ \mu\text{m}$, $V_I = 5$ V, $A = 12 \times 10^{-9}$, $C_c = 1.0$ pF, $R_e = 10\ \Omega$, $C_e = 0.15$ pF, $L_e = 1$ nH, $V_{input} = 0.4$ V, and $R_c = 10$, 20, 30, and 40 Ω.

As can be seen from Fig. 1.35, as the chip parasitic resistance R_c changes from 10 to 40 Ω, the impulse-response time increases from 0.37 to 1.67 ns, and the pulse waveform is gradually distorted. The reason for this phenomenon is that the parasitic resistance of the chip increases, and the stronger the blocking effect on the carrier, the greater the RC time-response constant and the longer the carrier transmission time in the photodetector.

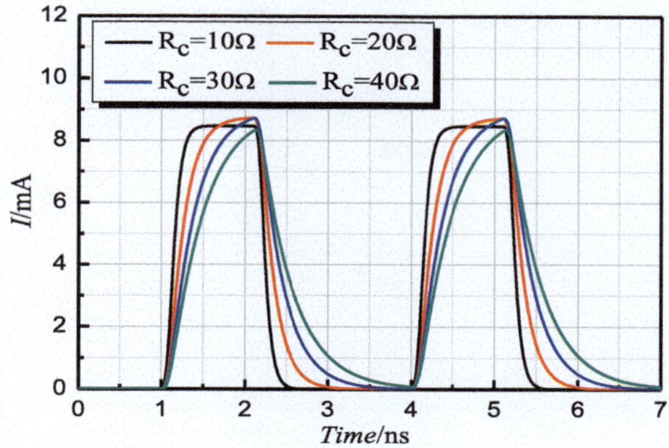

Fig. 1.35 Impulse-response curves under different chip parasitic resistances R_c

From the above analysis, it can be seen that an excessive chip parasitic resistance R_c has an adverse impact on the impulse response of the detector. Therefore, it is possible to improve the impulse-response characteristics of the PIN photodetector and suppress the waveform distortion by reducing the value of the chip parasitic parameter R_c.

Figure 1.36 shows the influence of the PIN photodetector on its impulse-response characteristics under different chip parasitic capacitances, C_c. The initial conditions are as follows: $W_i = 4.2\ \mu\text{m}$, $V_I = 5$ V, $A = 12 \times 10^{-9}$, $R_c = 5\ \Omega$, $R_e = 10\ \Omega$, $C_e = 0.15$ pF, $L_e = 1$ nH, $V_{input} = 0.4$ V, and $C_c = 0.1$, 8, 18, and 30 pF.

As can be seen from Fig. 1.36, as the chip parasitic capacitance C_c changes from 0.1 pF to 30 pF, the impulse-response time increases from 0.22 ns to 1.76 ns, and the output current decreases from 8.5 mA to 4.4 mA. Therefore, with an increase in the chip parasitic capacitance C_c, the impulse-response time increases, the output current decreases, and the detector can no longer respond well to the incident pulse.

The decrease in the output current is due to the shunt effect of the parallel parasitic capacitance on the current generated in the active region. The increase in response

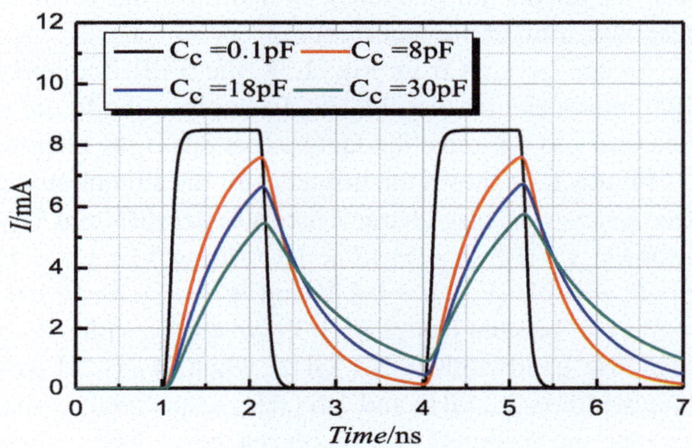

Fig. 1.36 Impulse-response curves under different chip parasitic capacitances C_c

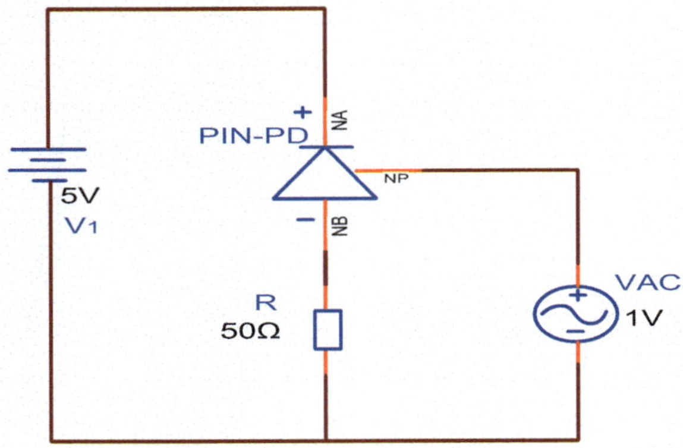

Fig. 1.37 AC small-signal simulation circuit of a PIN photodetector

time is due to the increase in the RC time constant with the increase in chip parasitic capacitance; the transmission time of the carriers in the photodetector becomes longer, and there is a certain response delay. Therefore, reducing the chip parasitic capacitance C_c can suppress the pulse-waveform distortion.

1.3.3 Frequency-Response Characteristics of a PIN-PD

The frequency–response characteristics of the photodetector were studied in the frequency domain, and the influence of package parasitic parameters on the frequency–response characteristics of the laser was analyzed. Using the circuit in Fig. 1.37, the frequency–response characteristics of the PIN photodetector were simulated using the PSpice AC small-signal analysis. In the peripheral circuit, V_1 is a 5-V reverse-bias voltage and *VAC* is a 1-V AC modulation voltage source.

The frequency–response characteristics of the PIN photodetector can be obtained using the simulation circuit in Fig. 1.37, as shown in Fig. 1.38. The horizontal axis represents the frequency ω, and the longitudinal axis represents the frequency response (output photocurrent I).

As can be seen from Fig. 1.38, the 3-dB frequency–response bandwidth of the PIN photodetector was 3 GHz. To improve the frequency–response bandwidth, it is necessary to optimize the factors that affect the response speed.

Figure 1.39 shows the influence of the PIN photodetector on its AC small-signal frequency–response characteristics under different W_i. The initial conditions are as follows: $V_1 = 5$ V, $A = 12 \times 10^{-9}$, $R_c = 5\ \Omega$, $C_c = 1.0$ pF, $R_e = 10\ \Omega$, $C_e = 0.15$ pF, $L_e = 1$ nH, $V_{input} = 0.4$ V, and $W_i = 3$ μm, 10 μm, 50 μm, and 100 μm.

As can be seen from Fig. 1.39, when the width W_i is 3 μm, the 3-dB frequency–response bandwidth is 3.1 GHz; when W_i is 50 μm and 100 μm, the response bandwidth is 2.2 GHz and 1.6 GHz, respectively; when the width W_i of region I is 10 μm, the response bandwidth reaches 4.1 GHz; hence, an optimal W_i of 10 μm maximizes the frequency–response bandwidth.

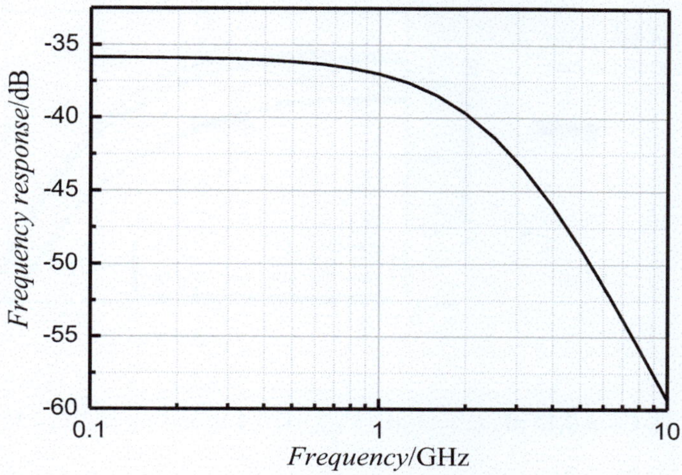

Fig. 1.38 Small-signal frequency–response characteristics of a PIN photodetector

Fig. 1.39 Frequency–response curves under different widths W_i

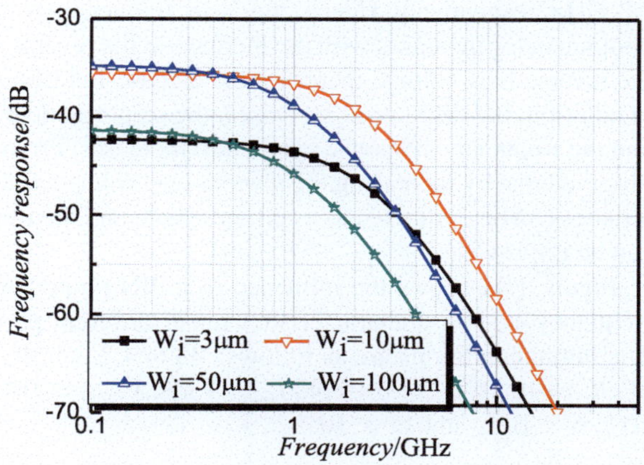

The increase in the width of region I first reduces the junction capacitance and the RC time constant. Second, a longer carrier drift time in region I increases the transit time. The interaction between the two results first increases the frequency–response bandwidth of the photodetector and then decreases it. From the above analysis, it can be seen that by selecting the appropriate W_i, the frequency–response characteristics can be optimized, and the frequency–response bandwidth can be improved, which is consistent with the time-domain simulation results.

Figure 1.40 shows the influence of the PIN photodetector on its AC small-signal frequency–response characteristics under different reverse-bias voltages, V_I. The initial conditions are as follows: $W_i = 4.2$ μm, $A = 12 \times 10^{-9}$, $R_c = 5\ \Omega$, $C_c = 1.0$

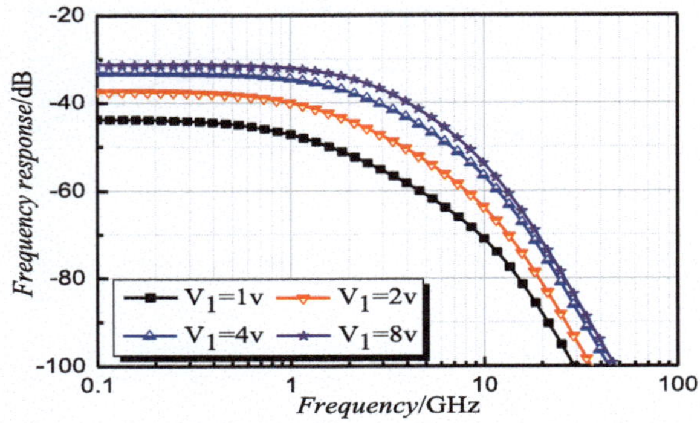

Fig. 1.40 Frequency–response curves under different reverse-bias voltages V_I

pF, $R_e = 5\ \Omega$, $C_e = 0.15$ pF, $L_e = 1$ nH, $V_{input} = 0.4$ V, and $V_I = 2$ V, 4 V, 6 V, and 8 V.

As can be seen from Fig. 1.40, when the reverse-bias voltage V_I is 2 V and 4 V, the 3-dB frequency–response bandwidth is 4.6 GHz and 5.6 GHz, respectively; when the reverse-bias voltage V_I is 6 V and 8 V, the response bandwidth is 8.9 GHz and 9.3 GHz, respectively. Hence, the 3-dB frequency–response bandwidth of the PIN photodetector increases with the increase in the reverse-bias voltage. The increase in the reverse-bias voltage increases the electric-field force of the carriers in region I; in addition, the drift speed increases, the response delay to incident light decreases, and the frequency–response bandwidth increases. From the above analysis, it can be observed that by increasing the reverse-bias voltage, the frequency–response characteristics can be optimized to a certain extent, and the frequency–response bandwidth can be improved.

Figure 1.41 shows the influence of a PIN photodetector on its AC small-signal frequency–response characteristics under different photosensitive surface areas, *A*. The initial conditions are as follows: $W_i = 4.2\ \mu$m, $V_I = 5$ V, $R_c = 5\ \Omega$, $C_c = 1.0$ pF, $R_e = 10\ \Omega$, $C_e = 0.15$ pF, $L_e = 1$ nH, $V_{input} = 0.4$ V, and $A = 6 \times 10^{-7}$, 12×10^{-8}, 6×10^{-8}, and 12×10^{-9}.

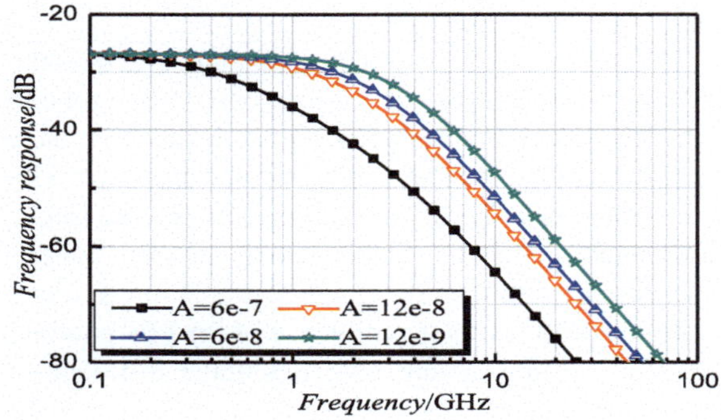

Fig. 1.41 Frequency–response curves under different photosensitive surfaces A

As can be seen from Fig. 1.41, when the photosensitive surface area A is 6 $\times$ 10^{-7}, the 3-dB frequency–response bandwidth is 2.1 GHz; when A is 12 $\times$ 10^{-8}, the response bandwidth is 4.9 GHz; when A is 6 $\times$ 10^{-8} and 12 $\times$ 10^{-9}, the response bandwidth is 5.8 GHz and 7.2 GHz, respectively; hence, the 3-dB frequency–response bandwidth of a PIN photodetector increases with the decrease of photosensitive surface A. When the photosensitive surface increases, the junction capacitance increases, and the PN-junction capacitance's effect on the photodetector becomes stronger, which delays the photodetector's response to the incident light and reduces the response bandwidth. Therefore, reducing the photosensitive surface A of the photodetector can improve the frequency–response bandwidth.

Figure 1.42 shows the influence of the PIN photodetector on its AC small-signal frequency–response characteristics under different chip parasitic resistances, R_c. The initial conditions are as follows: $W_i = 4.2$ μm, $V_I = 5$ V, $A = 12 \times 10^{-9}$, $C_c = 1.0$ pF, $R_e = 10$ Ω, $C_e = 0.15$ pF, $L_e = 1$ nH, $V_{input} = 0.4$ V, and $R_c = 10$, 20, 30, and 40 Ω.

As can be seen from Fig. 1.43, when the resistance $R_c = 40$ Ω, the 3-dB frequency–response bandwidth is 4.1 GHz; when $R_c = 30$ Ω, the response bandwidth is 4.9 GHz; when $R_c = 20$ Ω, the response bandwidth is 5.5 GHz; when $R_c = 10$ Ω, the response bandwidth is 6.7 GHz. That is, the 3-dB frequency–response bandwidth decreases with the increase of the chip parasitic resistance.

The increase in the chip parasitic resistance in the photodetector increases the RC time constant, which increases the transmission time of the photogenerated carriers in the detector, and lowers the frequency–response bandwidth. Therefore, the frequency–response bandwidth can be improved by reducing the chip parasitic resistance.

Figure 1.43 shows the influence of the PIN photodetector on its AC small-signal frequency–response characteristics under different chip parasitic capacitances, C_c. The initial conditions are as follows: $W_i = 4.2$ μm, $V_I = 5$ V, $A = 12 \times 10^{-9}$, $R_c = 5$ Ω, $R_e = 10$ Ω, $C_e = 0.15$ pF, $L_e = 1$ nH, $V_{input} = 0.4$ V, and $C_c = 0.1$ pF, 8 pF, 18 pF, and 30 pF.

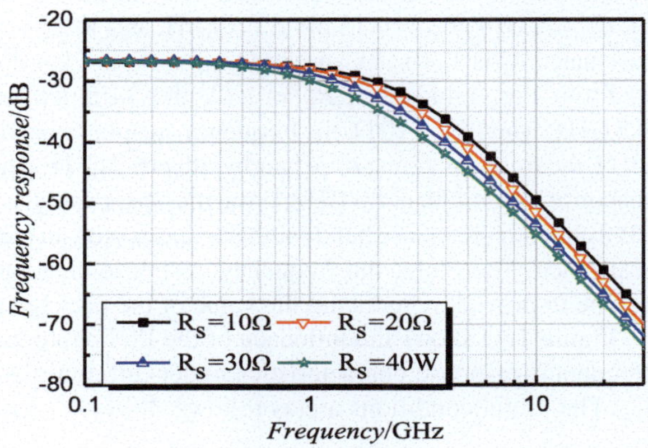

Fig. 1.42 Frequency–response curves under different chip parasitic resistances RC

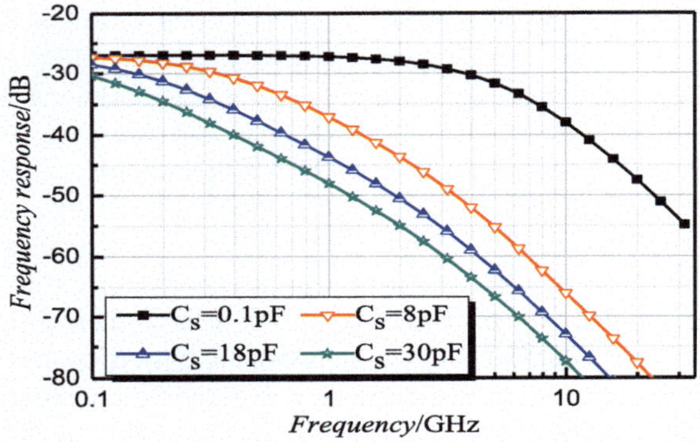

Fig. 1.43 Frequency–response curves under different parasitic chip capacitances C_c

From Fig. 1.43, when the parasitic chip capacitance $C_c = 30$ pF, the frequency–response bandwidth is 0.6 GHz; when $C_c = 18$ pF, the frequency–response bandwidth is 0.6 GHz; when $C_c = 8$ pF, the frequency–response bandwidth is 1.65 GHz; and when $C_c = 0.1$ pF, the frequency–response bandwidth is about 14 GHz.

It can be seen from the above that the 3-dB frequency–response bandwidth of the photodetector increases with the decrease of the chip parasitic capacitance. This is because with the increase in chip parasitic resistance, the RC time constant in the detector increases, the time between the generation and output of the photogenerated carriers increases, and the response bandwidth decreases. Hence, reducing the parasitic capacitance of the chip can obtain better frequency–response characteristics.

The internal parameters play a decisive role in the frequency–response and pulse-response characteristics of the PIN photodetector. In addition, the parasitic parameters of the detector chip in the packaging process significantly affect its performance. The influence of the package parasitic parameters (R_e, C_e, and L_e) on the frequency–response characteristics of the detector is analyzed below.

Figure 1.44 shows the influence of the PIN photodetector on its AC small-signal frequency–response characteristics under different package parasitic resistances, R_e. The initial conditions are as follows: $W_i = 4.2$ μm, $V_I = 5$ V, $A = 12 \times 10^{-9}$, $R_c = 5$ Ω, $C_c = 1.0$ pF, $C_e = 0.15$ pF, $L_e = 1$ nH, and $V_{input} = 0.4$ V. The package parasitic resistance R_e is 1, 10, 20, and 30 Ω.

From Fig. 1.44, when $R_e = 1$ Ω, the 3-dB frequency–response bandwidth is 9.8 GHz; when $R_e = 10$ Ω, the frequency–response bandwidth is 8.3 GHz; when $R_e = 20$ Ω, the frequency–response bandwidth is 6.2 GHz; when $R_e = 30$ Ω, the frequency–response bandwidth is 4.8 GHz. From the above analysis, it can be observed that the 3-dB frequency–response bandwidth decreases with an increase in the package parasitic resistance. Therefore, the frequency–response characteristics of the photodetector can be improved by reducing the value of the package parasitic resistance R_e.

Figure 1.45 shows the influence of the PIN photodetector on its AC small-signal frequency–response characteristics under different package parasitic capacitances, C_e. The initial conditions are as follows: $W_i = 4.2$ μm, $V_I = 5$ V, $A = 12 \times 10^{-9}$,

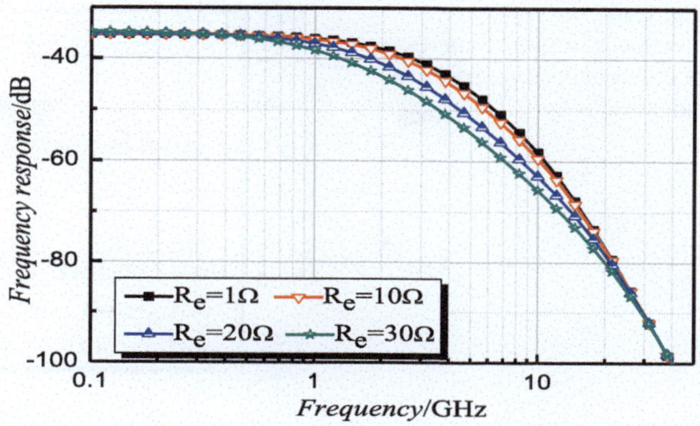

Fig. 1.44 Frequency–response curves under different package parasitic resistances R_e

$R_c = 5\ \Omega$, $C_c = 1.0$ pF, $R_e = 10\ \Omega$, $L_e = 1$ nH, and $V_{input} = 0.4$ V. The package parasitic capacitance C_e is taken separately: 0.15 pF, 0.35 pF, 0.75 pF, and 1 pF.

As can be seen from Fig. 1.45, when $C_e = 0.15$ pF, the 3-dB frequency–response bandwidth is approximately 13.1 GHz; when $C_e = 0.35$ pF, the response bandwidth is approximately 11.8 GHz; when $C_e = 0.75$ pF, the response bandwidth is 10.1 GHz; and when $C_e = 1$ pF, the response bandwidth is 9.2 GHz. It can be seen from the above analysis that the frequency–response bandwidth of the photodetector decreases with the increase of the package parasitic capacitance. Therefore, the frequency–response characteristics of the photodetector can be improved by reducing the value of the package parasitic capacitance C_e.

Figure 1.46 shows the effect of the PIN photodetector under different package parasitic inductances L_e on its AC small-signal frequency–response characteristics. The initial conditions are as follows: $W_i = 4.2\ \mu$m, $V_1 = 5$ V, $A = 12 \times 10^{-9}$, $R_c = 5\ \Omega$, $C_c = 1.0$ pF, $R_e = 10\ \Omega$, $C_e = 0.15$ pF, and $V_{input} = 0.4$ V. The package parasitic inductance L_e is 0.1 nH, 1 nH, 5 nH, 10 nH, 20 nH, and 30 nH.

Figure 1.46 shows a resonance peak in the response curve, which is caused by the parasitic oscillation, which is caused by the parasitic parameters (R_e, C_e, and L_e) of the package constituting the resonant circuit. When the inductance L_e is 0.1 nH and 1 nH, the 3-dB frequency–response bandwidth is 3.1 GHz and 3.9 GHz, respectively;

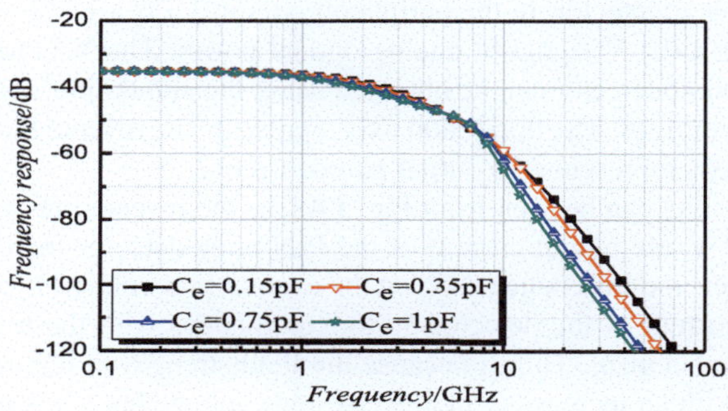

Fig. 1.45 Frequency–response curves of different package parasitic capacitances C_e

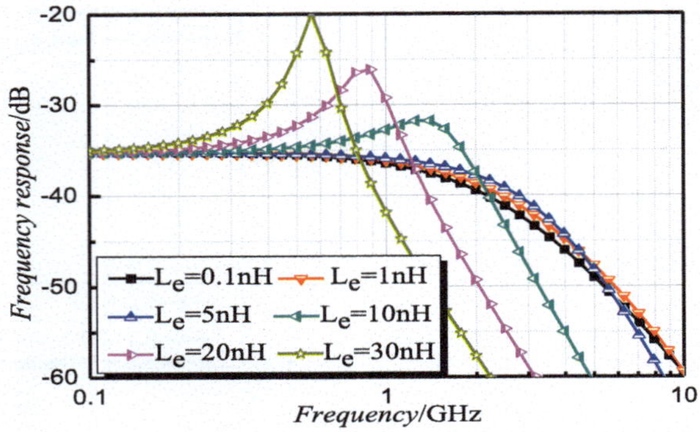

Fig. 1.46 Frequency–response curves of different package parasitic inductances L_e

when the inductance L_e is 5 nH and 10 nH, the frequency–response bandwidth is 4.2 GHz and 2.2 GHz, respectively; when the inductance L_e is 20 nH and 30 nH, the response bandwidth is 1.3 GHz and 0.41 GHz, respectively.

From the above analysis, it can be seen that with an increase in inductance L_e, the 3-dB frequency–response bandwidth does not always decrease; it first increases and then decreases. Therefore, an optimal parasitic inductance value exists that maximizes the frequency–response bandwidth. Therefore, the frequency–response bandwidth of the photodetector can be compensated by controlling the length of the gold lead wire and using the resonance effect.

1.3.4 DC Characteristics of a PIN-PD

Under reverse-bias voltage V_I, the PIN photodetector converts the incident optical power into a photocurrent, which is the most basic working characteristic of a detector. Therefore, it is essential to simulate basic DC characteristics. This section describes the simulation of the V-I characteristics (relationship between the reverse-bias voltage and output current).

The circuit in Fig. 1.47 was used to simulate the V-I characteristics of the PIN photodetector. In the peripheral circuit, V_I is a 1-V reverse-bias voltage, and V_{input} is a 0.4-V signal of analog optical power. The V-I characteristic curve of the PIN photodetector can be obtained using the simulation circuit in Fig. 1.47, as shown in Fig. 1.48. The horizontal axis represents the reverse-bias voltage V_I and the vertical axis represents the output photocurrent I.

As can be seen from Fig. 1.48, as the reverse-bias voltage V_I changes from 0 to 5 V, the output current of the PIN photodetector increases with an increase in the reverse-bias voltage. When the reverse-bias voltage V_I is greater than 5 V, the output current of the detector no longer changes with the reverse-bias voltage, but tends to stabilize. This is because the field strength in region I does not always increase with an increase in the reverse-bias voltage but tends to saturate. By increasing the

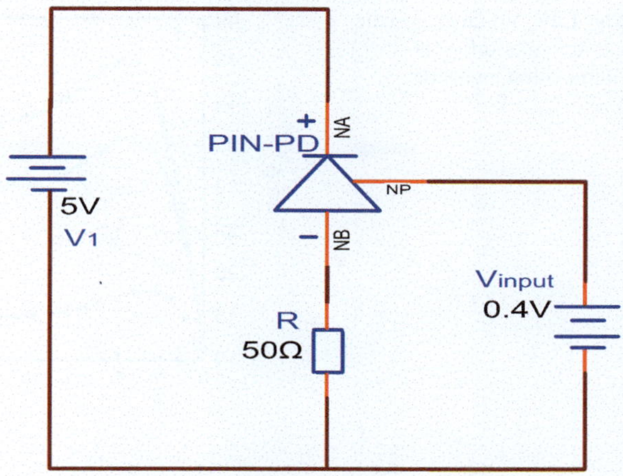

Fig. 1.47 DC-characteristic simulation circuit of a PIN-PD

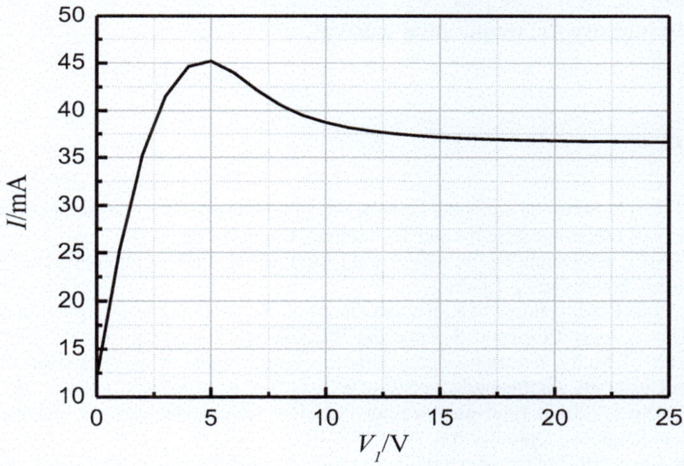

Fig. 1.48 V-I characteristics of a PIN photodetector

reverse-bias voltage V_I, the output current of the detector can be increased to a certain extent.

Figure 1.49 shows the influence of the PIN photodetector on its V-I characteristics under different illumination intensities, V_{input}. The initial conditions are as follows: $W_i = 4.2\ \mu m$, $V_I = 5$ V, $A = 12 \times 10^{-9}$, $R_c = 5\ \Omega$, $C_c = 1.0$ pF, $R_e = 10\ \Omega$, $C_e = 0.15$ pF, $L_e = 1$ nH, and $V_{input} = 0.1$ V, 0.2 V, 0.3 V, and 0.4 V.

As can be seen from Fig. 1.49, when the illumination intensity V_{input} is constant and the reverse-bias voltage increases, the output photocurrent increases and tends to saturate; when the reverse-bias voltage V_I is constant, the greater the illumination intensity, the greater the output current, in a linear relationship. This is because the

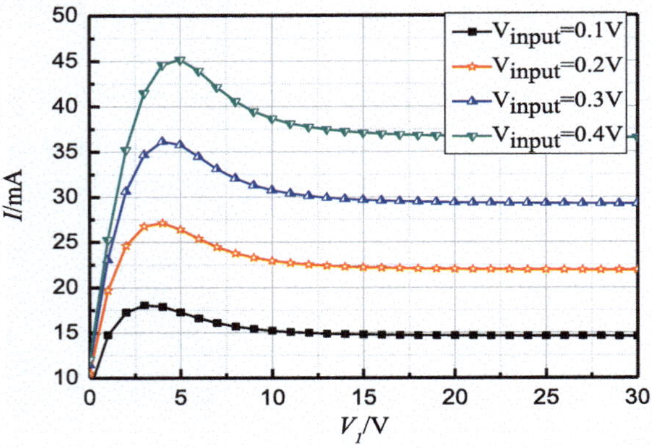

Fig. 1.49 V-I characteristic curves under different illumination intensities V_{input}

greater the light intensity, the more photons are involved in the stimulated absorption, and the more electron–hole pairs will be produced. The input optical power can be increased to improve the output photocurrent.

References

1. Chen M (2011) Study on equivalent circuit model of semiconductor laser and photodetector. University of Electronic Science and Technology, Chengdu, pp 42–43
2. Zhao J (2015) Equivalent circuit model and experimental study of photodetector. Chongqing University, Chongqing, pp 19–26
3. Xu W (2008) Establishment of equivalent circuit model and parameter extraction of photodetector. University of Electronic Science and Technology, Chengdu, pp 16–17
4. Duan H (2010) Study on response characteristics of semiconductor laser based on rate equation. Yanshan University, Qinhuangdao, pp 58–59
5. Wang Q, Xu L (2016) Equivalent circuit model of semiconductor laser and its parameter extraction. Semicond Optoelectron 37(2):186–189.
6. Chen W, Liu S (1991) Dynamic modulation analysis model of laser diode and its application. J Semicond 12(7):416–422
7. Chen W (2001) Optoelectronic device model and OEIC simulation. National Defense Industry Press, Beijing, pp 47–243
8. Xiao Y, Chen F (2010) Simulation analysis of packaging parasitic parameters on frequency response of semiconductor laser. Prog Laser Optoelectron 47(12):94–97
9. Qiu K, Mei K (1992) Study on high speed and ultra-high speed modulation mechanism of semiconductor laser. J Univ Electron Sci Technol 21(S1):53–57
10. Xie J (1992) Response characteristics of high speed semiconductor lasers. Opt Commun Res 5(3):42–48
11. Chen J, Zeng J (2008) Study on modulation characteristics of injection locked semiconductor laser. Semicond Optoelectron 29(2):173–176
12. Zhang M, Lin Z (1995) Study on high frequency semiconductor laser module and its characteristics. Acta PHOTONICA Sinica 24(1):25–29
13. Liao X (1990) Discussion on modulation bandwidth of ultrahigh speed semiconductor laser module. Opt Commun Res (2):36–46
14. Qiu K, Liu X (1993) Study on high frequency response characteristics of semiconductor lasers. J Univ Electron Sci Technol 22(4):395–399

15. Xiao Y (2011) Semiconductor laser equivalent circuit model simulation and parameter extraction. University of Electronic Science and Technology, Chengdu, pp 25–26
16. Chen W, Liu B, Liu S (1994) Research on PIN photodiode circuit model. Acta electronica Sinica 22(11):95–97
17. Tao Q (2001) 1.3 μm high speed PIN photodiode. Semicond Optoelectron 22(4):271–274
18. Li Q, Li W, Qi L et al (2018) 10 Gbit/s mesa InGaAs/InP pin high speed photodetector. Semicond Technol 43(4):280–284

Chapter 2
Analysis of PPM-Coded Modulated Signals

In this chapter, according to the characteristics of pulse-position modulation (PPM) signals, a mathematical model of a PPM signal is established, and the power spectrum of a single-pulse position modulation signal is analyzed in detail. In addition, a design scheme for a PPM time-slot synchronization system is proposed.

2.1 Basic Principle and Mathematical Model of PPM

There are three main forms of pulse-position modulation [1, 2]: single pulse, differential pulse, and multipulse. Single-pulse position modulation is used to map a group of binary n-bit data groups into a single-pulse signal at a certain time slot in a time period composed of 2^n time slots. This modulation method is essentially phase modulation. We denote the n-bit data group as $M = (m_1, m_2, \ldots, m_n)$, which can be expressed using the following formula [3]:

$$\Phi\text{:L} = \mathrm{m}_1 + 2\mathrm{m}_2 + \cdots + 2^{\mathrm{n}-1}\mathrm{m}_\mathrm{n} \quad \in (0, 1, 2, \ldots, 2^{n-1}), \tag{2.1}$$

where L represents the position number of the time slot, Φ represents the mapping relationship, and $L = \Phi(m_1, ..., m_n)$. The mapping determined by Eq. (2.1) is a one-to-one mapping that satisfies the uniqueness of the modulation.

$$S_n(t) = \begin{cases} P_C \; LT_C < t < (L+1)T_C \\ 0 \qquad \text{other time}, \end{cases} \tag{2.2}$$

where $S_n(t)$ represents the modulated signal, P_C represents the pulse power, and T_C represents the time-slot length.

For example, for a four-PPM modulation pulse:

X. Ke, *Coding Theory in Optical Wireless Communication Systems*, Optical Wireless Communication Theory and Technology, https://doi.org/10.1007/978-981-19-9837-9_2

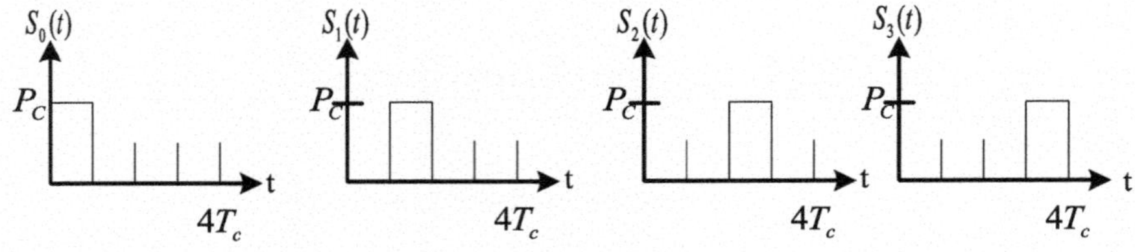

Fig. 2.1 Schematic diagram of an L-PPM modulation pulse

If M = (0, 0), then L = 0; if M = (1, 0), then L = 1;
If M = (0, 1), then L = 2; if M = (1, 1), then L = 3.

Here, 0, 1, 2, and 3 correspond to slot positions, as shown in Fig. 2.1. It can be seen that the mapping relationship determined by Eq. (2.1) is a one-to-one linear relationship, which satisfies the uniqueness requirement of modulation.

2.1.1 L-PPM Signal Power-Spectral Density (PSD) Analysis

In an actual atmospheric laser-communication system, there is a minimum time interval between adjacent light pulses emitted by a pulsed laser; this is the minimum delay time needed by a laser reconstructing an "inverted particle beam." This interval is called a guard slot (T_D). To understand the performance of L-PPM modulation for atmospheric laser communication, we analyzed the power spectrum of a random train of single-pulse position-modulation waveforms [4–6].

From Eq. (2.1), L-PPM coding maps n-bit information into $L = 2^n$ time slots and adds guard time slots to form an L-PPM frame. The L-PPM frame structure is shown in Fig. 2.2. The guard time slot $T_D = D\tau$ (τ is the length of a time slot, D is the number of guard time slots), and the information period is $T_S = L\tau$. Then, the length of one frame $T_f = T_D + T_S = (L + D)\tau = M\tau$.

In each frame, an optical pulse is located in the ith ($i = 0, 1, 2\ldots.$ L − 1) time slot, carrying n bits of information. After a photoelectric conversion is performed by the avalanche photodetector (APD) at the receiving end, a series of pulse sequences

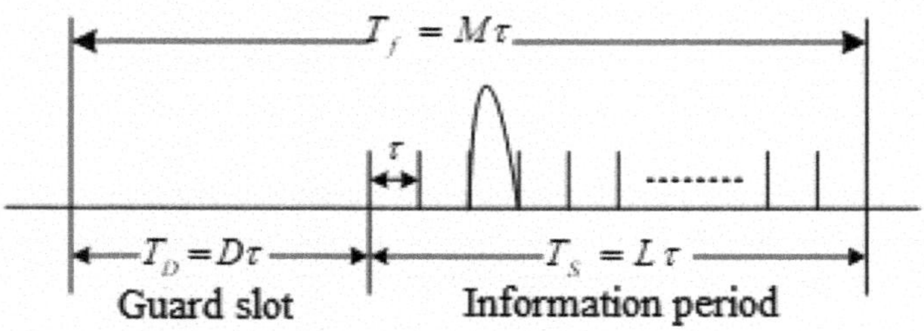

Fig. 2.2 L-PPM frame

is obtained. Mathematically, this L-PPM pulse sequence can be expressed as [7]

$$x(t) = \sum_n p(t - nM\tau - c_n\tau - e_n),$$

where p is the PPM pulse function, τ is the length of the time slot, c_n is an integer from 0 to M, representing the position of the PPM random pulse whose set is independent and identically distributed (IID), and e_n represents the random jitter of the pulse position, which is caused by the random noise of the receiver; this set is also IID.

It is known from the literature [8] that the received L-PPM pulse is not a stationary random process; therefore, its power spectrum cannot be obtained by the Fourier transform of its autocorrelation function. However, for any definite signal $x(t)$, its power-spectral density can be expressed as [9]

$$S(\omega) = \lim_{T\to\infty} \frac{|X_T(\omega)|^2}{T}, \tag{2.3}$$

where $X_T(\omega)$ is the spectral function of the truncation function $x_T(t)$ of $x(t)$ at time $[-T, T]$, defined as

$$X_T(\omega) = \int_{-T}^{T} x(t)e^{-j\omega t}dt. \tag{2.4}$$

Let $T = NM\tau$; then,

$$X_T(\omega) = \sum_{n=-N}^{N} P(\omega)e^{-j\omega(nM\tau + c_n\tau + e_n)}, \tag{2.5}$$

where $P(\omega)$ is the Fourier transform of the pulse-waveform function $p(t)$. The power-spectral density represented by Eq. (2.3) becomes

$$S(\omega) = \lim_{T\to\infty} \frac{E|X_T(\omega)|^2}{T}. \tag{2.6}$$

$$E\{|X_T(\omega)|^2\} = |P(\omega)|^2 \sum_{m,n=-N}^{N} e^{j\omega M\tau(m-n)} \cdot E\{e^{j\omega\tau(c_m - c_n)}\}E\{e^{j\omega(e_m - e_n)}\} \tag{2.7}$$

$$= |P(\omega)|^2[(2N+1) + \left|E\{e^{j\omega\tau c_n}\}\right|^2 \cdot \left|E\{e^{j\omega e_n}\}\right|^2 \sum_{m\neq n} e^{j\omega M\tau(m-n)}] \tag{2.8}$$

$$= |P(\omega)|^2 \left[(2N+1)(1 - |C(\omega)|^2 |M_e(\omega)|^2) + |C(\omega)|^2 |M_e(\omega)|^2 \left| \sum_{n=-N}^{N} e^{-j\omega M \tau n} \right|^2 \right]. \tag{2.9}$$

$C(\omega)$ and $M_e(\omega)$ are the characteristic functions of c_n and e_n, respectively, and can be expressed as

$$C(\omega) \equiv E\{e^{j\omega\tau c_n}\} = \frac{1}{M} \sum_{n=0}^{M-1} e^{j\omega\tau n} \tag{2.10}$$

$$M_e(\omega) \equiv E\{e^{j\omega e_n}\}. \tag{2.11}$$

Then, the power-spectral density can be expressed as

$$S(\omega) = \frac{|P(\omega)|^2}{2} \left[\frac{1}{M\tau}(1 - |C(\omega)|^2 |M_e(\omega)|^2) + \frac{2\pi}{(M\tau)^2} |M_e(\omega)|^2 \sum_{k=-\infty}^{\infty} \delta(\omega - 2\pi k/\tau) \right]. \tag{2.12}$$

The first part of Eq. (2.12) is related to noise, and the second part is related to discrete frequency components, which are composed of the fundamental frequency of the slot clock and its harmonic components. Therefore, the power-spectral density of noise can be expressed as

$$N(\omega) = \frac{|P(\omega)|^2}{2M\tau}(1 - |C(\omega)|^2 |M_e(\omega)|^2).$$

The total power of the slot-clock base frequency is

$$P_s = \frac{2}{(M\tau)^2} |M_e(\omega)|^2 |P(\omega)|^2.$$

The time-slot frequency component is affected neither by the probability distribution nor modulation sensitivity. The time-slot frequency component is only affected by the pulse shape. If the pulse width is equal to the time-slot length, the time-slot frequency component will not exist [10]; this provides a theoretical basis for establishing time-slot synchronization. Figure 2.3 shows the relationship between the duty cycle of the pulse (square wave) in the time slot and the frequency component of the time slot.

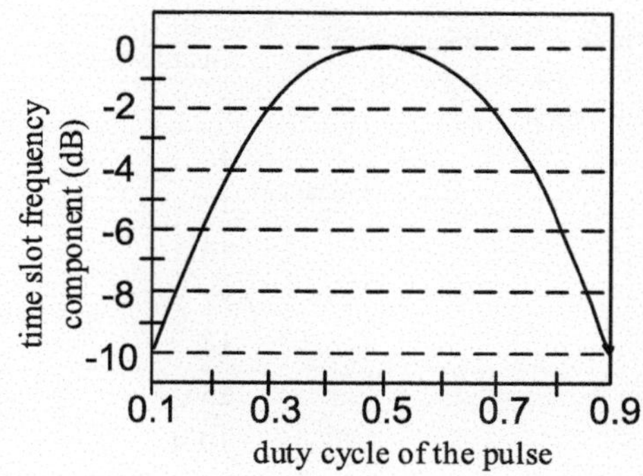

Fig. 2.3 Relationship between the pulse width and time-slot frequency component [10]

2.1.2 Transmission Efficiency of L-PPM Signals [11]

In a communication system, the modulation and demodulation of information play a key role in communication performance. To maintain reliability, the selected modulation method should have a high transmission rate (amount of information transmitted per unit time) and high transmission efficiency (amount of information transmitted per unit of energy). The requirements for a high rate and high efficiency may conflict. In an actual system, a certain technical index can be pursued according to actual needs. In the field of wireless laser communication, under the premise of ensuring a certain communication code rate, the efficiency must be as high as possible, or the transmission power must be as low as possible.

Optical pulse-position modulation is considered a suitable signaling method [21]. The ratio of the pulse power to the average power of this signaling method can be particularly high. Compared with the usual optical-intensity on–off keyed modulation (OOK), PPM can reduce the average power of optical radiation, and low radiation power can prolong the working life of the laser. Optical PPM is especially suitable for occasions that require low average power to transmit information; for example, submarine communication and indoor computer infrared (IR) wireless communication. The wireless local area network (LAN) standard, IEEE 802.11, has stipulated that the IR physical layer must adopt PPM.

The frame format of L-PPM is shown in Fig. 2.2. After the calculation, the data-transmission rate R of an L-PPM signal is

$$R = \log_2 L / T_f = \log_2 L / (L + D)\tau \text{ bit/s}. \tag{2.13}$$

The energy-utilization ratio of L-PPM and OOK modulation is

$$S_{ratio} = \log_2 L. \tag{2.14}$$

As shown in Table 2.1, when the number of guard time slots D is greater than 32, the power-utilization rate of the PPM is higher than that of OOK. As long as L

Table 2.1 Comparison of PPM and OOK power utilization

Guard slots D	Time slots L	S_{ratio}
0	3	0.396
4	6	0.646
12	10	0.83
16	11	0.865
32	17	1.022
64	28	1.202
256	78	1.571
1024	232	1.9645
1024	256	2

is greater than 16, the energy-utilization rate of PPM is higher than that of OOK; therefore, PPM is called a power-efficient modulation. PPM is especially suitable for communication systems with limited pulse-repetition intervals or communication environments where the bandwidth is abundant and power is precious.

2.2 PPM Ideal Channel-Capacity Analysis

The amount of information in an atmospheric channel is an ongoing concern. Hence, we analyzed the ideal channel capacity of PPM. Since Shannon proposed his formula in 1954, information theory has received increasing attention and has become an important and basic part of communication theory.

2.2.1 *Shannon's Formula*

Shannon's formula is a combination of probability theory and communication theory. When a symbol x is input into the channel, the conditional probability of the output symbol y is denoted as $W(y/x)$. The source output, with a $P_X(x)$ probability distribution in the channel, is used as the input, and the output Y of the channel is a probability variable whose distribution P_Y is given by $P_Y(y) = \sum_{x=\zeta} W(y|x)P_X(x)$. In this formula, ζ is the channel input letter. C is the channel capacity [12]:

$$C = \max_{P_X} I(X;Y). \tag{2.15}$$

In the formula, $I(X;Y)$ is mutual information, which means that $\max_{P_X} I(X;Y)$ is the maximum value of all possible probability distributions $I(X;Y)$ regarding input X.

From probability theory, Shannon derived that the average entropy is the difference between the average uncertainty $H(X)$ of the sender (source) and the uncertainty degree and measure of the source after the receiver receives the message Y (that is, the amount of information lost, owing to channel interference) $H(X|Y)$[13]:

$$I(X;Y) = H(X) - H(X|Y). \tag{2.16}$$

When Shannon studied the channel, he found that the noise in the channel is mainly Gaussian white noise; that is, white noise whose power-spectral density function follows a Gaussian distribution. A Gaussian channel contains Gaussian white noise. Shannon's formula can be derived for a Gaussian channel:

$$C = B\ln(1 + S/N) = B\ln(1 + S/(N_0 \times B)), \tag{2.17}$$

where N_0 is the signal-to-noise ratio per unit frequency, B is the bandwidth, and S/N is the signal-to-noise ratio (SNR). According to Shannon's formula, the amount of information C is related to both the bandwidth and the signal-to-noise ratio S/N. When the bandwidth B tends to infinity, the amount of information in the Gaussian channel tends to its limit:

$$C = \lim_{B\to\infty} B\ln(1 + S/(N_0 \times B)) \approx 1.44S/N_0(bits/s). \tag{2.18}$$

2.2.2 PPM Ideal Channel-Capacity Analysis

The channel capacity of L-PPM is given by [14]

$$C = (1 - e^{-\lambda_s\tau})\ln L \text{ nats/channel use}, \tag{2.19}$$

where λ_s is the photon density, τ is the time-slot length, and a nat is a natural unit of information. The average number of photons contained in each information frame is $\lambda_s\tau$, and the channel capacity can be transformed into

$$C_T = \frac{C}{L\tau} = \frac{1 - e^{-\lambda_s\tau}}{\tau}\frac{\ln L}{L} \text{ nats/s} \tag{2.20}$$

$$C_{ph} = \frac{C}{\lambda_s\tau} = \frac{1 - e^{-\lambda_s\tau}}{\lambda_s\tau}\ln L \text{ nats/photon}. \tag{2.21}$$

For a specific communication system, the time-slot length τ is fixed. C_{ph} is a function of the modulation number L, so a maximum value L^* of L exists, which maximizes the channel capacity C_{ph} of each photon. When the differential value of C_{ph} with respect to L is zero, we can obtain the value of L^*. However, it should be noted that if C_T and τ are both fixed values, a change in L will inevitably lead to a change in λ_s; that is, λ_s depends on L, as shown in Eq. (2.21):

$$\lambda_s = -\frac{1}{\tau}\ln(1 - \frac{C_T \tau L}{\ln L}). \tag{2.22}$$

Substituting Eq. (2.22) into Eq. (2.21) and differentiating it, we obtain

$$\frac{Z(L^*)}{1 - Z(L^*)}\left[1 - \frac{1}{\ln L^*}\right] + \ln[1 - Z(L^*)] = 0, \tag{2.23}$$

Where

$$Z(L) = \frac{C_T \tau L}{\ln L}. \tag{2.24}$$

let $\alpha \triangleq C_T\tau$(same as the channel capacity of each PPM slot),The value of L^* is the solution to Eq. (2.24).

When the solution given by Eq. (2.24) is not an integer, the nearest integer can be used as the actual value of L^*. This difference is important only when the value of L^* is relatively small. As shown in Eq. (2.21), when L takes L^*, the channel capacity reaches the maximum value C^*. From Eq. (2.24), when optimizing L, the relational expression is satisfied for any α:

$$\frac{Z^*}{1 - Z^*}\left[1 - \frac{1}{\ln L^*}\right] + \ln[1 - Z^*] = 0. \tag{2.25}$$

We define

$$Z^* = \frac{C_T \tau L^*}{\ln L^*} = \frac{C^*}{\ln L^*}. \tag{2.26}$$

Then, Eq. (2.26) becomes

$$C^* = f(Z^*), \tag{2.27}$$

where

$$f(Z) = \frac{Z^2}{Z + (1 - Z)\ln(1 - Z)} \tag{2.28}$$

is obtained by (27). Hence,

$Z = 1 - e^{-\lambda_s \tau}$.

Therefore, it can be seen that $0 < Z < 1$. In fact, $Z = 0$ indicates that the pulse energy is infinite, and $Z = 1$ indicates that the pulse energy is zero. Now, we need to determine the threshold value of $f(Z)$. We define $g(Z) = Z + (1 - Z)\ln(1 - Z)$;

then,

$$f(Z) = \frac{Z^2}{g(Z)}.$$

We can obtain

$$f'(Z) = \frac{2Z^2 + Z(2 - Z)\ln(1 - Z)}{g^2(Z)}. \tag{2.29}$$

We use the Taylor-series expansion to obtain
$\ln(1 - Z) < -Z - \frac{Z^2}{2} - \frac{Z^3}{3} - \frac{Z^4}{4}$.
For $0 < Z < 1$, we obtain

$$2Z^2 + Z(2 - Z)\ln(1 - Z) < -\frac{Z^4}{6}\ln(1 + Z - \frac{3}{2}Z^2) < 0. \tag{2.30}$$

For all $0 < Z < 1$, Eqs. (2.29) and (2.30) indicate $f\prime(Z) < 0$; therefore, $f(Z)$ varies monotonically with Z. For all $Z, f(Z) < f(0_{+}) = 2$ and $f(Z)$ tends to 2, as Z tends to 0.

We know that for an ideal PPM channel, when the bandwidth tends to infinity, the channel capacity tends to infinity, which is obviously contradictory to $C^* < 2$, because α decreases, L^* increases, Z^* tends to 0, and the energy of each pulse tends to 0.

Hence, $C_{ph}^* = \frac{C^*}{\lambda_s^* \tau}$, not because C^* and C_{ph}^* tend to infinity, but because $\lambda_s^* \tau$ tends to 0 and C_{ph}^* tends to infinity. The finiteness of C^* is similar to the Shannon capacity limit of the additive white Gaussian-noise channel (AWGN) [15]. For AWGN channels, in the case of a reliable transmission (e.g., zero probability error), the ratio of energy E to the unilateral noise power-spectral density N_0 is limited by $E/N_0 < 1$.

2.3 Structure and Working Principle of a PPM Modulation and Demodulation System

2.3.1 Structure and Composition of a PPM System

As shown in Fig. 2.4, in the transmitter and receiver, the sampling, processing, synchronization, decision, and decoding of the received signal, are all performed by the sampling digital signal processor (DSP) and related logic in the transmitter and receiver. With the emergence of high-speed digital signal processors and flexible and convenient field-programmable logic-gate arrays (FPGAs), people have been able to implement high-performance all-digital PPM transmitters and receivers with

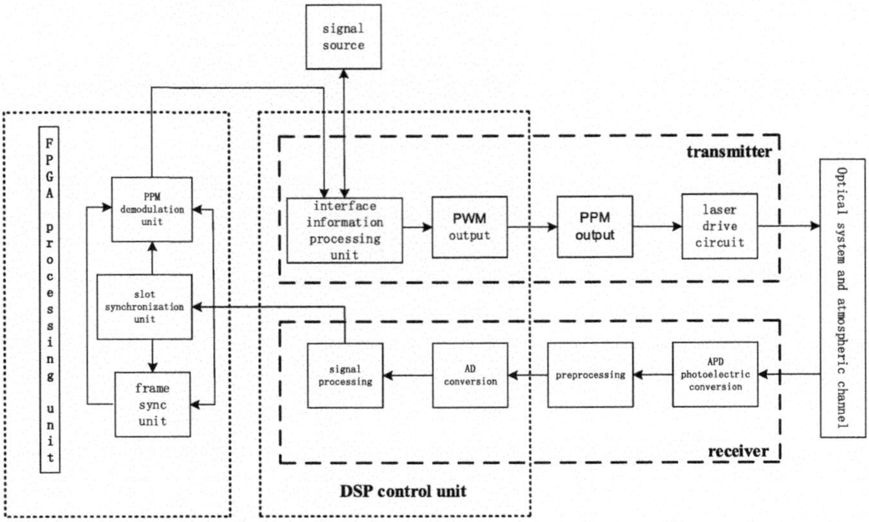

Fig. 2.4 PPM modulation and demodulation system

software on a hardware platform built using general-purpose DSP processors and FPGAs.

As shown in Fig. 2.4, the PPM system consists of a DSP control unit, FPGA processing unit, and peripheral processing unit. The DSP handles the pulse-width modulation (PWM) coding of binary signals, system control, communication with the host, signal processing, and digitization of analog signals. The FPGA extracts the time-slot and frame synchronization signals and decodes the PPM signal. The peripheral processing unit includes a PPM output unit, laser-modulation drive circuit, optical system, APD photoelectric converter, and preprocessing unit.

2.3.2 *PPM Modulation Principle*

The functions of the PPM system are shown in Fig. 2.4, and are described as follows.

(1) How the transmitter works

The information sent by the signal source is processed by the interface information-processing unit into a binary data stream. According to the requirements of the PPM modulation system, L-bit binary information is encoded into a PWM pulse, and the modulated PWM pulse sequence is sent to the PPM output unit. It triggers the generation of a narrow PPM pulse sequence. The narrow PPM pulse modulates the laser through its driving circuit, and the signal emitted by the laser is sent to an atmospheric channel through the optical system.

(2) How the receiver works

After the optical system captures and receives the optical pulse signal from the atmospheric channel, the APD photodetector converts the optical signal into an electrical signal. Because of interference from atmospheric-channel and circuit noise, the electrical signal is attenuated; therefore, preprocessing is required before demodulation. Preprocessing functions include signal amplification, signal filtering, and signal pulse-width processing. The processed signal is sent to the AD conversion module of the DSP, which converts the analog signal into a digital signal and sends it to the FPGA through three channels.

One channel is used to extract and recover the time-slot clock synchronization signal, the other is used to recover and extract the frame-synchronization clock signal, and the third is sent to the demodulation module. The demodulation of the PPM is based on the establishment of time-slot and frame synchronization. The information after PPM demodulation is sent to the interface information-processing unit of the DSP and then sent to the receiver.

(3) Modulator implementation principle

A block diagram of the modulator hardware is shown in Fig. 2.5. The system has four interfaces for exchanging information with the outside world: an asynchronous serial interface (UART), universal serial-bus interface (USB), telephone audio interface (PCM), and fiber-optic interface.

An asynchronous serial interface acts as the communication interface between the computer and DSP. The modulation system we designed adopts a hexadecimal PPM code; therefore, the DSP needs to code every four bits of the computer's binary information into a PPM pulse. Unfortunately, it is very complicated to directly generate PPM pulses from binary information; therefore, the binary information is first transformed during modulation; that is, the binary information is compiled into PWM waves [28], and then, PPM waves are generated from the PWM waves through hardware circuits. The conversion of the PWM to PPM waves is shown in Fig. 2.6. Therefore, the design of the PPM modulator is converted into the design of a PWM modulator, and the DSP chip is convenient for PWM modulation.

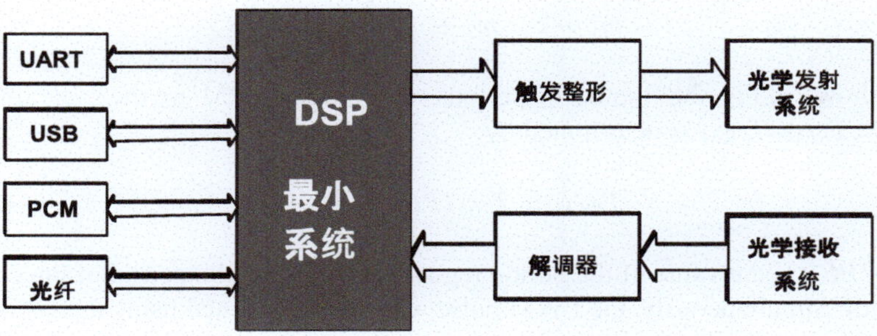

Fig. 2.5 Block diagram of the system-hardware composition

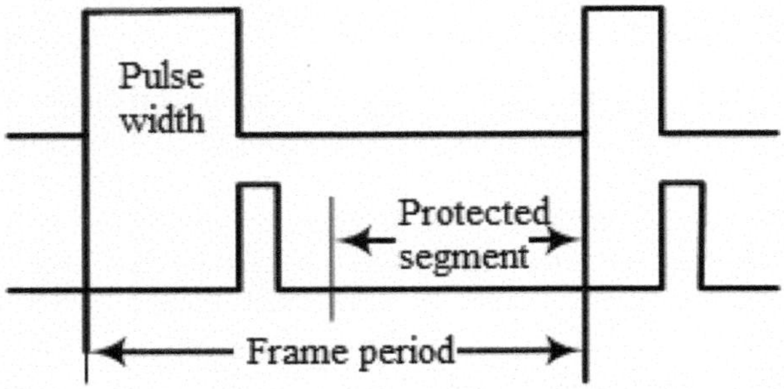

Fig. 2.6 Schematic diagram of PWM to PPM conversion

To generate a PWM wave, two key values need to be determined: one is the period of the PWM wave, and the other is the pulse width of the PWM wave. The cycle and pulse width of the PWM wave are determined by the data sent by the upper computer, and are also related to the number of PPM modulations.

We denote the n-bit data as $M = (m_1, m_2, ...m_n)$, and set the base number of the PPM modulation system to be L; that is, in a PPM frame period, the information segment has a total of L time slots. The time of each time slot is denoted as τ, and the guard time-slot time is denoted as T_D. The frame period of the system is

$$T_{ppm} = L \times \tau + T_D. \tag{2.31}$$

The frame rate of the system is

$$f_R = (L \times \tau + T_D)^{-1}. \tag{2.32}$$

The transmission rate of the system is

$$R = (\log_2 L)/(L \times \tau + T_D). \tag{2.33}$$

We denote the time-slot position corresponding to each group of data as l; then, the mapping relationship between l and the data is

$$l = m_1 + 2 \times m_2 + \cdots + 2^{n-1} * m_n. \tag{2.34}$$

As shown in Fig. 2.8, the frame periods of PWM and PPM are the same; that is, $T_{PWM} = T_{PPM}$. T_{PWM} is determined by

$$T_{PWM} = 2 \times TPER \times T_{OSC}, \tag{2.35}$$

where $TPER$ is the value of the period register and T_{OSC} is the period of the crystal oscillator. Simultaneously, the PWM pulse width is determined using the following formula:

$$T_{ON} = 2 \times (TPER - CMPR). \quad (2.36)$$

Here, *CMPR* is the value of the comparison register. *TPER* and *CMPR* are values that can be set by software; so as long as the value of T_{PPM} is known, the value of *TPER* can be inferred, and the value of *CMPR* is calculated according to the data to be encoded. From these two values, T_{ON} can be deduced. Thus far, the two key problems with generating PWM waves have been solved.

If the modulation and demodulation system is designed with a hexadecimal PPM, that is, L = 16, the repetition frequency of the laser is 1 MHz, the guard time slot can be set to $T_D = 100\mu s$, the clock frequency is 30 MHz, and $T_{OSC} = 33$ ns. In the following analysis, let a frame contain 16 guard slots and 16 communication slots, and each slot $T_s = 1/0.16\mu s$. From this, we can determine:

1. The slot frequency $f_s = 0.16MHz$;
2. The frame length $T_f = (16 + 16) \times T_s = 200\mu s$ and the frame rate is 5000 nat/s;
3. One frame represents one data point, which is represented by the position of the optical pulse (there is only one optical pulse per frame) in 16 communication time slots;
4. One frame (T_f) transmits a four-bit number, so the communication rate of this optical communication system is $\frac{1s}{T_f} \cdot 4bits = 0.02Mbits/s$.

The difficulty with the modulator design is generating the narrow PPM pulses and the precise delay of the pulses. We take advantage of the convenient and accurate characteristics of PWM waves to complete the conversion from PWM to PPM through hardware circuits, thereby solving the key problem of PPM modulation.

2.3.3 Maximum-Likelihood Demodulation of PPM Signals

In an atmospheric laser-communication system, the optical pulse emitted by the laser must be transmitted through the atmosphere, detected by the receiver, demodulated, and judged, before the transmitted information can be recovered. In the entire channel, atmospheric-channel and receiver noise will affect the transmitted signal light pulse [16, 17].

When studying the maximum-likelihood demodulation of a PPM signal, the mathematical model of noise is generally established according to the following two situations: one is to regard the atmospheric channel as a Poisson probability-distribution model independent of the signal [14], and assume that the receiver noise also follows the Poisson distribution; thus, the common influence of the two on the signal light pulse can be calculated according to the Poisson distribution model. In the other situation, with strong background-light interference, the noise superimposed on the signal is regarded as an additive Gaussian white-noise distribution [13–16]. We studied the maximum-likelihood demodulation of PPM signals for the two different noise models, mainly around the decision metric and the demodulation-error (correct) probability of the maximum-likelihood demodulation.

(1) Maximum-likelihood demodulation of PPM signals under a poisson noise-distribution model

When we study the maximum-likelihood demodulation of a PPM signal, we regard the atmospheric channel as a Poisson probability-distribution model that is independent of the signal [14], and assume that the receiver noise also obeys the Poisson distribution. Hence, the difference between the two on the signal optical pulse can be calculated according to the Poisson distribution model. The output current of the photodetector at the receiving end is

$$i_d(t) = \sum_{j=1}^{k(t)} h(t - t_j), \tag{2.37}$$

where $h(t)$ is the impulse response of the photodetector, t_j is the moment of photoelectron generation, and $k(t)$ is the number of photoelectrons generated in the interval $[0, t)$, using is a Poisson process in which the counting intensity is modulated by random transmission data.

In the digital PPM receiver, the output signal of the photodetector should be sampled first; the discretization process of $i_d(t)$ is shown in Fig. 2.7.

Mathematically, it can be expressed as

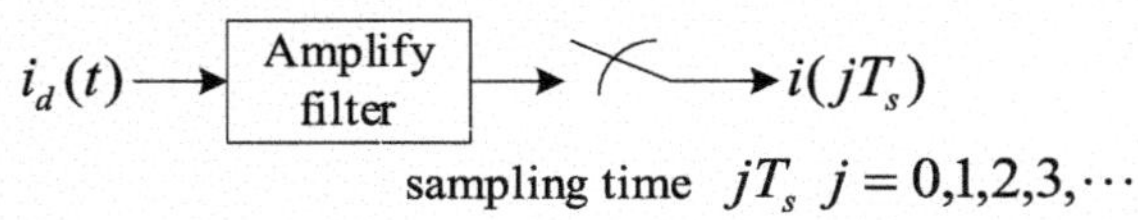

Fig. 2.7 Discretization of the output current of the photodetector

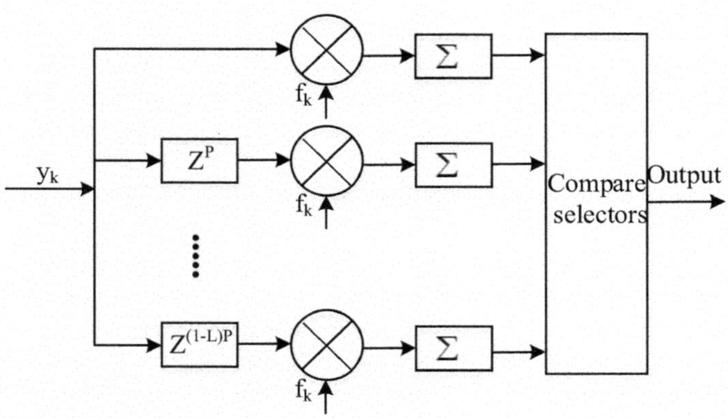

Fig. 2.8 L-PPM receiver block diagram with P samples per slot

$$i(jT_s) = \int_{jT_s-\tau}^{jT_s} Ai_d(t)dt = eAk(j), \tag{2.38}$$

where T_s is the sampling interval, τ is the response time of the amplification filter circuit (it has been set to be much larger than the response time of the photodetector), e is the electron charge, A is the circuit-amplification factor, and $k(j)$ is the photoelectron generated in the $[jT_s - \tau, jT_s)$ period. The number, subject to a Poisson distribution, is a sufficient statistic for the received signal. If the number of samples in one slot is P, the number of samples in one frame is PM (A signal frame consists of M time slots). The photoelectron-count data vector sampled by the receiver in one frame can be expressed as

$$K = [k(0), k(1), k(2), \ldots, k(PM - 1)]. \tag{2.39}$$

Each of the components $k(j)(j = 0, 1, 2, \cdots, PM - 1)$ follows a Poisson distribution and is independent of the others. The count intensity is $\lambda_s f(j-N)+\lambda_b$, where λ_s is the signal intensity; λ_b is the noise intensity (the influence of noise on the signal light pulse is calculated according to the Poisson distribution model); and $f(j - N)$ is the signal-pulse waveform normalized by the maximum value. N is the position of the first sample value of the signal light pulse, which is a random variable modulated by the source, and estimated at $\{0, P, 2P, 3P, \cdots, (M - 1)P\}$.

The conditional probability that the observed photoelectron-count vector is under the conditions of $N = n$ (that is, the optical pulse is known to be emitted in n time slots) can be expressed as

$$P_r(K|N = n) = \prod_{j=0}^{PM-1} \frac{\{[\lambda_s f(j-n) + \lambda_b]\tau\}^{k(j)}}{k(j)!} \exp\{-[\lambda_s f(j-n) + \lambda_b]\tau\} \tag{2.40}$$

$$= \prod_{j=0}^{PM-1} \frac{(\lambda_b\tau)^{k(j)}}{k(j)!} \exp\{-[\lambda_s f(j-n) + \lambda_b]\tau\} \prod_{j=0}^{PM-1} [1 + \frac{\lambda_s}{\lambda_b} f(j-n)]^{k(j)}$$

$$= c \prod_{j=0}^{PM-1} [1 + \frac{\lambda_s}{\lambda_b} f(j-n)]^{k(j)} \tag{2.41}$$

$$c = \prod_{j=0}^{PM-1} \frac{(\lambda_b\tau)^{k(j)}}{k(j)!} \exp\{-[\lambda_s f(j-n) + \lambda_b]\tau\}.$$

In Eq. (2.41), if there is no intersymbol interference, that is, $f(j - n)$ has a value only in the interval $[nP, (n + 1)P]$, then, c is a constant, independent of n, and the likelihood function of the photoelectron-count data vector K can be obtained as

$$L(n) = \prod_{j=0}^{PM-1} [1 + \frac{\lambda_s}{\lambda_b} f(j-n)]^{k(j)}. \tag{2.42}$$

Taking the logarithm of Eq. (2.42), letting $x = \frac{\lambda_s}{\lambda_b}$, and omitting the constant term, the log-likelihood function can be obtained as

$$\ln L(n) = \sum_{j=0}^{PM-1} k(j) \ln[1 + xf(j-n)]. \tag{2.43}$$

Maximum-likelihood demodulation is used to calculate the above log-likelihood function for M possible pulse positions n; hence, the pulse position where the log-likelihood function reaches a maximum is judged as the position of the emitted light pulse. The maximum-likelihood estimate $\hat{n}$ of the pulse position n can be expressed as:

$$\hat{n} = \arg\max \ln L(n), \tag{2.44}$$

$$n \in \{0, P, 2P, \cdots, (M-1)P\}.$$

For a rectangular pulse signal with a pulse width of LT_s,

$$f(j-n) = \begin{cases} 1, j = n, n+1, \ldots, n+L-1 \\ 0, \text{ otherwise.} \end{cases} \tag{2.45}$$

We substitute Eq. (2.45) into Eq. (2.43) to obtain

$$\ln L(n) = \ln(1+x) \sum_{j=n}^{n+L-1} k(j). \tag{2.46}$$

Because $\ln(1+x)$ is unrelated to n, the decision metric for n is as follows:

$$l(n) = \sum_{j=n}^{n+L-1} k(j). \tag{2.47}$$

Equation (2.47) accumulates the L count samples of the sending pulse of the n time slot and judges the time slot with the maximum accumulated count as the sending-pulse time slot; this is called the maximum accumulated-count judgment criterion. When sampling once per time slot, namely $P = 1$, Eq. (2.41) becomes $l(n) = k(n)$, and the decision criterion at this time is the common maximum count criterion.

Assuming that all Q symbols are transmitted equally, the average symbol error probability can be expressed as

$$P_e = \frac{1}{Q}\sum_{n=0}^{Q-1} P_e(n) \tag{2.48}$$

where $P_e(n)$ represents the probability of misjudging the remaining $Q - 1$ symbols when the transmitted symbol is n. The exact calculation of $P_e(n)$ is difficult and only its upper bound is calculated here:

$$P_e(n) \le P[\bigcup_{\substack{m=0\\m\ne n}}^{Q-1} \{l(n) \le l(m)\}|n]. \tag{2.49}$$

Equation (2.49) represents a pessimistic estimate of an incorrect decision when the two decision metrics are equal. Using the union bound of the probability, we can deduce that

$$P\left[\bigcup_{\substack{m=0\\m\ne n}}^{Q-1} l(n) \le l(m)|n\right] \le \sum_{\substack{m=0\\m\ne n}}^{Q-1} P[l(n) \le l(m)|n]. \tag{2.50}$$

We reuse the Bhattacharyya circle to obtain

$$P[l(n) \le l(m)|n] \le \sum_{K} [P_r(K|m)P_r(K|n)]^{\frac{1}{2}}. \tag{2.51}$$

From Eq. (2.51), we find

$$\begin{aligned}\sum_{K}[P_r(K|m)P_r(K|n)]^{\frac{1}{2}} &= \prod_{j=0}^{PQ-1}\sum_{k=0}^{\infty}\left\{[(\lambda_s f(j-m)+\lambda_b)\tau]^k \exp[-(\lambda_s f(j-m)+\lambda_b)\tau]/k!\right\}^{\frac{1}{2}}\\ &= \left\{[(\lambda_s f(j-n)+\lambda_b)\tau]^k \exp[-(\lambda_s f(j-n)+\lambda_b)\tau]/k!\right\}^{\frac{1}{2}}\\ &= \exp\left\{-\frac{\tau\lambda_b}{2}\sum_{j=0}^{PQ-1}\left[\sqrt{1+xf(j-n)}-\sqrt{1+xf(j-m)}\right]^2\right\}.\end{aligned} \tag{2.52}$$

Using Eq. (2.52), the upper bound of the error probability of various signal waveforms can be calculated under the condition of unsigned interference. For the orthogonal PPM symbol set ($L \le P$) without slot interference, we obtain

$$P_e \le (Q-1)\exp\left\{-\tau\lambda_b\sum_{j=0}^{L-1}\left[\sqrt{1+xf(j)}-1\right]^2\right\}. \tag{2.53}$$

For rectangular signals,

$$P_e \leq (Q-1)\exp\left[-\tau\lambda_b L\left(\sqrt{1+x}-1\right)^2\right]. \tag{2.54}$$

The error-probability bound under the condition of no background noise ($\lambda_b = 0$) is

$$P_e \leq (Q-1)\exp(-\tau\lambda_s L). \tag{2.55}$$

At this time, the channel is a symmetrical Q meta-erased channel, and its error probability is

$$P_e = \frac{Q-1}{Q}\exp(-\tau\lambda_s L). \tag{2.56}$$

(2) Maximum-likelihood demodulation of PPM signals under Gaussian white noise [18]

In the case of strong background-light interference, the noise superimposed on the signal can be regarded as an additive white Gaussian-noise distribution [13–16]. Similar to the maximum-likelihood demodulation of a PPM signal under the Poisson noise-distribution model, the output current of the photodetector is sampled first. Using T_{sa} as the sampling interval to sample P times per time slot, the M-element PPM signal is sampled PM times in one frame. The photoelectron-count observation vector can be expressed as

$$Y = [y_1, y_2, \ldots, y_{PM}], \tag{2.57}$$

where y_i is the ith sampled value. When the transmit pulse signal $f(t)$ is in the jth time slot,

$$\begin{cases} y_i = f_i + n_i, \; i = (j-1)P+1, \ldots, jP \\ y_i = n_i, \qquad \textit{otherwise}, \end{cases} \tag{2.58}$$

where $f_i = f\{[i-(j-1)P]T_{sa}\}$ is the sampling value of the optical pulse signal and n_i is the sampling value of the Gaussian white noise, which are independent Gaussian random variables. Let their mean value be 0 and their variance be σ^2. If $f(t)$ is a rectangular-pulse signal with a pulse width of WT_{sa}, it has the following form:

$$\begin{cases} f(mT_{sa}) = \text{non - zero constant}, \; m = 1, 2, \ldots, W \\ f(mT_{sa}) = 0, \qquad \text{otherwise}. \end{cases} \tag{2.59}$$

OPPM (overlapping PPM; we will discuss OPPM modulation in Chap. 7) is not considered, so $W \leq P$. Assuming that the system is precisely synchronized and there is no intersymbol interference, the probability-density function of observation vector Y, when pulse $f(t)$ is located in the jth slot, can be expressed as

$$
\begin{aligned}
\Lambda_j(y) = & \prod_{i=1, i\neq j}^{M} \prod_{k=1}^{P} \exp\left[-\frac{y_{(i-1)P+k}^2}{2\sigma^2}\right] \\
& \times \prod_{l=1}^{W} \exp\left\{-\frac{\left[y_{(j-1)P+l} - f(lT_{sa})\right]^2}{2\sigma^2}\right\} \times \prod_{l=W+1}^{P} \exp\left[-\frac{y_{(j-1)P+l}^2}{2\sigma^2}\right] \\
= & \frac{\prod_{i=1}^{M}\prod_{k=1}^{P} \exp\left[-\frac{y_{(i-1)P+k}^2}{2\sigma^2}\right]}{\prod_{k=1}^{P} \exp\left[-\frac{y_{(j-1)P+k}^2}{2\sigma^2}\right]} \times \prod_{l=1}^{W} \exp\left\{-\frac{\left[y_{(j-1)P+l} - f(lT_{sa})\right]^2}{2\sigma^2}\right\} \\
& \times \prod_{l=W+1}^{P} \exp\left[-\frac{y_{(j-1)P+l}^2}{2\sigma^2}\right] \\
= & \frac{\prod_{i=1}^{M}\prod_{k=1}^{P} \exp\left[-\frac{y_{(i-1)P+k}^2}{2\sigma^2}\right]}{\prod_{k=1}^{W} \exp\left[-\frac{y_{(j-1)P+k}^2}{2\sigma^2}\right] \times \prod_{k=W+1}^{P} \exp\left[-\frac{y_{(j-1)P+k}^2}{2\sigma^2}\right]} \\
& \times \prod_{l=1}^{W} \exp\left\{-\frac{\left[y_{(j-1)P+l} - f(lT_{sa})\right]^2}{2\sigma^2}\right\} \times \prod_{l=W+1}^{P} \exp\left[-\frac{y_{(j-1)P+l}^2}{2\sigma^2}\right] \\
= & \prod_{i=1}^{M}\prod_{k=1}^{P} \exp\left[-\frac{y_{(i-1)P+k}^2}{2\sigma^2}\right] \times \frac{\prod_{l=1}^{W} \exp\left[-\frac{\left[y_{(j-1)P+l} - f(lT_{sa})\right]^2}{2\sigma^2}\right]}{\prod_{l=1}^{W} \exp\left[-\frac{y_{(j-1)P+l}^2}{2\sigma^2}\right]}.
\end{aligned}
\tag{2.60}
$$

Because $\prod_{i=1}^{M}\prod_{k=1}^{P} \exp\left[-\frac{y_{(i-1)P+k}^2}{2\sigma^2}\right]$ is unrelated to the pulse $f(t)$ in the jth slot, it can be ignored when calculating the likelihood function. The likelihood function of the observation vector can be expressed as

$$
L(j) = \frac{\prod_{l=1}^{W} \exp\left[-\frac{\left[y_{(j-1)P+l} - f(lT_{sa})\right]^2}{2\sigma^2}\right]}{\prod_{l=1}^{W} \exp\left[-\frac{y_{(j-1)P+l}^2}{2\sigma^2}\right]}.
\tag{2.61}
$$

Taking the logarithm of Eq. (2.61), the log-likelihood function can be obtained as follows:

$$
\ln L(j) = \sum_{k=1}^{W} \frac{y_{(j-1)P+k}^2 - \left[y_{(j-1)P+k} - f_k\right]^2}{2\sigma^2}.
\tag{2.62}
$$

We simplify and remove the constant term to obtain

$$\ln L(j) = \sum_{k=1}^{W} y_{(j-1)P+k} f_k = \sum_{k=1}^{W} y\left[(j-1)P + k\right] T_{sa} f(kT_{sa}). \tag{2.63}$$

The decision metric can be defined as

$$\xi_j = \sum_{k=1}^{W} y_{(j-1)P+k} f_k. \tag{2.64}$$

If the optical pulse signal $f(t)$ is regarded as a rectangular pulse signal with a pulse width of WT_{sa}, the influence of f_k on the decision metric can be ignored, and the decision metric can be simplified as

$$\xi_j = \sum_{k=1}^{W} y_{(j-1)P+k}. \tag{2.65}$$

For white Gaussian noise, the maximum-likelihood demodulation method is similar to the Poisson noise-distribution model, in which the log-likelihood function is calculated for M possible pulse positions j, which maximizes the log-likelihood function. The pulse position is determined as the position at which the light pulse was emitted. The maximum-likelihood estimate $\hat{j}$ of the pulse position j can be expressed as

$$\hat{j} = \arg\max\left(\xi_j\right),\ j = 1, 2, \ldots, M. \tag{2.66}$$

We use $P_{f_j}\left(f_j\right)$ to represent the probability that the receiver makes a correct decision when the transmitted optical pulse is located in the jth time slot. (Here, the correct rate is used to measure the performance of the receiver because the derivation of the correct rate is more accurate than the demodulation error.) Then, the average correct-symbol rate of this receiver can be expressed as

$$P_c = \frac{1}{M} \sum_{j=1}^{M} P_{f_j}\left(f_j\right). \tag{2.67}$$

Because $P_{f_j}\left(f_j\right)$ is equal to $j = 1, 2, \ldots, M$, we obtain

$$\begin{aligned} P_c = P_{f_j}\left(f_j\right) &= P(\xi_j \geq \xi_1, \xi_j \geq \xi_2, \cdots, \xi_j \geq \xi_{j-1}, \xi_j \geq \xi_{j+1}, \ldots, \xi_j \geq \xi_M) \\ &= \left[P(\xi_j \geq \xi_1)\right]^{M-1}\ j \neq 1, \end{aligned} \tag{2.68}$$

where $\xi_j = \sum_{k=1}^{W} y_{(j-1)P+k} f_k$ and $\xi_1 = \sum_{k=1}^{W} y_k f_k$ are two random variables. The probability-density function of ξ_j is studied first.

$$\xi_j = \sum_{k=1}^{W} f_k y_k = \sum_{k=1}^{W} f_k (f_k + n_k) = \sum_{k=1}^{W} (f_k^2 + f_k n_k) = \sum_{k=1}^{W} f_k^2 + \sum_{k=1}^{W} f_k n_k$$

$$E(\xi_j) = E\left(\sum_{k=1}^{W} f_k^2 + \sum_{k=1}^{W} f_k n_k\right) = \sum_{k=1}^{W} f_k^2 + \sum_{k=1}^{W} f_k E(n_k) = \sum_{k=1}^{W} f_k^2$$

$$D(\xi_j) = D\left(\sum_{k=1}^{W} f_k^2 + \sum_{k=1}^{W} f_k n_k\right) = \sum_{k=1}^{W} f_k^2 D(n_k) = \sigma^2 \sum_{k=1}^{W} f_k^2$$

So, $\xi_j \sim N\left(\sum_{k=1}^{W} f_k^2, \sigma^2 \sum_{k=1}^{W} f_k^2\right)$

$$E(\xi_i) = 0, i \neq j$$

$$D(\xi_j) = \sum_{k=1}^{W} f_k^2 \sigma^2 = E\sigma^2$$

$$D(\xi_i) = E\sigma^2;$$

then,

$$P_c = \int_{-\infty}^{\infty} [P(\xi_i \prec z)]^{L-1} f_{\xi_j}(z) dz. \tag{2.69}$$

Among the above,

$$P(\xi_i \prec z) = \int_{-\infty}^{Z} \frac{1}{\sqrt{2\pi E\sigma^2}} \exp\left(-\frac{u^2}{2E\sigma^2}\right) du$$

$$f_{\xi_j}(z) = \frac{1}{\sqrt{2\pi E\sigma^2}} \exp\left(-\frac{(z-E)^2}{2E\sigma^2}\right).$$

2.3.4 Implementation of the Maximum-Likelihood Demodulation of PPM Signals

In the L time slots of an L-PPM signal, a signal pulse $f(t)$ is sent in one of the time slots. The laser does not send pulses in the other time slots. The output current of the photodetector at the receiving end is [19]

$$i_d(t) = \sum_{j=1}^{k(t)} h(t - t_j), \tag{2.70}$$

where $k(t)$ represents the number of photoelectrons generated in the interval [0,t], which is a Poisson process whose counting intensity is modulated by random transmission data, $h(t)$ is the impulse response of the photodetector, and t_j is the moment of photoelectron generation. The discretization of $i_d(t)$ is

$$i(jT_s) = \int_{jT_{sa}-\tau}^{jT_{sa}} Ai_d(t)dt = eAk(j). \tag{2.71}$$

Here, T_{sa} is the sampling interval, τ is the response time of the amplification filter circuit (it has been set to be much larger than the photodetector response time), e is the electronic charge, A is the circuit magnification, and $k(j)$ is the $[jT_{sa} - \tau, jT_{sa}]$ period. The number of generated photoelectrons follows the Poisson distribution, which is a sufficient statistic of the received signal. If the number of samples in a time slot is P (P $= T_s/T_{sa}$; the length of the time slot is T_s, and the sampling time is T_{sa}), then the number of samples in one frame is LP, and the number of photons sampled by the receiver in one frame is expressed as a data vector:

$$Y = \left[y_1, y_2, \cdots, y_{LP}\right]. \tag{2.72}$$

When the transmit pulse signal is in the jth time slot,

$$y_i = f_i + n_i\,,\; i = (j-1)P + 1, \cdots, jP$$

$$f_i = f((i-(j-1)P)T_{sa})$$

$$y_i = n_i \quad otherwise.$$

Among them, n_i is the sampling value of the white Gaussian-noise process; therefore, ($i = 1, 2, \ldots$, LP) are independent Gaussian random variables, with a mean value of 0 and variance σ^2. If signal $f(t)$ is a waveform of width W, it has the following form:

$$\begin{aligned} &f(jT_{sn}) \neq 0 \; j = 1, \ldots, W \\ &f(jT_{sa}) = 0 \quad otherwise. \end{aligned}$$

Because we do not consider O-PPM (overlapping PPM), W $\leq$ P. If the system is precisely synchronized and there is no intersymbol interference, then when the pulse $f(t)$ is located in the j th time slot, where $j = (1, 2, \ldots$, L), the probability density of the observation vector Y can be expressed as

$$\Lambda_j = \prod_{i=1, i\neq j}^{L} \prod_{k=1}^{P} \exp\left[-\frac{y_{(i-1)P+k}^2}{2\sigma^2}\right]$$

$$\times \prod_{l=1}^{W} \exp\left[-\frac{\left[y_{(j-1)P+l} - f(lT_{sa})\right]^2}{2\sigma^2}\right] \times \prod_{l=W+1}^{P} \exp\left[-\frac{y_{(j-1)P+l}^2}{2\sigma^2}\right]$$

$$= \prod_{i=1}^{L}\prod_{k=1}^{P} \exp\left[-\frac{y_{(j-1)P+k}^2}{2\sigma^2}\right] \times \frac{\prod_{l=1}^{W} \exp\left[-\frac{\left[y_{(j-1)P+l} - f(lT_{sa})\right]^2}{2\sigma^2}\right]}{\prod_{l=1}^{W} \exp\left[-\frac{y_{(j-1)P+l}^2}{2\sigma^2}\right]}. \tag{2.73}$$

Because the number of $\prod_{i=1}^{M}\prod_{k=1}^{P} \exp\left[-\frac{y_{(j-1)P+k}^2}{2\sigma^2}\right]$ time slots j is irrelevant, it can be ignored when calculating the likelihood function. Thus, the jth log-likelihood function can be obtained as

$$\ln(\Lambda_j) = \sum_{k=1}^{W} \frac{y_{(j-1)P+k}^2 - \left(y_{(j-1)P+k} - f_k\right)^2}{2\sigma^2}. \tag{2.74}$$

We simplify and remove the constant to obtain

$$\ln(\Lambda_j) = \sum_{k=1}^{W} y_{(j-1)P+k} f(kT_{sn}) = \sum_{k=1}^{W} y((j-1)P+k)T_{sn} f(kT_{sa}). \tag{2.75}$$

For the white Gaussian noise, the maximum-likelihood judgment of the PPM signal sampled P times per time slot is determined in this manner:

We calculate ln (Λ j) for $j = 1, 2, \ldots,$ L; then, the signal $f(t)$ located in the jth time slot is estimated to be

$$\hat{j} = \arg\left(\max\left(\ln \Lambda_j\right)\right) \quad (j \in \{1, 2, \cdots, L\}). \tag{2.76}$$

A structural block diagram corresponding to the maximum-likelihood demodulation of the L-PPM signal is shown in Fig. 2.8.

2.3.5 *Maximum-Likelihood Demodulation Bit-Error-Rate Analysis*

When the transmitted pulse signal is in the jth time slot, $y_k = f_k + n_k$ is in the jth time slot. If $y_k = n_k$ in the other time slots, $\xi_j = \sum_{k=1}^{W} f_k y_k$ is the value when the signal is in the jth time slot. If $\xi_j \geq \xi_i$ ($i = 1, 2, \ldots,$ L, $i \neq j$), the pulse is judged to be located in the jth time slot [20]. Let $p_{f_j}(f_j)$ represent the probability of a correct decision when the signal is located in the jth time slot; then, the average correct symbol rate

$p_c = \frac{1}{L}\sum_{j=1}^{L} p_{f_j}(f_j)$ of this receiver, because $p_{f_j}(f_j)$ is equal to $j = 1, 2, \ldots$, L; therefore,

$$E(\xi_j) = \sum_{k=1}^{W} f_k^2 = E$$

$$\begin{aligned} p_c &= p_{f_j}(f_j) = p(\xi_j \geq \xi_1, \xi_j \geq \xi_2, \cdots, \xi_j \geq \xi_{j-1}, \xi_j \geq \xi_{j+1}, \cdots \xi_j \geq \xi_m) \\ &= \left[p(\xi_j \geq \xi_1)\right]^{L-1} j \neq 1 \end{aligned} \tag{2.77}$$

$$\begin{aligned} E(\xi_i) &= 0, i \neq j \\ D(\xi_j) &= \sum_{k=1}^{W} f_k^2 \sigma^2 = E\sigma^2 \\ D(\xi_i) &= E\sigma^2. \end{aligned}$$

Then,

$$p_c = \int_{-\infty}^{\infty} \left[p(\xi_i \prec z)\right]^{L-1} f_{\xi_j}(z) dz. \tag{2.78}$$

Among them,

$$\begin{aligned} p(\xi_i \prec z) &= \int_{-\infty}^{Z} \frac{1}{\sqrt{2\pi E\sigma^2}} \exp\left(-\frac{u^2}{2E\sigma^2}\right) du, \\ f_{\xi_j}(z) &= \frac{1}{\sqrt{2\pi E\sigma^2}} \exp\left(-\frac{(z-E)^2}{2E\sigma^2}\right). \end{aligned}$$

2.3.6 Demodulator Structure and Realization Principle

The demodulation system used the Verilog hardware-description language to design each module. A block diagram of the demodulator implementation is shown in Fig. 2.9. It is mainly composed of a likelihood operator, data latch, data comparator, counter, and register [34]. PPM_IN represents the data sampled by the AD module in the DSP and sent to the likelihood operator of the FPGA through the data line of the DSP. The clock output by the synchronization subsystem has a slot clock and frame clock, which are sent to the demodulation system to control the counter, comparator, and register. The working principle of each module is described in detail below [21–25].

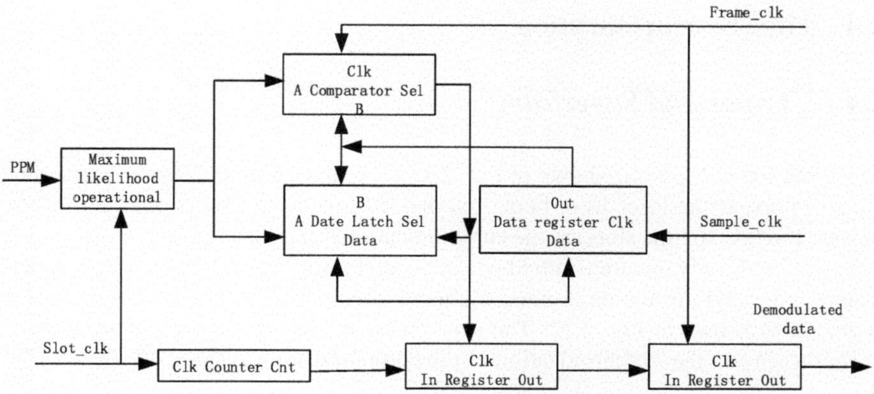

Fig. 2.9 Block diagram of the demodulator FPGA implementation

1. The likelihood operator receives the data sampled by the AD, calculates the maximum-likelihood value of each time slot, and sends the maximum-likelihood value to the data latch and comparator at the time-slot frequency.
2. The port equation of data latch U1 is SEL = (A > B) & ($/T_{frame}$); that is, when A > B and $T_{frame} = 0$, SEL = 1, which allows the PPM_IN output of the AD converter to be loaded into the data latch. Therefore, the value of the data latch is always the maximum-likelihood operation value in a PPM frame and the state of the data register is updated when each PPM frame arrives.
3. The clock of register U4 is the output of comparator A > B. Only when the SLOT_CLK pulse arrives in the state of A > B is the state of the data register updated; that is, at the position of the largest time slot in each PPM frame.
4. The clock of register U5 is the frame-section clock, which outputs the position of the pulse output by U4 in the frame when the frame clock arrives.

The experimental results for the demodulator are presented in Fig. 2.10. It can be seen [26] that the circuit can satisfy the predetermined functional requirements.

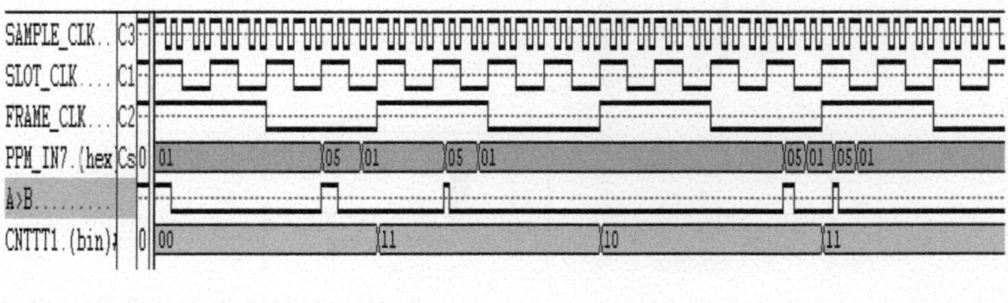

Fig. 2.10 Demodulator test results [26–28]

2.4 Time Synchronization

2.4.1 Frame and Superframe

The PPM frame format is shown in Fig. 2.11. The first 16 time slots are guard time slots to prevent the laser from being overloaded, owing to excessive transmission power. The last 16 time slots are the communication time slots. Each communication time slot holds only one transmitted pulse. Several information frames form a superframe, and each information frame contains a protection segment and an information segment, as shown in Fig. 2.12. The superframe is preceded by a synchronization header to ensure the synchronization of the communication system.

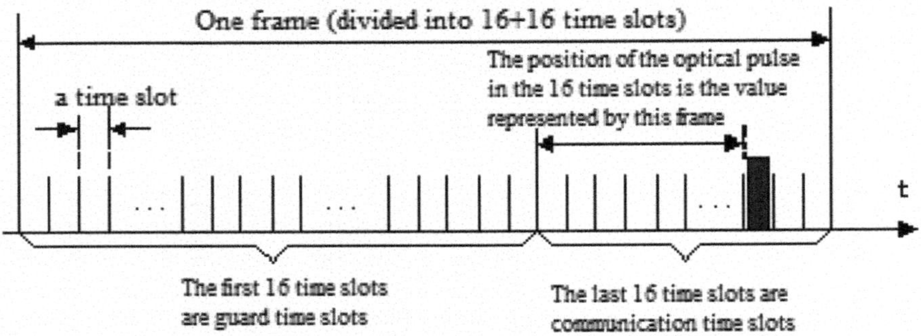

Fig. 2.11 PPM frame format [27, 28]

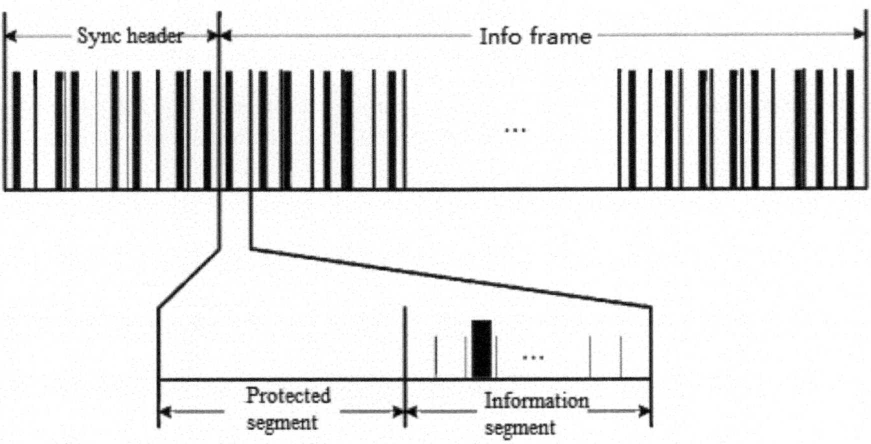

Fig. 2.12 PPM superframe structure [27, 28]

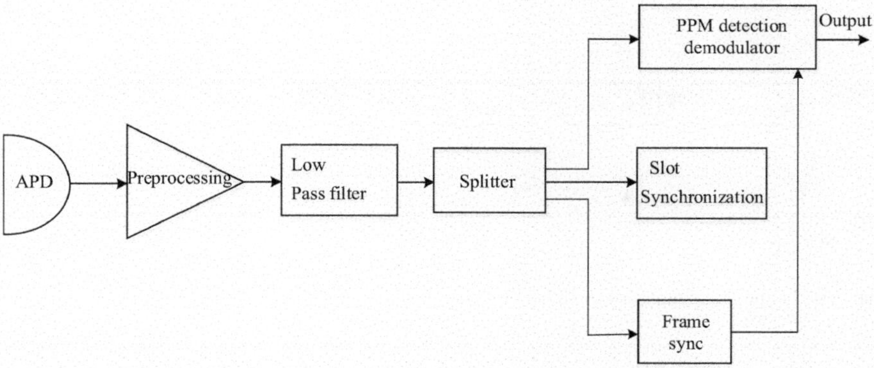

Fig. 2.13 Receiver composition block diagram [27, 28]

2.4.2 Frame Synchronization

To ensure that the communication system can correctly demodulate the information in the PPM frame, the information-detection part of the receiver and the superframe must be strictly synchronized in time. Figure 2.13 shows a schematic diagram of the receiver. The output of the avalanche photodetector (APD) is a series of unipolar narrow-pulse signals, which correspond to light pulses output by the laser.

Because the optical pulse signal is doped with many noise signals after passing through the atmospheric channel, and the signal is also attenuated significantly, the signal output by the APD is first processed by a preprocessor, and then a low-pass filter is used to filter out high-frequency signals. The splitter sends the filtered signal to the demodulation-detection unit, time-slot–synchronization unit, and frame-synchronization unit.

2.4.3 Slot Synchronization [21–25]

The guard time slot of the PPM signal designed by the modulation and demodulation system is half of the frame period. Regardless of the time slot that the PPM pulse is in, the position of the leading edge of the pulse in the time slot remains unchanged. A digital phase-locked loop (DPLL) provides the conditions for extracting the synchronization signal. The time-slot synchronization subsystem we designed receives the time-slot synchronization signal extracted by the DPLL. A block diagram of the time-slot synchronization subsystem is shown in Fig. 2.14.

As shown in Fig. 2.16, the signal output by the splitter is first sent to the amplitude comparator to remove low-amplitude noise. The width comparator can remove noise signals from the narrow pulse and the pulse stretcher adjusts the duty cycle of the pulse signal to 50% of the output-level signal. Subsequently, after A/D sampling, the analog

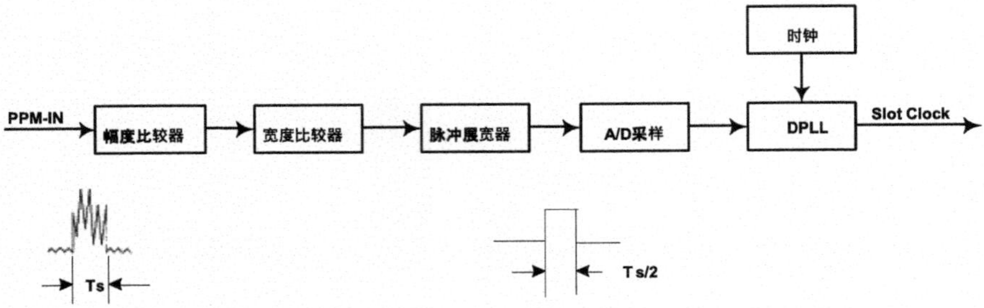

Fig. 2.14 Time-slot synchronizer [27, 28]

signal is converted into a digital signal, which is sent to the digital phase-locked loop to extract the synchronous time-slot clock.

The time-slot synchronization subsystem extracts the time-slot clock signal using the phase-locked loop. The principle of the digital phase-locked loop is shown in Fig. 2.15, and the DPLL was designed using the Verilog hardware description language. An exclusive-or (XOR)-gate phase detector was used in the design. The XOR-gate phase detector compares the phase difference $\Phi_e = \Phi_{in} - \Phi_{out}$ between the phase of the input signal F_{in} and the phase of the output signal F_{out}. It outputs the error signal S_e as the counting-direction signal of the K variable-mode up–and-down counter. When the loop is locked, S_e is a square wave with a duty ratio of 50%, and the absolute phase difference is 90°. Therefore, the phase-difference limit of the phase detector of the XOR gate is ± 90°.

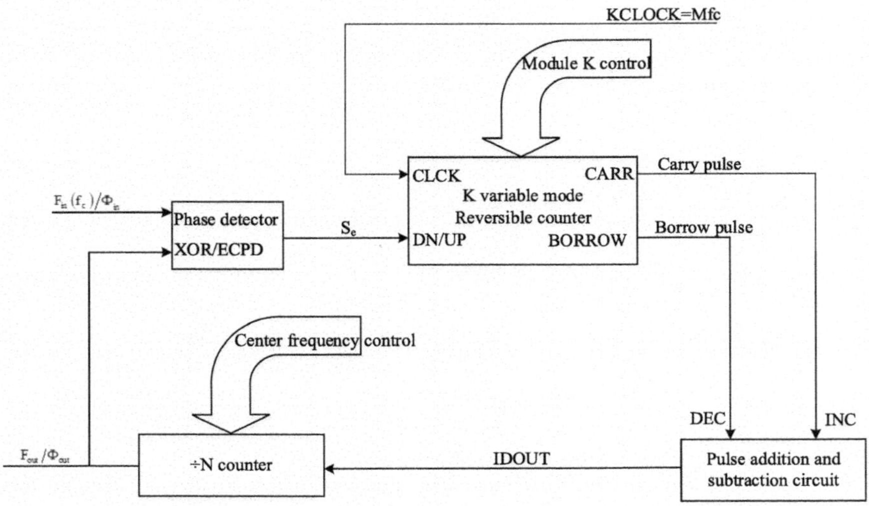

Fig. 2.15 Structure diagram of the digital phase-locked loop [27, 28]

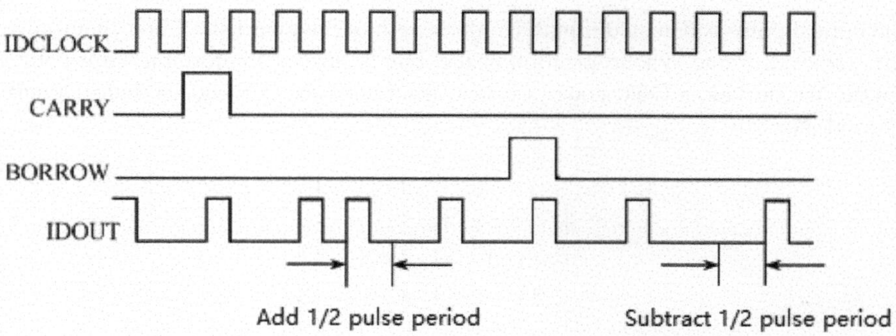

Fig. 2.16 Working waveform of the pulse addition-and-subtraction circuit [27, 28]

The K variable-mode up-and-down counter performs addition and subtraction operations according to the phase-difference signal S_e. When S_e is low, the counter performs an addition operation. If the addition result reaches the preset modulus value, a carry pulse signal CARRY is output to the pulse addition-and-subtraction circuit. When S_e is high, the counter performs a subtraction operation. If the result is zero, then a borrow pulse signal BORROW is output to the pulse addition-and-subtraction circuit.

The Verilog design code for the K variable-mode reversible counter can be found in [27, 28].

The pulse addition-and-subtraction circuit tracks and adjusts the frequency and phase of the input signal, and finally locks the output signal to the frequency and signal of the input signal. The working waveform is shown in Fig. 2.16.

The pulse addition-and-subtraction circuit completes the frequency and phase adjustment of the loop, which is called a numerically controlled oscillator. When there is no carry/borrow pulse signal, the external reference clock is divided by two. When there is a CARRY pulse signal, half of the pulse is inserted into the output divide-by-two signal to increase the frequency of the output signal. When there is a BORROW pulse signal, half of the pulse is subtracted from the output two-frequency signal to reduce the frequency of the output signal. The Verilog design code can be found in [27, 28].

2.4.4 Frame-Synchronization Code Detection [27, 28]

During the demodulation process, the receiver must calculate the position of the optical pulse in the 16 communication time slots and use it as the data transmitted in one frame [27, 28]. To obtain the position of the optical pulse in the communication time slot, we must first determine the starting position of the frame. For example, if we increase a counter by one for each time slot and reset it to 0 at the beginning of the frame, the data transmitted in this frame is equal to the counter value when the optical pulse is detected minus 16 (the length of the guard slot). To synchronize the frames, a frame-synchronization header can be inserted before the information frame. We provide a Gold code (a type of pseudo-noise (PN) code) with a length (period) N = 54:

$$\mathrm{pn} = \begin{bmatrix} 1 & -1 & 1 & 1 & 1 & 1 \\ 1 & -1 & 1 & 1 & -1 & 1 \\ -1 & 1 & -1 & 1 & 1 & 1 \\ -1 & -1 & -1 & 1 & -1 & 1 \\ 1 & -1 & 1 & -1 & 1 & 1 \\ 1 & -1 & 1 & 1 & 1 & 1 \\ 1 & -1 & -1 & -1 & 1 & -1 \\ -1 & -1 & -1 & 1 & -1 & 1 \\ -1 & 1 & 1 & -1 & -1 & -1 \end{bmatrix}.$$

It exhibits good autocorrelation and cross-correlation properties. Here, we make another sequence equal to

$$y = \left[\mathrm{sign}(\mathrm{randn}(1,\ 1000))\mathrm{PN}\ \mathrm{sign}(\mathrm{randn}(1,\ 1000))\right].$$

That is, sequence y contains a segment of the PN code, and the 1000 points before and after the PN code are random numbers. When we designed the "frame-synchronization header," we used the cross-correlation characteristics of the 54-bit Gold code given above.

Without frame synchronization, a receiver cannot determine the exact position of the light pulse in the frame because the starting position of the frame is unknown. The only information available to the receiver is the number of time slots between the two optical pulses; therefore, the frame-synchronization code should be associated with this number of time slots.

As shown in Fig. 2.17, the receiving end calculates the number of time slots between two optical pulses, represented by C_1, C_2, ... Note that C_i includes the guard time-slot period. If the guard time slot in C_i is subtracted, then the number of communication timeslots S_i between the two optical pulses is obtained. Here, we set the guard timeslot length as 16. Then,

$$\begin{aligned} S_1 &= C_1 - \text{Number of guard slots} = C_1 - 16 \\ S_2 &= C_2 - \text{Number of guard slots} = C_2 - 16 \\ &\ldots \\ S_i &= C_i - \text{Number of guard slots} = C_i - 16. \end{aligned}$$

If the length of the communication time slot is M, the value range of S_i is $1 \le S_i \le 3\,M - 1$. In the modulation and demodulation system, the length of the

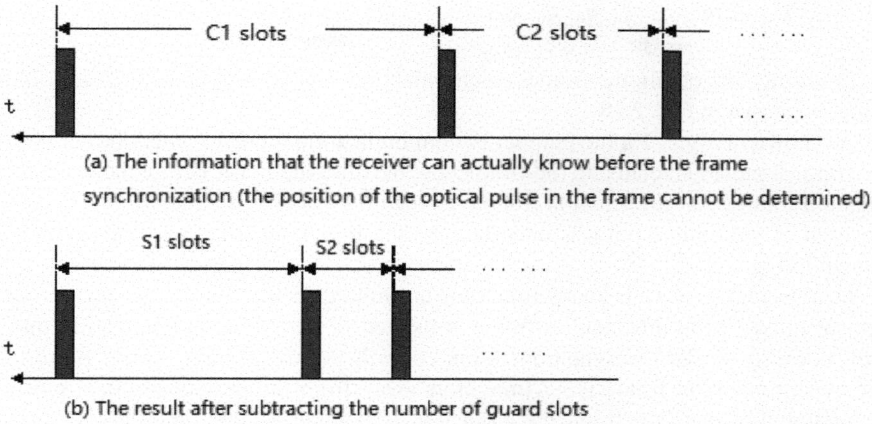

Fig. 2.17 Schematic diagram of the information actually obtained by the receiver [27, 28]

communication time slot is 16, and $1 \le S_i \le 47$. It can be observed that S_i can use a six-bit binary-number representation ($47 = 10\ 1111$ B).

Depending on the number of time slots between optical pulses, one frame can obtain six-bit data S_i. The length of the Gold code used for frame synchronization is 54, so the frame-synchronization Gold code can only be represented by $54 \div 6 = 9$ frames. We replace the "− 1" in the frame-synchronization Gold code designed earlier with "0" to obtain

$$\mathrm{x}' = \begin{bmatrix} 1\,0\,1\,1\,1\,1 \\ 1\,0\,1\,1\,0\,1 \\ 0\,1\,0\,1\,1\,1 \\ 0\,0\,0\,1\,0\,1 \\ 1\,0\,1\,0\,1\,1 \\ 1\,0\,1\,1\,1\,1 \\ 1\,0\,0\,0\,1\,0 \\ 0\,0\,0\,1\,0\,1 \\ 0\,1\,1\,0\,0\,0 \end{bmatrix}.$$

The total length of x′ is still 54; we divide it into nine lines with each line containing six numbers; one line represents a six-bit binary number. Then,

$$x' = \left[47\ 45\ 23\ 5\ 43\ 47\ 34\ 5\ 24\right],$$

where $47 = 101{,}111$ B (first row of x′), $45 = 101{,}101$ B (second row of x′), and so forth. The nine-element frame-synchronization code corresponds to the Gold code as follows:

$$[S_1, S_2, S_3, S_4, S_5, S_6, S_7, S_8, S_9] = x' = [47, 45, 23, 5, 43, 47, 34, 5, 24].$$

From this, we obtain the frame-synchronization header designed using the Gold code, as shown in Fig. 2.18.

As shown in Fig. 2.17, the number of communication time-slot intervals between two optical pulses is related to the Gold code. Because the frame before the synchronization frame is an information frame, the data it transmits are random; that is, the position of the optical pulse is uncertain, so the following situation may occur, as shown in Fig. 2.18.

In a special case (the guard time slot is omitted in the figure), when the last optical pulse of the information frame is located in the 16th communication time slot, because we set the synchronization code $S_1 = 47$, the first optical pulse of the synchronization frame is 47 time slots away from the previous optical pulse, exceeding it by one. The frame has a range of 16 communication time slots, so there is no optical pulse in the first synchronization frame. This is inconsistent with the requirements of "only one optical pulse per frame" required by PPM modulation, and that it should only be represented by nine frames. The frame-synchronization header now requires more than nine frames to be expressed.

We set the position of the optical pulse in the positioning frame as the third time slot of the 16 communication time slots, which ensures that there is only one optical pulse in the next nine synchronization frames. In addition, the optical pulse in the synchronization frame has a uniquely determined location. As shown in Fig. 2.19, this problem is caused by the uncertainty of the position of the optical pulse in the last information frame.

We set the position of the optical pulse in the positioning frame as the third time slot of the 16 communication time slots, which ensures that there is only one optical pulse in the next nine synchronization frames. In addition, the optical pulse in the synchronization frame has a uniquely determined location. In summary, the structure of our "frame-synchronization header" is shown in Fig. 2.20, and the position of the optical pulse in each frame is shown in Fig. 2.21.

The receiving end can detect the number of time slots C_i between two adjacently received light pulses and then calculate $S_i = C_i - 16$. After obtaining S_i, the receiver

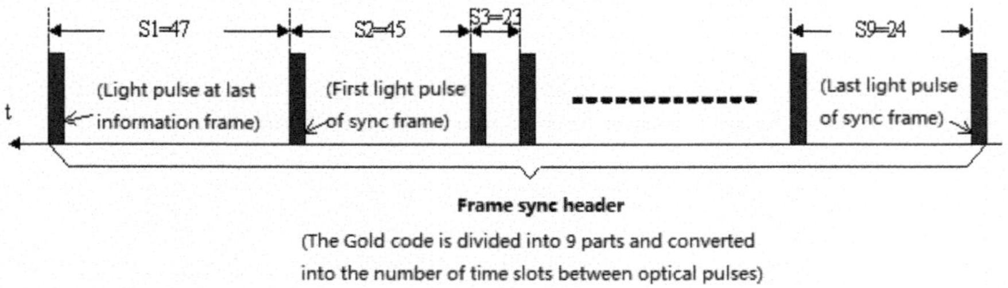

Fig. 2.18 Frame-synchronization header [27, 28]

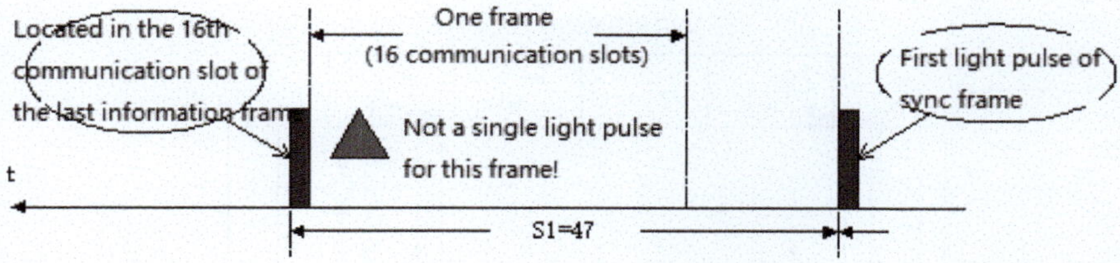

Fig. 2.19 Problem caused by the uncertainty of the position of the optical pulse in the last information frame [27, 28]

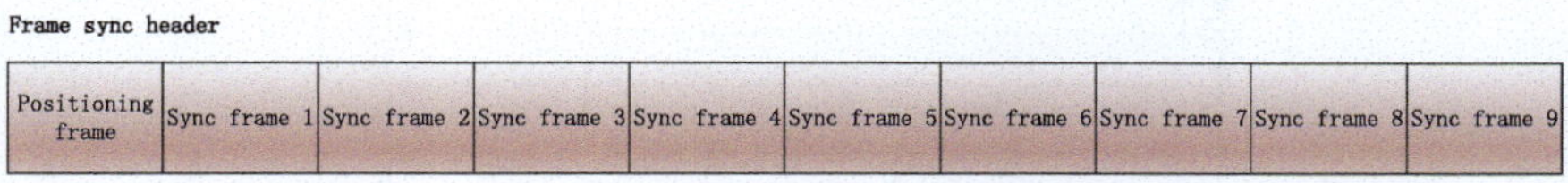

Fig. 2.20 Frame-synchronization header structure [27, 28]

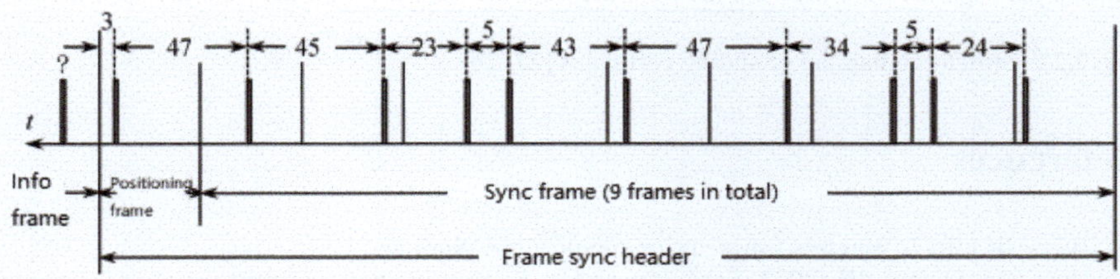

Fig. 2.21 Position of the optical pulse in the communication time slot in each frame of the frame-synchronization header [27, 28]

uses the "sliding correlation method" to detect the frame-synchronization header in the superframe. The structure of the correlator is illustrated in Fig. 2.22.

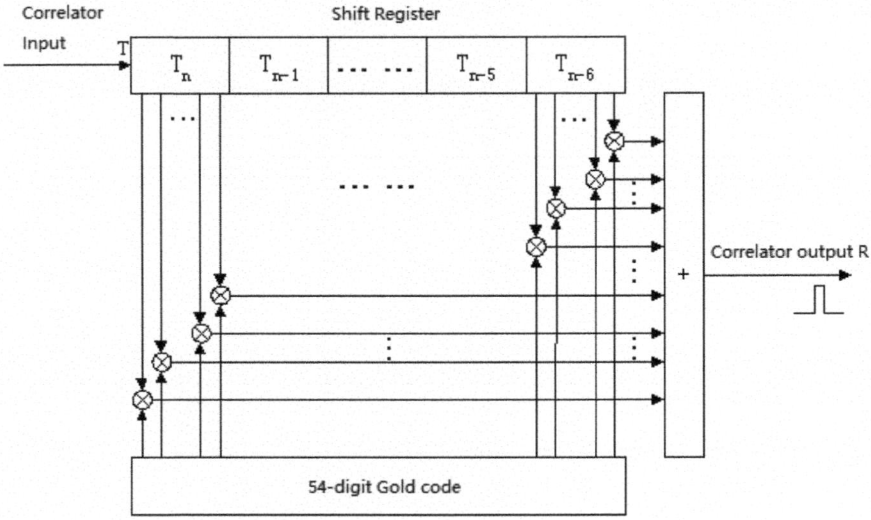

Fig. 2.22 Structural diagram of sliding correlator [27, 28]

References

1. Barry JR (1997) Wireless Infrared Communications. Springer, US
2. Haas SM, Shapiro JH (2002) Space-time codes for optical wireless communications. Eurasip J Adv Signal Process 2002(3):211–220
3. Liang Z, Daosu Y (1998) Basic analysis of code modulation of multi-pulse PPK. J Univ Electron Sci Technol China 27(5):51–521
4. Karp S, Gagliardi RM (1970) The design of a pulse-position modulated optical communication system. IEEE Trans Commun Technol 17(6):670–676
5. Otte R, Jong LD (2002) Wireless optical PPM telemetry and the influence of lighting flicker. IEEE Trans Instrum Meas 47(1):51–55
6. Leeson MS (2004) Pulse position modulation for spectrum-sliced transmission. IEEE Photonics Technol Lett 16(4):1191–1193
7. Chen CC, Gardner C (1986) Performance of PLL synchronized optical PPM communication systems. IEEE Trans Commun 34(10):988–994
8. Davidson FM, Sun X (1989) Slot clock recovery in optical PPM communication systems with avalanche photodiode photodetectors. IEEE Trans Commun COM-37(11):1164–1172
9. Fan C, Zhan D, Xu B et al (1995) Principles of communication Fourth Edition. National Defense Industry Press, 10
10. Elmirghani J, Cryan RA, Clayton FM (1995) On the spectral estimation and synchronisation of the Cyclostationary optical fibre PPM process. IEEE Trans Commun 43(234):1001–1012
11. Chuanyun Z (2000) Transmission efficiency of optical PPM. J Guilin Insti Electron Technol 20(4):53–55
12. Iwaki H (2002) Information transmission and coding theory. Science Press, pp 51–52
13. Zuo Y, Tang D (2000) Research on the amount of information in optical communication. Opt Commun Technol 24(2):160–162
14. Lesh J (2003) Capacity limit of the noiseless, energy-efficient optical PPM channel. IEEE Trans Commun 31(4):546–548
15. Massey J (2003) Capacity, cutoff rate, and coding for a direct-detection optical channel. IEEE Trans Commun 29(11):1615–1621

16. Liu H (2002) TMS320LF240xDSP structure, principle and application. Published by Beihang University Society 29(04):31–33+42
17. Du A, Ke X (2006) Study on the influence of atmospheric channel on laser PPM signal. Laser J 2006(01):73–74
18. Pu D, Zou C (2002) Maximum likelihood detection of sampled PPM signal under white Gaussian noise. J Guilin Inst Electron Technol 22(3):5–8
19. Zou C, Huang X, Ao F, Zhang D (1999) Maximum likelihood demodulation of optical PPM sampling signal. J Univ Electron Sci Technol China 28(3):247–250
20. Otte R, de Jong LP, van Roermund AHM (1996) A novel PPM demodulator for wireless optical transmission systems. In: Proceedings of the Electronics'96, Sozopol, Bulgaria
21. Ling G, Gagliardi R (1986) Slot synchronization in optical PPM communications. IEEE Trans Commun 34(12):1202–1208
22. Otte R, De Jong LP, Van Roermund AHM (1998) Slot synchronization by reducing the PPM pulse width in wireless optical systems. IEEE Trans Circ Syst II Analog & Dig Signal Process 45(7):901–903
23. Wang F, Huang B, He D (1995) Laser underwater PPM communication time slot synchronizer. J Huazhong Univ Sci Technol 23(10): 52–56
24. Zhang J, Zheng J, Wan X (1994) Phase locking technology. Xi'an University of Electronic Science and Technology Press, 6
25. Vilnrotter VA, Rodemich ER, Tan HH (2009) A synchronization technique for optical PPM signals. In: Military communications conference-communications-computers: teamed for the 90s. IEEE, 2009
26. Ding D, Ke X (2005) Design and simulation of PPM modulation and demodulation system for atmospheric laser communication. Opt Commun Technol 29(1):50–52
27. Ke X, Zhao L, Ding D (2007) A realization of time slot synchronization and frame synchronization in atmospheric laser communication. Semicond Optoelectron, 5
28. Ding D (2005) Design of PPM modulation and demodulation system for atmospheric laser communication. Xi'an University of Technology

Chapter 3
MPPM Coded Modulation Signal Analysis

This chapter first describes the basic principles of multipulse position modulation (MPPM), defines its performance parameters, such as transmission capability, transmission efficiency, and bandwidth utilization, and compares the performance of MPPM and pulse-position modulation (PPM) in terms of bandwidth and power utilization. Subsequently, an expression of optical MPPM channel capacity in the protected period is derived, and the channel capacity, information transmission rate, and channel-capacity energy efficiency of MPPM and PPM are compared.

3.1 MPPM Fundamentals

MPPM is a modulation mode that maps a continuous binary bitstream of length L into the MPPM symbol of a light pulse appearing simultaneously in p time slots in an information frame composed of m time slots, denoted as (m, p) MPPM [1–3], and $\binom{m}{p} \geq 2^L$. . For three-pulse MPPM, denoted as (m, 3) MPPM, $P = 3$, and its frame structure is shown in Fig. 3.1.

Compared to on–off keying (OOK) modulation, PPM, differential PPM (DPPM), and MPPM have higher transmission rates and optical power utilization [4, 5]. Compared with PPM and DPPM, MPPM can improve both bandwidth efficiency and power efficiency [6]. Therefore, a further analysis of MPPM modulation is crucial for wireless laser communication.

X. Ke, *Coding Theory in Optical Wireless Communication Systems*, Optical Wireless Communication Theory and Technology, https://doi.org/10.1007/978-981-19-9837-9_3

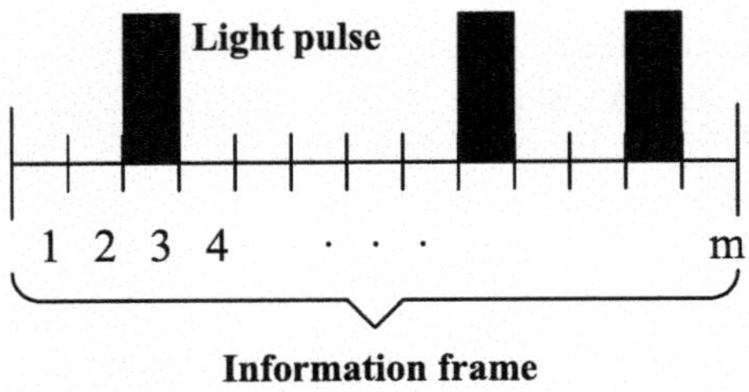

Fig. 3.1 (m, 3) MPPM frame structure

3.2 Analysis of MPPM Performance Parameters

3.2.1 *Signaling Ability*

MPPM can be divided into multipulse combinations and multipulse arrangements, depending on the combination and arrangement of the pulses. The type of arrangement or combination represents the MPPM transmission ability. For a multipulse arrangement, each pulse should have different characteristics, such as a different pulse width, or through the subcarrier frequency modulation, phase modulation. Because multipulse alignment is strictly limited, this aspect has not been introduced at present. We only discuss the multipulse combination method; hence, the multipulse mentioned in this chapter refers to the multipulse combination form.

Suppose that a combinative relation of MPPM with k pulses in 2^m time slots is $NC_k(m)$; then,

$$NC_k(m) = \frac{2^m!}{k!(2^m - k)!}. \tag{3.1}$$

According to Eq. (3.1), the transmission capacity of an MPPM can be easily calculated, as listed in Table 3.1 [5].

It can be proven that $NC_k(m)$ reaches a maximum value when $k = 2^{m-1}$, and it can also be seen that ordinary OOK can be equivalent to MPPM mode $k = 2^m$. As shown in Table 3.1, the transmission capacity of MPPM increases sharply with an increase in k. For example, if $m = 4$, that is, an information frame has 16 time slots, PPM ($k = 1$) can only transmit four bits of information. The two-pulse, three-pulse, and

Table 3.1 Comparison of MPPM transmission capability

m	1	2	3	4	5	6	7	8
$NC_1(m)$	2	4	8	16	32	64	128	256
$NC_2(m)$	1	6	28	120	496	2016	8128	32,640
$NC_3(m)$	0	4	56	560	4960	41,664	341,376	2,763,520
$NC_4(m)$	0	1	70	1820	35,960	635,376	10,668,000	174,792,640

four-pulse combinations can transmit six ($2^6 < 120 < 2^7$), nine ($2^9 < 560 < 2^{10}$), and 10 ($2^{10} < 1820 < 2^{11}$) bits of information, respectively.

3.2.2 MPPM Signal-Space Analysis

Suppose that an information frame has 2^m time slots, and the number of time slots where the pulse is located is recorded as l: $l \in \{0, 1, 2, \cdots, 2^m - 1\} = Z_{2^m}$. Therefore, the signal constellation of PPM can be expressed as an integer set Z_{2^m}. However, if the signal point distance $d(l_1, l_2)$ is defined at this time, it is

$$d(l_1, l_2) = |l_1 - l_2|. \tag{3.2}$$

It does not reflect the energy characteristics or corresponding anti-interference characteristics of the pulse signal. The band-limited signal $d(l_1, l_2)$ reflects only the crosstalk between pulses to some extent. However, because of the characteristics of PPM, for stationary-noise interference, each signal has the same energy; therefore, if the crosstalk between pulses is not considered, each signal point has the same anti-interference performance.

For the two-pulse MPPM modulation mode, the time-slot numbers of the two pulse positions are l_1 and l_2, where $l_1, l_2 \in \{0, 1, 2, \cdots, 2^m - 1\} = Z_{2^m}$. However, because the two pulses cannot overlap, the two-pulse MPPM modulation-signal constellation S_C is a subset of $Z_{2^m} \times Z_{2^m}$, as shown in Fig. 3.2.

$$S_C = \{(l_1, l_2) \mid l_1 > l_2, \ l_1, \ l_2 \in Z_{2^m}\} \tag{3.3}$$

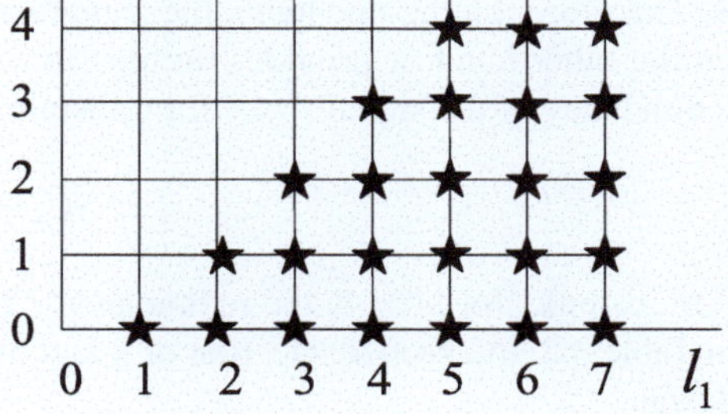

Fig. 3.2 (8, 2) MPPM constellation (m = 3, ★ is a constellation point)

Similarly, owing to the characteristics of MPPM, it is of no practical significance to define the distance between the signal points of two-pulse MPPM using the similar distance definition of quadrature-amplitude modulation (QAM). Therefore, the distance between the signal points of two-pulse MPPM is defined as

$$d((l_1, l_2), (l_1', l_2')) = \min\{|l_1 - l_1'|, |l_2 - l_2'|\}. \tag{3.4}$$

By analogy, the signal constellation of k-pulse MPPM is $S_C(k)$ and the distance between signal points in the corresponding k-pulse MPPM constellation is $d(l,\ l')$.

$$S_C(k) = \{(l_1,\ \cdots, l_k) \mid l_1 > l_2 > \cdots > l_k,\ l_i \in Z_{2^m},\ i = 1,\ 2,\ \cdots,\ k\} \tag{3.5}$$

$$d(l,\ l') = d((l_1, \cdots, l_k),\ (l_1', \cdots l_k')) = \min\{|l_1 - l_1'|, \cdots,\ |l_k - l_k'|\} \tag{3.6}$$

It is easy to prove that, for any signal point $l,\ l', l''$, the following two relations always hold:

$$d(l,\ l') = d((l', l) \tag{3.7}$$

$$d(l, l') + d((l', l'') \geq d((l', l^n) \geq 0. \tag{3.8}$$

Thus, when defining the distance, the signal constellation of a k-pulse MPPM forms a distance space.

3.2.3 Analysis of MPPM Signal Redundancy

For MPPM with 2^m time slots and k pulses, if b bits are to be transmitted, the total capacity of the signal constellation S is required:

$$|S| = NC_K(m) \geq 2^b. \tag{3.9}$$

When Eq. (3.9) is established, the required signal-point set S_O is a true subset of S. Therefore, on the one hand, the signal space can have considerable redundancy; on the other hand, it provides the possibility for error-control coding design. The redundancy of the signal space for transmitting b bits is defined as r_s:

$$r_s = NC_K(m)/2^b. \tag{3.10}$$

For example, for $m = b$, the redundancy $r_s(k)$ is shown in Table 3.2 [5]. As shown in Table 3.2, the proper selection of k and m can effectively control the redundancy design.

Table 3.2 Redundancy of k-pulse MPPM (m = b)

m	1	2	3	4	5	6	7	8
$r_s(2)$	0.5	1.5	4.5	8.5	15.5	31.5	64.5	128.5
$r_s(3)$	0	1	7	35	155	651	2667	10,795
$r_s(4)$	0	0.25	8.75	114.75	1124.75	9928.75	83,344.75	682,784.75

3.2.4 Transmission Efficiency

The number of bits per photon transmitted is typically defined as the transmission efficiency, which is expressed in bits/photon. For PPM, let each information frame be divided into N_{PPM} equal time slots. At the transmitting end, one of the N_{PPM} time slots generates a light pulse, the position of which indicates the message being transmitted. Therefore, a message frame can transmit $\log_2^{N_{PPM}}$ bits of information.

At the receiving end, a photon-counting circuit composed of photodiodes counts the number of photons in the light pulse. When λ is used to represent the mean number of photons contained in an optical pulse, the transmission efficiency of the PPM is defined as

$$\xi_{PPM} = \frac{\log_2^{N_{PPM}}}{\lambda} (bits/photon). \tag{3.11}$$

It can be seen from Eq. (3.11) that the smaller λ is, the higher the transmission efficiency ξ_{PPM} of the PPM. For MPPM, light pulses are transmitted in multiple time slots of an information frame. If an information frame contains N_{MPPM} time slots and K light pulses, $\binom{N_{MPPM}}{K}$ MPPM symbols will be generated, and $\log_2\binom{N_{MPPM}}{K}$ bits of information will be transmitted in each information frame. The transmission efficiency of the MPPM can be expressed by Eq. (3.12).

$$\xi_{MPPM} = \frac{\log_2\binom{N_{MPPM}}{K}}{K\lambda} (bits/photon). \tag{3.12}$$

As can be seen from Eq. (3.12), the smaller λ is, the larger the MPPM transmission efficiency ξ_{MPPM} is.

We assume that the time length of the information frame is T and the number of time slots contained is N. The width of the light pulse τ (the light pulse fills the entire time slot, so it is also called the time-slot width) can be expressed by Eq. (3.13).

$$\tau = \frac{T}{N}(s) \tag{3.13}$$

As can be seen from Eqs. (3.11), (3.12), and (3.13), regardless of whether PPM or MPPM is used, theoretically, ξ can increase infinitely with the increase in N, and the width of the light pulse τ (time-slot width) becomes increasingly narrower.

3.2.5 Bandwidth Utilization

We set the occupied bandwidth as B; then,

$$B = \frac{N}{T}\ (Hz). \tag{3.14}$$

It can be seen from Eqs. (3.13) and (3.14) that B is the reciprocal of τ, and B is proportional to N. With an increase in N, τ will decrease and B will increase. Therefore, when the time length of the information frame is fixed, an increase in N leads to an expansion of the bandwidth. If the bit rate is R, the bandwidth utilization is $\frac{R}{B}$.

$$\frac{R}{B}_{PPM} = \frac{\log {}^{N_{PPM}}_{2}}{N_{PPM}} \tag{3.15}$$

As can be seen from Eq. (3.15), increasing N leads to a rapid decline in the bandwidth utilization of PPM.

$$\frac{R}{B}_{MPPM} = \frac{\log_2 \binom{N_{MPPM}}{K}}{N_{MPPM}} \tag{3.16}$$

As shown in Fig. 3.3, when three pulses are contained in an information frame, that is, K = 3, the bandwidth utilization of MPPM decreases rapidly with an increase in time slots N_{MPPM}. As shown in Fig. 3.4, when N_{MPPM} is fixed, the bandwidth utilization of the MPPM increases rapidly with an increase in pulse numbers K.

3.3 MPPM Characteristics

3.3.1 Comparison of MPPM and PPM Bandwidth Usage

The following compares the bandwidths of PPM and MPPM with the same transmission efficiencies. Assume that the time length of both information frames is T. The information frame of PPM contains N' time slots, whereas the information frame

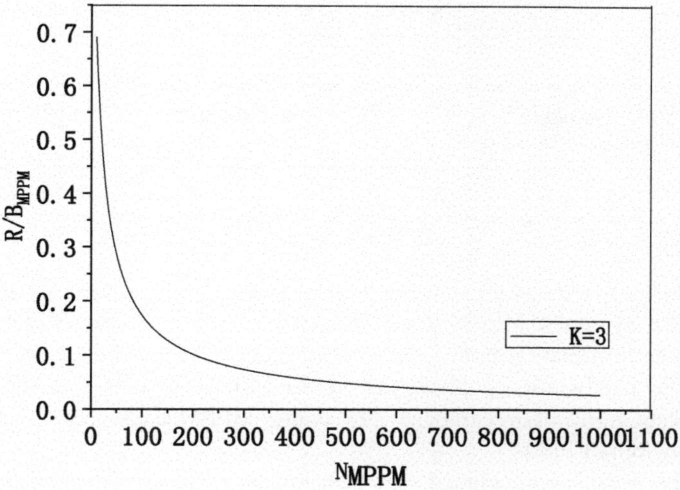

Fig. 3.3 Relationship between MPPM bandwidth usage R/B and the number of time slots N_{MPPM} in an information frame

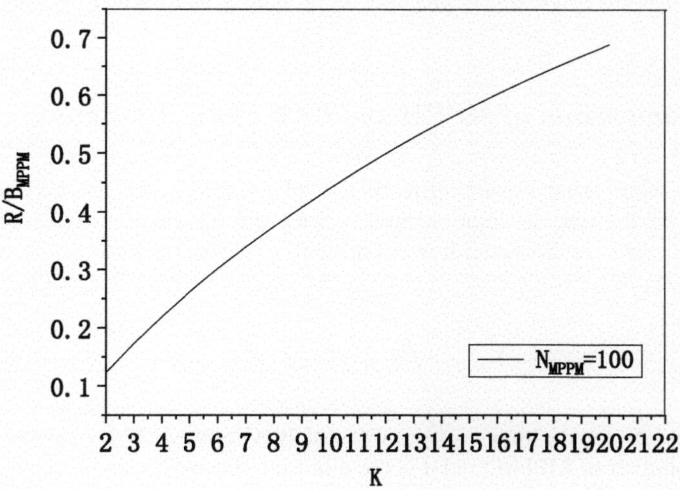

Fig. 3.4 MPPM bandwidth utilization $\frac{R}{B}$ and the number of pulses K contained in the information frame

of MPPM contains N time slots and the pulse number is two. By substituting the defined parameters into Eqs. (3.11) and (3.12), the transmission efficiency of PPM and two-pulse MPPM can be calculated, and Eq. (3.17) can be obtained if they are equal.

$$\left(\frac{N'}{2}\right)^2 = \frac{N(N-1)}{2} \tag{3.17}$$

Generally, N > > 1; hence, both sides of Eq. (3.17) can use the quadratic power at the same time to obtain Eq. (3.18).

$$\frac{N'}{\sqrt{2}} = N(N >> 1) \tag{3.18}$$

Equation (3.18) indicates that under the same transmission efficiency, the time slot required by a two-pulse MPPM in an information frame is $\frac{1}{\sqrt{2}}$ times that required by PPM in an information frame. Substituting Eq. (3.18) into Eq. (3.14), it can be found that the bandwidth occupied by two-pulse MPPM is $\frac{1}{\sqrt{2}} \approx 0.71$ PPM; that is, under the same transmission efficiency, two-pulse MPPM saves approximately half of the bandwidth of PPM.

Similarly, it can be concluded that with the same transmission efficiency, the bandwidth occupied by K-pulse MPPM is $\frac{K}{K\sqrt{K!}}$ PPM; namely, $\frac{N'}{\frac{K}{K\sqrt{K!}}} = N$. When K = 3, the bandwidth occupied by three-pulse MPPM is $\frac{1}{3/\sqrt[3]{3!}} \approx 0.61$ PPM. Thus, in terms of saving bandwidth, three pulses are more advantageous than two.

3.3.2 Comparison of MPPM and PPM Power Utilization

In general, comparing power utilization involves comparing the average power consumed by the two modulation modes when the information-transmission rate and bit-error rate are the same. It is not difficult to derive the power ratio of MPPM and PPM [7].

$$P_{MPPM} \Big/ P_{PPM} = 2k\sqrt{\frac{n}{2Nd}}, \tag{3.19}$$

where n and N are the time-slot numbers of PPM and MPPM, respectively; k is the pulse number of MPPM; and d is the minimum Hamming distance between the MPPM-modulated code words. When n = N, d = 2, and k = 1, the value of Eq. (3.19) is 1, and the MPPM degenerates to PPM. A proper selection of n, d, and k values can reduce the power required by MPPM to less than that of PPM. If MPPM and PPM are combined with convolution codes to form grid-coded modulation, MPPM has a lower power requirement.

3.3.3 Comparison of Anti-interference Performance Between MPPM and PPM

The deep-space channel is a photon-counting channel that follows a Poisson distribution. The probability of receiving m photons in a light-pulse time slot is

$$P(m) = \frac{e^{-\lambda}\lambda^m}{m!}, \tag{3.20}$$

where λ denotes the average number of photons contained in each light pulse. Thus, the probability that a light pulse cannot be detected is $\varepsilon = P(0) = e^{-\lambda}$. Suppose the time-slot number of MPPM is N, and each information frame is represented by a combination of k pulses. Then, the probability that w pulses cannot be detected is

$$P_w = \binom{k}{w}\varepsilon^w(1-\varepsilon)^{k-w}. \tag{3.21}$$

Using $N_k = \binom{N-k+w}{w}$, the total probability of a pulse being undetected, namely, the symbol-error probability, is

$$P(s) = \sum_{w=1}^{k}\frac{N_K - 1}{N_K}\binom{k}{w}\varepsilon^w(1-\varepsilon)^{k-w}. \tag{3.22}$$

When k = 2, the bit error rates of MPPM and PPM were compared. It was found that the anti-interference performance of MPPM is better than that of PPM; that is, when other parameters are the same, the bit error rates of MPPM are lower than those of PPM.

One of the reasons why MPPM has better anti-jamming performance than PPM is that MPPM has redundancy. For example, when $N = 7,\quad k = 2$, and $\binom{7}{2} = 21$. At this time, MPPM can only transmit four bits of information; $2^4 = 16$, so five pulse combinations are redundant. This redundancy can reduce the bit error rates of MPPM and the system to enhance the anti-interference performance. Therefore, if N and k are properly set, errors can be controlled. It should be noted that as the redundancy increases, the power efficiency and frequency efficiency of the system decrease.

3.4 MPPM Channel-Capacity Analysis

The channel is the signal-transmission medium and can be divided into wired and wireless channels. Typically, a generalized communication system can be divided into modulation and coding channels. The function of a channel is to transmit information in the form of signals. Studying a channel involves determining the maximum amount of information that it can transmit; that is, the channel capacity.

The MPPM channel model is given below. According to this model, the expression of the MPPM channel capacity in the protected period was derived, and the channel capacity, information-transmission rate, and channel-capacity energy efficiency of MPPM and PPM were analyzed and compared.

3.4.1 *Channel Model*

(1) MPPM symbol for the protection period

A transmission period has m information time slots, and p time slots each have a pulse, which is recorded as (m, p) MPPM, so that $Q = \binom{m}{p}$ pulse-combination forms can be obtained. In these Q combinations (later called Q symbols), with equal-probability transmission, the transmitted information is $\log_2 Q$ bits/ symbol.

Figure 3.5 shows the frame structure of a two-pulse optical MPPM symbol ((m, 2) MPPM) with a protected period. Symbolic time T is divided into M time slots, each of which has a width of τ (T/M)(s). The first D (in practice, D must be a natural number) time slots without information transmitted are called the protection period T_D, which is determined by the minimum repetition cycle of the pulsed laser. It is the minimum time required to reconstruct the particle-number inversion after each laser pulse is emitted.

The last m + D time slots are called the signal transmission period T_S, among which the D time slots are also the protection period, which is the same as T_D above. The m slot positions T_S correspond to Q symbols. During the time interval T_D of the

Fig. 3.5 (m, 2) MPPM frame structure

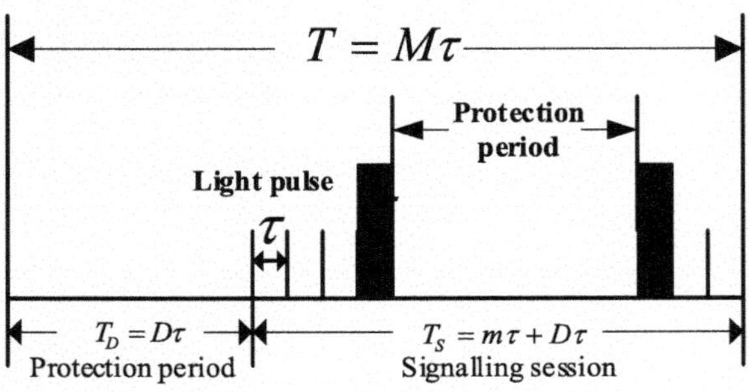

emitting laser source, two slots in transmission period T_S are opened (the rest are closed) and a symbol is sent. The receiver counts the photons of each information time slot of the transmission period to obtain m numbers $\{n_i\}(i = 1, \cdots, m)$. The two time slots with the maximum number of photons are then selected as the received symbol.

(2) Channel model of an (m, p) MPPM

For a noiseless channel (quantum noise without background noise, but with photon counting), the probability of detecting no photons in the (m-2) time slots when the laser is turned off is 1. In the time slot switched on by the laser, the photon count n_i follows a Poisson distribution. If the average photon-count rate per second is S, then the probability that the receiver detects no photons is

$$q = e^{-S\tau}. \tag{3.23}$$

The relation between the photon-counting rate S and the incident signal power P of the receiver is

$$S = [\eta/(hf)]P \tag{3.24}$$

η is the quantum efficiency of the photoelectric detector and hf is the photon energy with a working frequency of f [8, 9].

If no photon is detected in one or more of the p pulses in a time slot, the received signal is called an erasing frame and cannot be considered as the frame being transmitted. At this point, the (m, p) MPPM can be simulated using a Q-element discrete memory-free erasure channel, as shown in Fig. 3.6 [10].

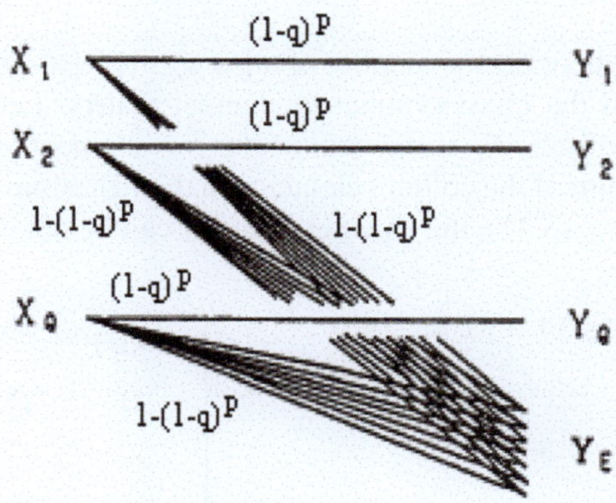

Fig. 3.6 Q-element erasing-channel model

3.4.2 Channel-Capacity Analysis

(1) Channel capacity

Let the average mutual information of a channel be I(X; Y), and the probability distribution of the input source be $P(x)$. Then, the channel capacity C of the channel is defined as

$$C = \max_{P(X)}\{I(X;Y)\} = \max_{P(X)}\{H(X) - H(X/Y)\}\ (bits/symbol), \tag{3.25}$$

where X and Y represent the input and output of the channel, respectively; H(X) represents the entropy of the channel input source X; and H(X/Y) represents the conditional entropy between the input and output symbols.

According to the channel model shown in Fig. 3.6, the corresponding channel matrix is shown in Fig. 3.7. The channel corresponding to this channel matrix is quasi-symmetric. The calculation formula for the channel capacity of the quasi-symmetric discrete channel is shown in Eq. (3.26) [11–13].

$$\begin{bmatrix} (1-q)^p & 0 & \cdots & 1-(1-q)^p \\ 0 & (1-q)^p & \cdots & 1-(1-q)^p \\ \cdots & \cdots & \cdots & \cdots \\ 0 & \cdots & (1-q)^p & 1-(1-q)^p \end{bmatrix}_{Q\times Q+1}$$

$$C = \log r - H(p_1', p_2', \cdots, p_s') - \sum_{k=1}^{n} N_k \log M_k, \tag{3.26}$$

where r is the number of input symbol sets and $(p_1', p_2', \cdots, p_s')$ is the row element in the quasi-symmetric channel matrix. Let the matrix be divided into n disjoint subsets. N_k is the sum of the row elements in the Kth submatrix Q_k and M_k is the sum of the column elements in the Kth submatrix Q_k. According to Fig. 3.7 showing Eq. (3.27), the channel capacity can be obtained as follows:

$$\begin{bmatrix} (1-q)^p & 0 & \cdots & 1-(1-q)^p \\ 0 & (1-q)^p & \cdots & 1-(1-q)^p \\ \cdots & \cdots & \cdots & \cdots \\ 0 & \cdots & (1-q)^p & 1-(1-q)^p \end{bmatrix}_{Q\times Q+1}$$

Fig. 3.7 Channel matrix

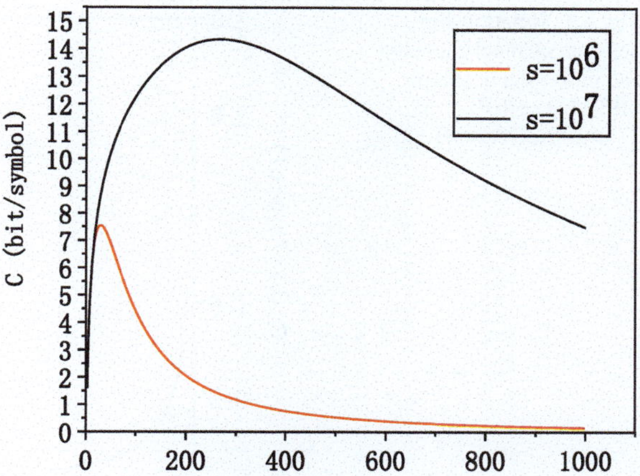

Fig. 3.8 Relationship between (m, 2) MPPM channel capacity and time slots ($T_s = 100\mu s$, $D = 8$)

$$C = (1-q)^p \log_2\binom{m}{p}$$
$$= (1-e^{-S\tau})^p \log_2\binom{m}{p} = (1-e^{-S\frac{T_S}{m+(p-1)D}})^p \log_2\binom{m}{p}(bit/symbol). \quad (3.27)$$

It can be seen from Eq. (3.27) that the channel capacity C decreases with an increase in the number of protected time slots D, so D = 8 in the following analysis. (In practice, D depends on specific requirements.)

In a channel where the average photon-counting rate per second (optical pulse power) S and symbol time T_S remain unchanged, C changes with m, as shown in Fig. 3.8. An optimal time-slot number m exists to maximize the channel capacity C. When $S = 10^6$, the optimal time-slot number is $m_s = 31$ and the corresponding maximum capacity is 7.5 bits. When $S = 10^7$, the optimal time-slot number is $m_s = 265$ and the corresponding maximum capacity is 14.3 bits. Thus, the larger the optical-pulse power, the larger m_s and the larger the corresponding channel capacity.

In addition, the channel capacities of MPPM and PPM were compared when the symbol time T_S and the average photon-count rate per second S were the same, as shown in Figs. 3.9, 3.10, and 3.11.

By comparing Figs. 3.9, 3.10, and 3.11, it can be seen that with the increase in S, the channel capacity of (m, 2) MPPM and (m, 3) MPPM is gradually better than that of PPM. It can also be obtained from Eq. (3.28) that if the time-slot number m is unchanged and T_S is given, the channel capacity C changes with the pulse number p, as shown in Fig. 3.12.

As can be seen from Fig. 3.12, when $S = 10^7$, C first increases and then decreases with the increase in pulse number p. If S increases again, C still increases and then decreases with p, and the value of C does not change significantly. Hence, increasing the number of pulses does not increase the channel capacity.

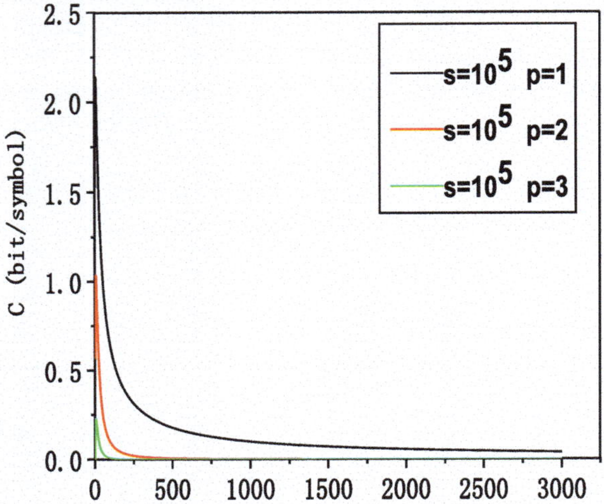

Fig. 3.9 Relationship between channel capacity and time slots in (m, p) MPPM ($s = 10^5$, $T_s = 100\mu s, D = 8$)

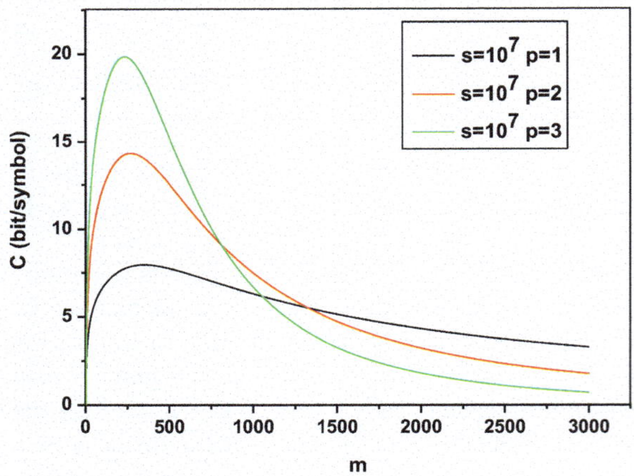

Fig. 3.10 Relationship between channel capacity and time slots in (m, p) MPPM ($s = 10^7$, $T_s = 100\mu s, D = 8$)

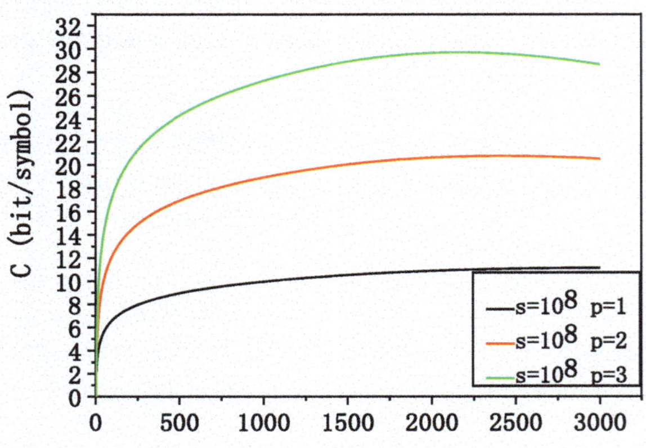

Fig. 3.11 Relationship between channel capacity and time slots in (m, p) MPPM ($s = 10^8$, $T_s = 100\mu s, D = 8$)

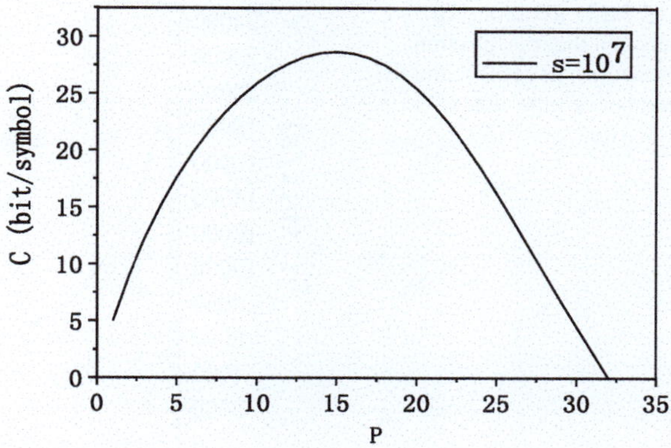

Fig. 3.12 Relationship between (m, p) MPPM channel capacity and pulse number ($T_s = 100\mu s$, $D = 8, m = 32$)

(2) Information-transmission rate [11, 14]

The information-transmission rate refers to the average amount of information transmitted in a channel per unit time. Because it takes T seconds to use each channel, the channel capacity C_t per second is

$$C_t = \frac{C}{T} = \frac{(1 - e^{-s\tau})^p \log_2 \binom{m}{p}}{pT_D + m\tau} = \frac{(1 - e^{-s\tau})^p \log_2 \binom{m}{p}}{(pD + m)\tau} (bits/s). \tag{3.28}$$

According to Eq. (3.28), when the average photon-counting rate per second S and the time-slot width τ (reciprocal of the system bandwidth) remain unchanged, C_t changes with m, as shown in Fig. 3.14. There is an optimal time-slot number m_s that maximizes the maximum information C_t (later called the information-transmission rate) transmitted per unit time.

By comparing Figs. 3.13, 3.14, and 3.15, it can be seen that the MPPM information-transmission rate C_t is better than that of PPM when the gap width τ is larger. However, with an increase in the time-slot width τ, the maximum information-transmission rate C_t gradually decreases.

(3) Channel-capacity energy efficiency

The average number of received photons used per channel is $ps\tau$, so the channel capacity (capacity energy efficiency) of C_p per received photon is

$$C_p = \frac{C}{ps\tau} = \frac{(1 - e^{-s\tau})^p \log_2 \binom{m}{p}}{ps\tau} = \frac{(1 - e^{-N_P})^p \log_2 \binom{m}{p}}{pN_P} (bits/photon). \tag{3.29}$$

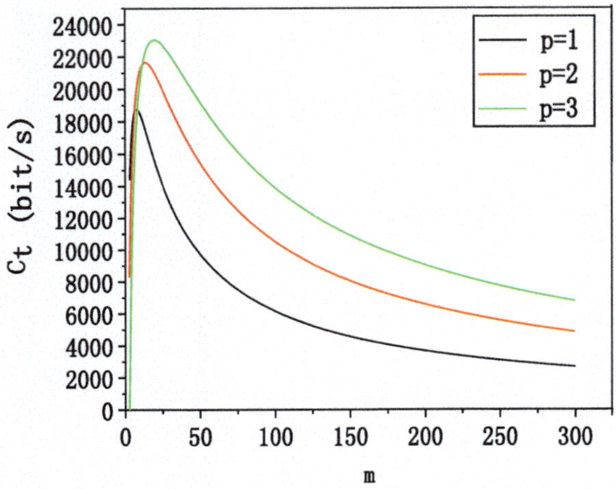

Fig. 3.13 Relationship between the transmission rate of (m, p) MPPM and the number of time slots ($D = 8$, $\tau = 10\mu s$, $s = 10^6$)

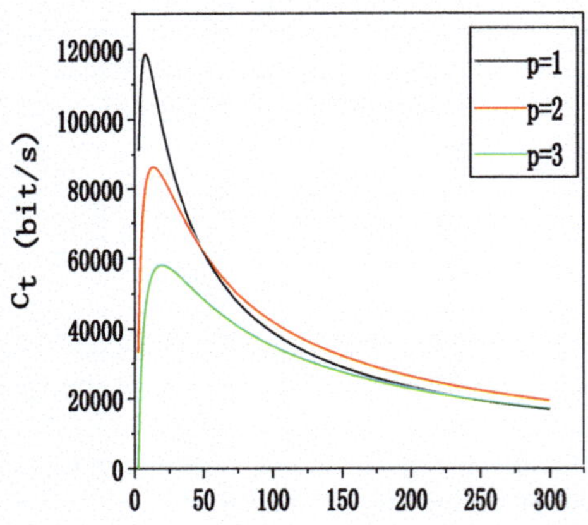

Fig. 3.14 (m, p) MPPM information-transmission rate and the number of time slots ($D = 8$, $\tau = 1\mu s$, $s = 10^6$)

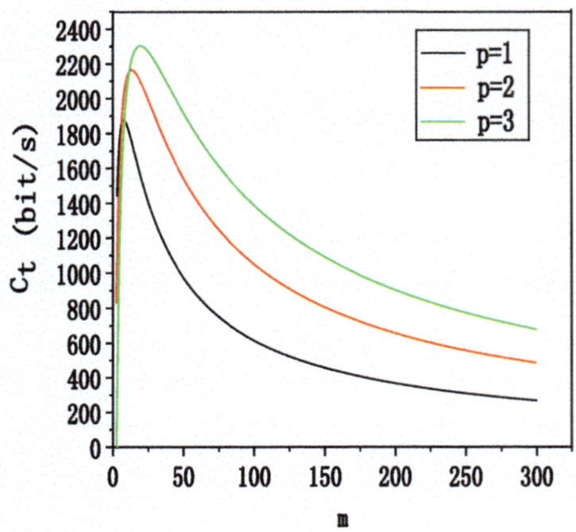

Fig. 3.15 Relationship between the transmission rate of (m, p) MPPM and the number of time slots ($D = 8$, $\tau = 100\mu s$, $s = 10^6$)

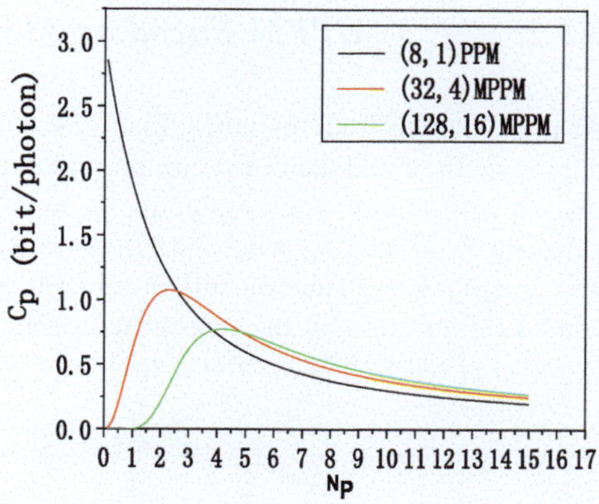

Fig. 3.16 Relationship between the energy efficiency of the (m, p) MPPM channel capacity and the number of photons contained in a single optical pulse

As shown in Fig. 3.16, when the number of photons contained in a single optical pulse is $N_P > 5$, the channel-capacity energy efficiency of MPPM is better than that of PPM.

3.5 Design of a Three-Pulse MPPM Coding and Decoding System

According to the three-pulse MPPM coding and decoding principle, an implementation method based on field-programmable gate arrays (FPGAs) is proposed below, and the hardware-implementation scheme and results are provided.

3.5.1 Three-Pulse MPPM ((M, 3) MPPM)

As shown in Table 3.1, when the time gap is $m \geq 3$, the transmission capacity of (m, 3) MPPM is significantly better than that of (m, 2) MPPM. As shown in Table 3.2, when $m > 2$, the redundancy of (m, 3) MPPM is significantly greater than that of (m, 2) MPPM. It can also be seen from Fig. 3.12 that when the length of the transmission period and the number of time slots it contains have fixed timing, the channel capacity of (m, 3) MPPM is better than that of (m, 2) MPPM. Therefore, it is very important to study three-pulse MPPM.

3.5.2 (M, 3) MPPM Encoding-Mapping Method

At present, only two encoding-mapping methods are available for (m, 3) MPPM: the list method and the constellation method. In (m, 3) MPPM coding, it is assumed that an information frame contains $m = 2^3$ time slots, and the values of the time slots are l_1, l_2, and l_3, where the three pulses are located within $\{0,\ 1,\ \cdots m-3\}$ and $l_1 \leq l_2 \leq l_3$. The constellation of (m, 3) MPPM is shown in Fig. 3.17, which can be represented by the two-pulse constellation $l_1 = 0,\ 1,\ \cdots, 5$, composed of l_2 and l_3 [14, 15]. The coding mapping of (m, 3) MPPM is denoted as [16]

$$\Phi(u) : (l_1, l_2, l_3). \tag{3.30}$$

The Φ value is defined from the two-pulse constellation of $l_1 = 0$ [17], and then ordered from the two-pulse constellation of $l_1 = 1,\ 2,\ \cdots, m-3$. When $l_1 = 0$, Φ is the value of the two-pulse constellation of $l_1 = 0$; when $l_1 = 1$, $l_1 = 1$ must include the two-pulse constellation l_2 and l_3 values, plus the $l_1 = 0$ two-pulse constellation state, for a total of C_{m-1}^2. When $l_1 = 2$, the values of l_2 and l_3 must be added to the total number of $l_1 = 0$ and $l_1 = 1$ in the two-pulse constellation of

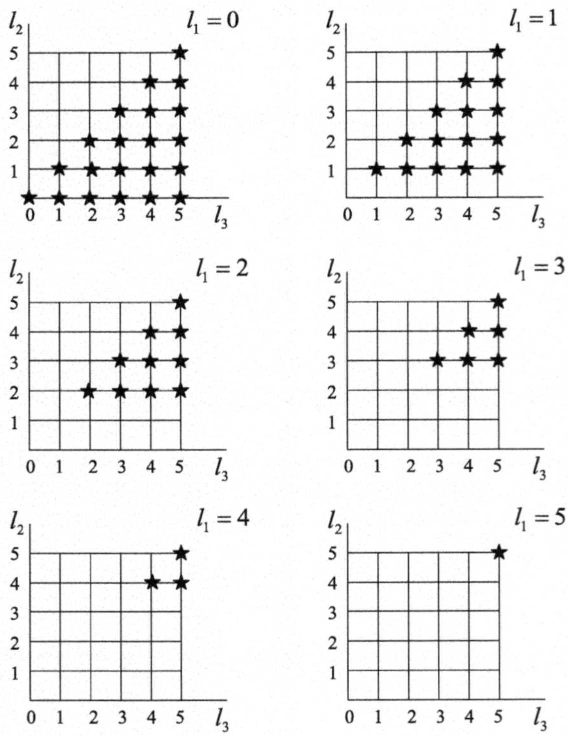

Fig. 3.17 (6, 3) MPPM constellation [17]

$l_1 = 2$ for $C_{m-1}^2 + C_{m-2}^2$. Thus, the coding and decoding relation of the (m, 3) MPPM modulation combination is

$$\Phi(u) = (l_2 - l_1) + C_{(l_3-l_1)+1}^2 + \sum_{i=0}^{l_1-1} C_{m-1-i}^2. \tag{3.31}$$

Thus, it can be seen that the decoding process of the three-pulse MPPM modulation with m time slots is given $l_1,\ l_2,\ l_3$ and solved with $\Phi(u)$. The decoding can be directly completed using Eq. (3.31). The coding process is given $\Phi(u)$ to solve $l_1,\ l_2,\ l_3$, and the constraint condition $l_1 \le l_2 \le l_3$ must be satisfied simultaneously. It is impossible to obtain an analytical solution using Eq. (3.31).

In summary, the constellation-diagram method can only decode; coding requires further study. Achieving $l_1 = l_2 = l_3$ is also very difficult. Figure 3.2 shows the frame structure of the (m, 3) MPPM. The MPPM symbol that maps the continuous binary bit stream of length L into the simultaneous occurrence of light pulses on three time slots in the information frame composed of m time slots, is denoted as (m, 3) MPPM.

It is assumed that the input length of the continuous binary bitstream is four, and the output symbol encoded by the three-pulse MPPM encoder is represented by (6, 3) MPPM. The time-slot numbers of the three pulses are denoted as l_1, l_2, and l_3, and the relationship between them is shown in Table 3.3.

Table 3.3 Relationship between input information, (6, 3) MPPM symbol, and the number of time slots in which the pulse is located

4 bits	(6, 3) MPPM	$(l_1,\ l_2,\ l_3)$
0000	111,000	(1, 2, 3)
0001	110,100	(1, 2, 4)
0010	110,010	(1, 2, 5)
0011	110,001	(1, 2, 6)
0100	101,100	(1, 3, 4)
0101	101,010	(1, 3, 5)
0110	101,001	(1, 3, 6)
0111	100,110	(1, 4, 5)
1000	100,101	(1, 4, 6)
1001	100,011	(1, 5, 6)
1010	011,100	(2, 3, 4)
1011	011,010	(2, 3, 5)
1100	011,001	(2, 3, 6)
1101	010,110	(2, 4, 5)
1110	010,101	(2, 4, 6)
1111	010,011	(2, 5, 6)

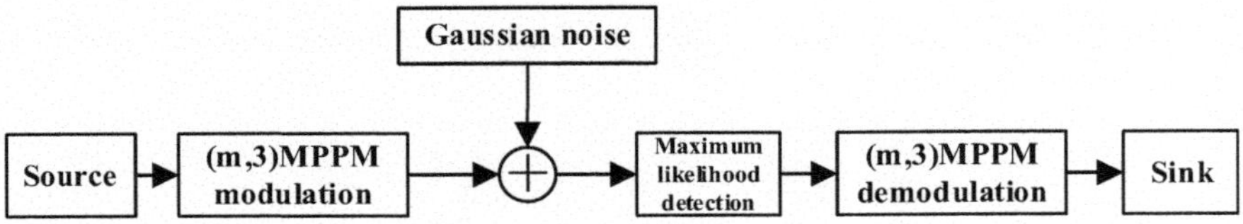

Fig. 3.18 Block diagram of the (m, 3) MPPM simulation system

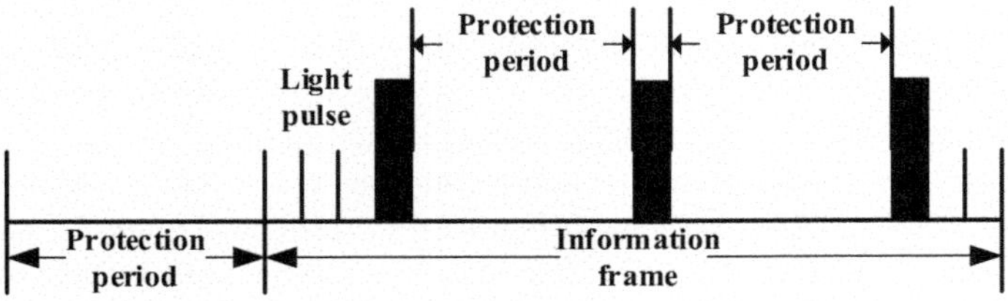

Fig. 3.19 (m, 3) MPPM frame structure with protection periods

3.5.3 (M, 3) MPPM Simulation Using Simulink

(1) (m, 3) MPPM simulation system

This implementation was simulated using Simulink. A block diagram of the simulation system is shown in Fig. 3.18.

As shown in Fig. 3.19, the source of this system emits a random binary sequence. In the (m, 3) MPPM modulation, four bits are grouped. After the (m, 3) MPPM modulation, it becomes an information frame which contains six time slots (assuming that each light pulse fills the entire time slot). Assuming that the minimum repetition period of the laser is t, a protection period of duration t is added between the two light pulses and the information frames.

(2) Simulation results of (m, 3) MPPM

Similar to the simulation system of (m, 2) MPPM, white Gaussian noise was used to simulate the channel noise of an optical transmission. At the receiving end, the maximum-likelihood detection method was used for demodulation [15, 16]. Figures 3.20, 3.21, 3.22, 3.23, and 3.24 show the results.

3.5.4 (m, 3) MPPM Function Simulation

(1) (m, 3) MPPM encoder

To verify the function of the (m, 3) MPPM encoder, it was simulated with Modelsim. Assuming that the rate of the input bitstream of the encoding module is 9.6 kbits/

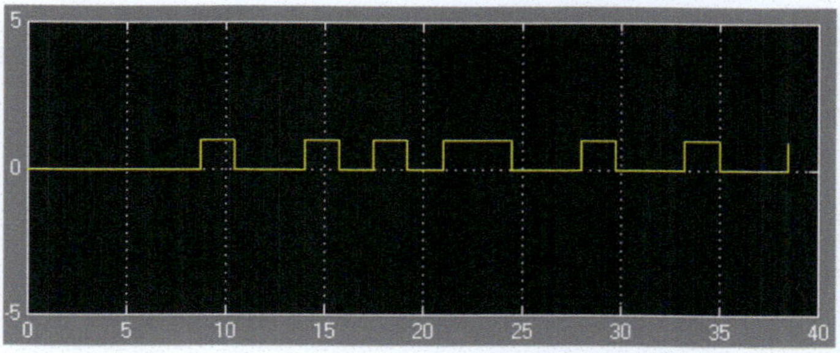

Fig. 3.20 Random binary source [18]

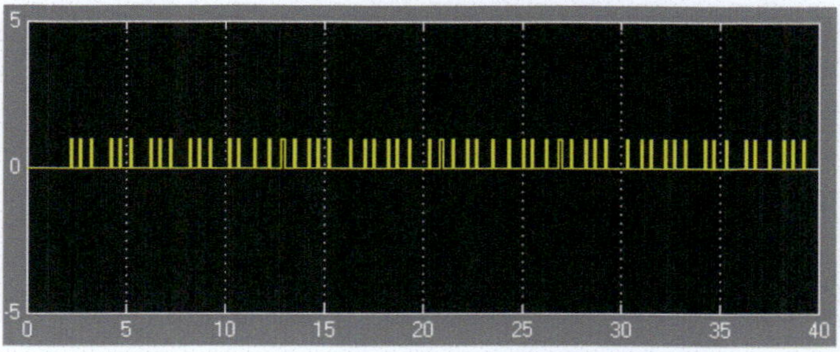

Fig. 3.21 Waveform after (m, 3) MPPM modulation [18]

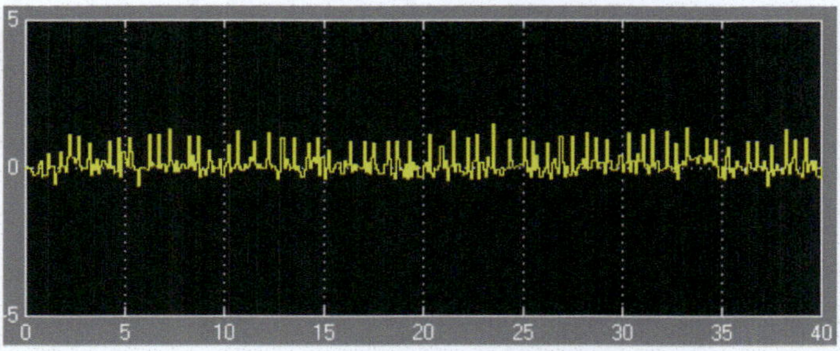

Fig. 3.22 Waveform after Gaussian channel [18] (SNR = 10 dB)

s and the encoding is carried out in groups of four bits, the length of an (m, 3) MPPM symbol is $\frac{4}{9.6\times10^3}s = 0.416$ ms. Considering the minimum time required to reconstruct the particle-number inversion after each laser pulse, i.e., the minimum repetition cycle of the pulsed laser, a protection period should be added after each pulse. In addition, to synchronize the coding and decoding modules, the beginning of each information frame also requires a frame-synchronization header. The structure of an (m, 3) MPPM symbol in the encoder is shown in Fig. 3.25.

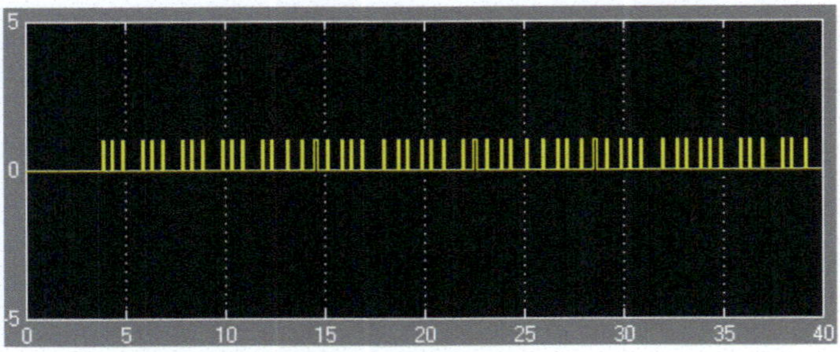

Fig. 3.23 Waveform after maximum-likelihood detection [18]

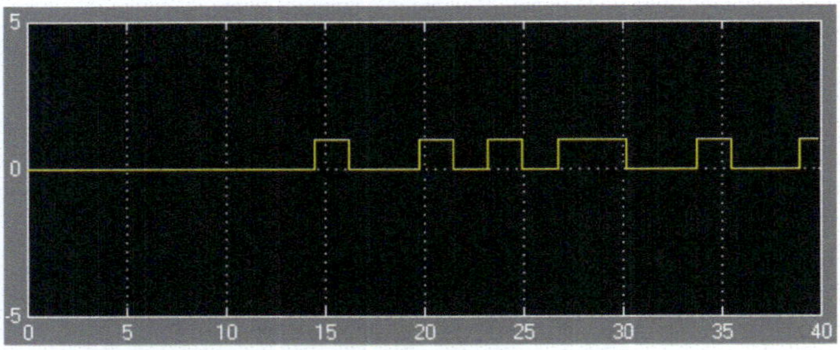

Fig. 3.24 Waveform after (m, 3) MPPM demodulation [18]

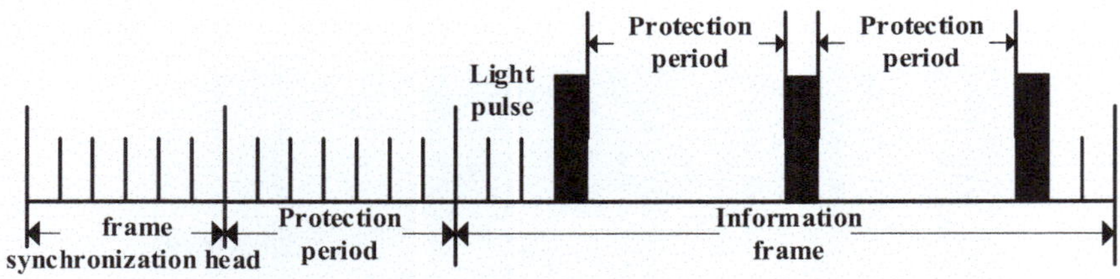

Fig. 3.25 (m, 3) MPPM symbol structure [18]

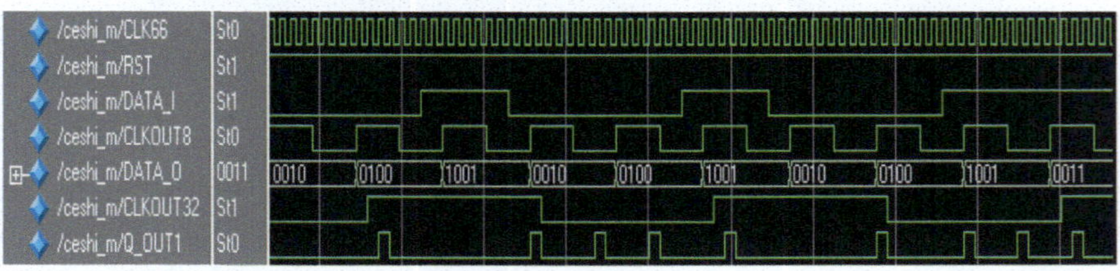

Fig. 3.26 Simulation waveform of the coding process [18]

For the three-pulse transmission of four-bit information, at least six time slots are required in the information frame, that is, m = 6 ($\binom{6}{3} > 2^4$). To distinguish the frame-synchronization head from the information frame, the number of time slots contained in the frame-synchronization head must be greater than the sum of the maximum number of time slots contained in the protection period and the number of time slots between two pulses.

In terms of transmitting four-bit information, the information frame contains six time slots, and the maximum number of time slots between two pulses is six. If the protection period contains four time slots, the number of time slots contained in the frame-synchronization head should be greater than $4 + 6 = 10$. Here, we take 13 and add a pulse indicating the beginning of a frame; the total number of time slots contained in the frame-synchronization head is 14. Thus, the (m, 3) MPPM symbol shown in Fig. 3.25 contains $14 + 3 \times 4 + 6 = 32$ time slots.

The duration of each time slot is $\frac{4}{9.6 \times 10^3 \times 32} = 0.013ms$, and the duration of the protected period is $4 \times 0.013 = 0.052ms$ (assuming that the protected period contains the four time slots above). A laser with a minimum repetition period of 0.05 ms was selected in the experiment, which meets the requirements (the duration of the protection period is 0.052 ms > 0.05 ms).

A (6, 3) MPPM symbol contains a total of 32 time slots. 0011 after coding, for example, received an MPPM (6, 3) symbol for 10,000,000,000,000,100,010,000,010, 000; the symbol 1 represents a pulse, and the frame-synchronization head of 0 with no pulse marks the beginning of an MPPM (6, 3) symbol. Table 3.4 shows the relationship between the input information, output (6, 3) MPPM symbol, and time-slot number of the three pulses of the coding module; the simulation waveform is shown in Fig. 30.

(2) (m, 3) MPPM decoder

In decoding, the frame-synchronization header is detected to determine the start of a frame, and the original input information, namely a four-bit binary number, is translated from the received frame signal according to Table 3.4. The decoder function is simulated using Modelsim, and the results are shown in Fig. 3.27.

3.5.5 FPGA Verification of the (m, 3) MPPM

We chose the Xilinx Company Spartan II series XC2S200 device for FPGA verification and used the ISE XST 6.2 integrated tool to simulate the synthesis, implementation, and wiring. Finally, the FPGA download file (.bit) was generated. The operation results of the (m, 3) MPPM encoder and decoder observed from the oscilloscope are shown in Fig. 3.28.

In the FPGA verification process, a set of M sequences was input, and the same set of M sequences was output after the (m, 3) MPPM coder and decoder, but with some

Table 3.4 Relationship between input information, (6, 3) MPPM symbol, and time-slot number of pulses [18]

4 bits	(6, 3) MPPM	$(l_1,\ l_2,\ l_3)$
0000	10,000,000,000,000,100,001,000,010,000,000	(1, 2, 3)
0001	10,000,000,000,000,100,001,000,001,000,000	(1, 2, 4)
0010	10,000,000,000,000,100,001,000,000,100,000	(1, 2, 5)
0011	10,000,000,000,000,100,001,000,000,010,000	(1, 2, 6)
0100	10,000,000,000,000,100,000,100,001,000,000	(1, 3, 4)
0101	10,000,000,000,000,100,000,100,000,100,000	(1, 3, 5)
0110	10,000,000,000,000,100,000,100,000,010,000	(1, 3, 6)
0111	10,000,000,000,000,100,000,010,000,100,000	(1, 4, 5)
1000	10,000,000,000,000,100,000,010,000,010,000	(1, 4, 6)
1001	10,000,000,000,000,100,000,001,000,010,000	(1, 5, 6)
1010	10,000,000,000,000,010,000,100,001,000,000	(2, 3, 4)
1011	10,000,000,000,000,010,000,100,000,100,000	(2, 3, 5)
1100	10,000,000,000,000,010,000,100,000,010,000	(2, 3, 6)
1101	10,000,000,000,000,010,000,010,000,100,000	(2, 4, 5)
1110	10,000,000,000,000,010,000,010,000,010,000	(2, 4, 6)
1111	10,000,000,000,000,010,000,001,000,010,000	(2, 5, 6)

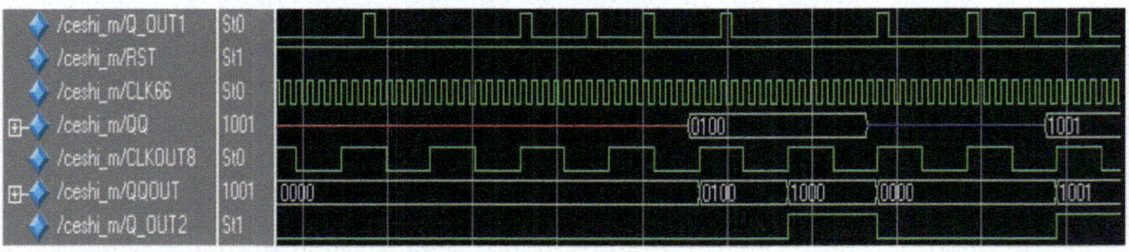

Fig. 3.27 Simulation waveform of the decoding process [18]

delay. This shows that the designed (m, 3) MPPM coder and decoder can complete the coding and decoding. A list of the main programs can be found in [18]. The hardware implementation of the three-pulse MPPM coding and decoding system can be found in [19].

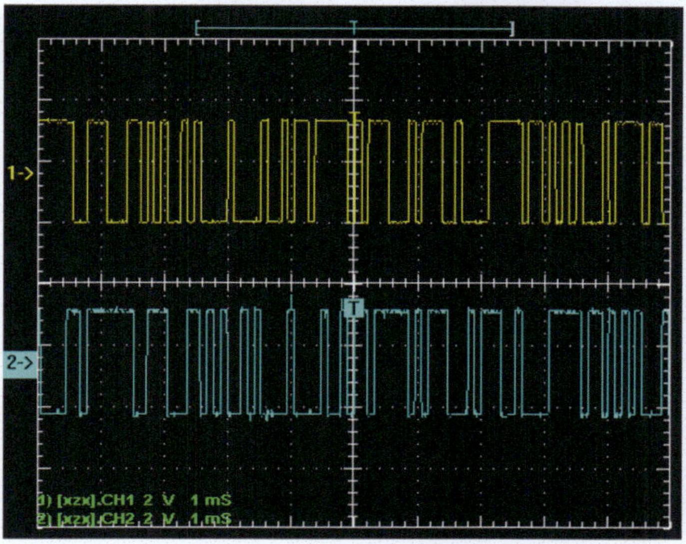

Fig. 3.28 Operation results of the (m, 3) MPPM codec [18]

References

1. Ke XZ, Xi XL (2004) Introduction to wireless laser communication. Beijing University of Posts and Telecommunications Press, Beijing, pp 5–20
2. Dong X, Li J, Orlik P (2008) Performance of a new transmitted reference pulse cluster system for UWB communications. IEEE Trans Veh Technol 57(05):3217–3224
3. Sugiyama H, Nosu K (1989) MPPM: a method for improving the band-utilization efficiency in optical PPM. IEEE J Lightwave Technol 7(03):465–472
4. Zhou L, Yin DS (1998) Basic analysis of multi-pulse PPK encoding modulation. J Univ Electron Sci Technol China 27(05):519–521
5. Otte R, Jong LD (2002) Wireless optical PPM telemetry and the influence of lighting flicker. IEEE Trans Instrum Meas 47(01):51–55
6. Ntogari G, Kamalakis T, Sphicopoulos T (2009) Performance analysis of space time block coding techniques for indoor optical wireless systems. IEEE J Sel Areas Commun 27(09):1545–1552
7. Guo JZ, Tan Y, Ai Y (2006) Research on modulation technology in satellite optical communication. Optical Commun Technol 35(04):45–46
8. Zou CY, Ao FL, Huang XF (2000) Analysis of optical PPM channel capacity without background noise. J Electron Sci 22(04):682–686
9. Qin L, Ke XZ (2007) Capacity analysis of optical MPPM channel without background noise. Opto-electron Eng 34(07):107–110
10. Le Bidan R, Leroux C, Jego C et al (2008) Reed-Solomon turbo product codes for optical communications: from code optimization to decoder design. EURASIP J Wirel Commun Netw 20(08):1–14
11. Fu ZY (2001) Information theory, 1st edn. Publishing House of Electronics Industry, Beijing, pp 51–72
12. Miao YI, Chen MS, Song LT et al (2009) Effect of imperfect slot synchronization on bit errors of MPPM in underwater laser communication. Laser Technol 33(06):596–597

13. Lee DCM, Kahn JM (1999) Coding and equalization for PPM on wireless infrared channels. IEEE Trans Commun 47(02):255–260
14. Ma HP (2003) Research on optical modulation and demodulation technology in mobile atmospheric laser communication. National University of Defense Technology, Changsha, pp 25–35
15. Piao DZ, Zou CY (2022) Maximum likelihood detection of sampled PPM signal under white Gaussian noise. J Guilin Univ Electron Technol. 22(03):5–8
16. Huang CF (2000) Encoding and decoding of multi-pulse pulse position keying modulation. J Chongqing Univ Posts Telecommun 12(03):7–9
17. Qin L, Ke XZ (2007) A two-pulse MPPM coding mapping method. J Xi'an Univ Technol 23(03):269–272
18. Qin L (2007) Design of multi-pulse modulation system in atmospheric laser communication. Xi'an University of Technology, Xi'an, pp 38–43
19. Qin L, Du Y, X, Ke XZ et al (2008) Implementation of three-pulse MPPM coding and decoding system in atmospheric laser communication. Semicond Optoelectron 29(03):403–433

Chapter 4
Analysis of DPPM-Coded Modulation Signals

Differential pulse-position modulation (DPPM) is a type of pulse-position modulation [1–4], similar to pulse-position modulation (PPM) and multipulse position modulation (MPPM). This chapter analyzes the bit-error rate characteristics of DPPM-coded modulation, introduces the implementation mode of DPPM-coded modulation, and finally provides a hardware-implementation scheme for coded modulation.

4.1 DPPM Coding

PPM is an orthogonal modulation mode [5]. In L-PPM, the input of a $\log_2^L$ bit corresponds to one of L different pulse-position patterns. Each form includes a high level of one time slot and a low level of the previous time slot. A pulse p(t) transmits information in the high-level timeslot. DPPM is an improvement over PPM. DPPM is obtained by deleting all low-level time slots after the high-level time slot of their corresponding PPM signals.

Let the signal form of L-DPPM be $S1_n(t)$ and the signal form of L-PPM be $S2_n(t)$, $0 \leq n \leq L-1$. A square wave is used to emit a laser signal [6–8], and $S1_n(t)$ and $S2_n(t)$ can be obtained:

$$S1n(t) = \begin{cases} 0 & t < nTc \\ Pc & nTc \leq t \leq (n+1)Tc \end{cases}$$

$$S2n(t) = \begin{cases} 0 \;\; t < nTc(n+1)Tc < Tc < t < (L-1)Tc \\ P_c \qquad\qquad nTc \leq t \leq (n+1)Tc. \end{cases} \tag{4.1}$$

P_c is the maximum energy and T_c is the time of each time slot. DPPM- and PPM-coding first convert the original information bit sequence into a time-slot sequence

X. Ke, *Coding Theory in Optical Wireless Communication Systems*, Optical Wireless Communication Theory and Technology, https://doi.org/10.1007/978-981-19-9837-9_4

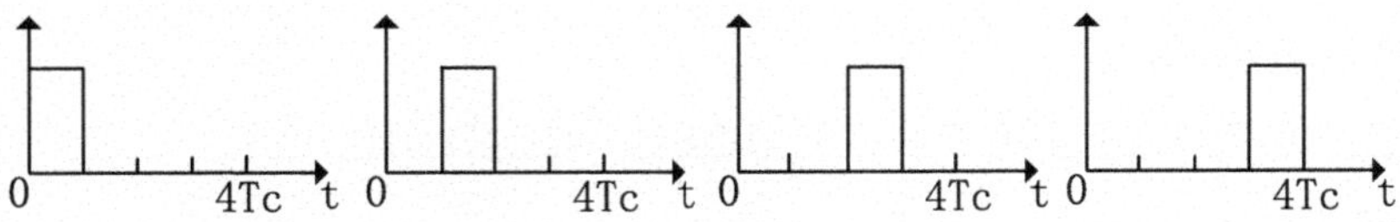

(a) Schematic diagram of a 4-PPM-modulated pulse position

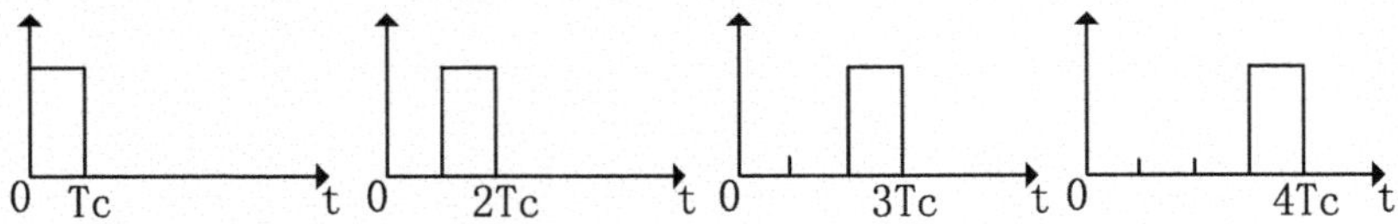

(b) Schematic diagram of 4-DPPM modulated pulse position

Fig. 4.1 Schematic diagrams of the difference in pulse-modulation position

according to the corresponding law, and then, the time-slot sequence modulates the laser through pulse $p(t)$ to form an optical signal and sends it out.

Both DPPM and PPM belong to position-mapping modulation. If the n-bit data group is written as $M = (m_1, m_2, \ldots, m_n)$ and the time-slot position is written as l, the mapping-coding relationship between the L-PPM and L-DPPM modulation can be written as

$$\begin{aligned} \phi : l &= m_1 + 2m_2 + \cdots + 2^{n-1} m_n \\ &\in \{0, 1, \ldots, n-1\}. \end{aligned} \tag{4.2}$$

For example, for 4-PPM and 4-DPPM modulation, if M = (0, 0), then $l = 0$; if M = (0, 1), then $l = 1$; if M = (1, 0), then $l = 2$; if M = (1, 1), then $l = 3$; 0, 1, 2, and 3 correspond to time-slot positions, respectively, as shown in Fig. 4.1.

One of the advantages of DPPM over PPM is synchronization [6]. DPPM does not have time synchronization, which is very important in PPM signal detection; however, it is not necessary in DPPM. DPPM's most important advantage is that it can achieve high energy and bandwidth utilization [7]. In L-PPM, with an increase in L, although the average energy utilization is improved, the bandwidth utilization is reduced, which is caused by the increase in the signal dimension and the energy peak-to-average ratio.

In DPPM, the low-level signal after the high level in the code group of a PPM modulation signal can be omitted without affecting the bit-error rate. For a fixed L, because the duty cycle of L-DPPM is higher than that of the corresponding L-PPM, its average power utilization will be lower than that of PPM. However, for a fixed average bit rate and fixed bandwidth, a higher L can be selected in DPPM; thus, a higher average power utilization is obtained.

With the same average bit rate, when the system adopts on–off keying (OOK) modulation mode, the required bandwidth is $W = R_b$, which is equal to the average bit rate; when the system adopts PPM modulation mode, the required bandwidth is $W = LR_b/\log_2 L$; when the system adopts DPPM modulation mode, the required

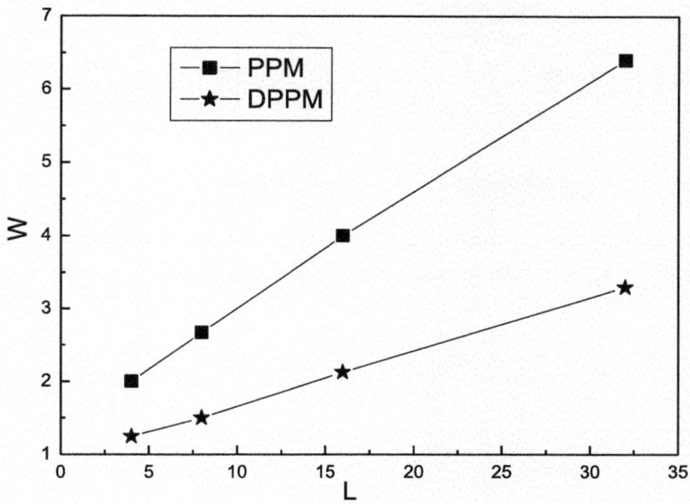

Fig. 4.2 Bandwidth required by the system

bandwidth is $W = (L + 1)\ R_b/2\log_2 L$. As shown in Fig. 4.2, the bandwidth required for different binary numbers of DPPM modulation modes increases gently with the increase of L. It can be seen that with the increase of L, the required bandwidth increases faster than that of DPPM modulation. Comparing the two lines, it can be seen that DPPM at the same L saves more bandwidth than PPM.

4.2 Bit-Error Characteristic Analysis

Because there is no fixed period for DPPM symbols, they are very difficult to analyze. If a time-slot signal goes wrong, it not only affects the bits close to the time slot, but also affects the subsequent bits, which makes the traditional definition of bit-error rate inapplicable to DPPM.

Figure 4.3b shows the continuous-time block diagram of a hard-decision decoding DPPM receiver. The received signal passes through the receiving filter $r(t) = p(-t)$, which is relative to $p(t)$ in Fig. 4.3a. A limiter in the output part of the filter detects whether the time slot is high or low. Figure 4.3c shows the discrete-time equalization system for obtaining the detected time-slot sequence. The discrete-time impulse response is $f_k = p_c p(t) \otimes h(t) \otimes r(t)|_{t=\mathrm{KT}}$ and $h(t)$ is the channel impulse response.

We considered a DPPM system through a continuous channel. Let P_0 be the probability that a low-level slot in a superframe is detected as a high-level slot and P_1 be the probability that a high-level slot in a superframe is detected as a low-level slot. Assuming that the input bit stream is a discrete and identically distributed Bernoulli process, the bit-error rate of a superframe containing F bits and N slots is

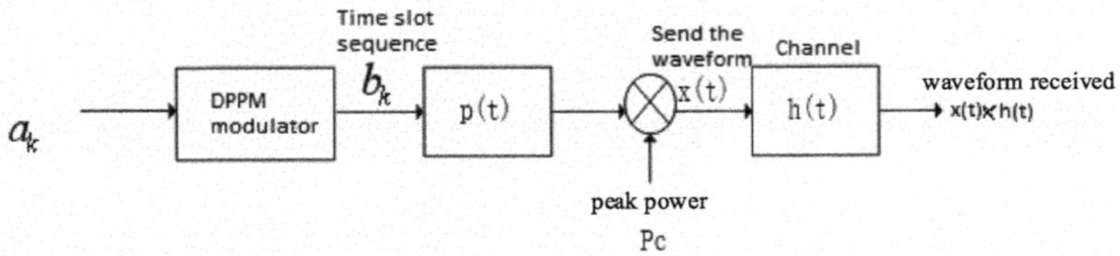

(a) Time-structure diagram for a DPPM transmitter. The unit-amplitude pulse p(t) can be generated in the high-level slot in the transmission sequence, and the pulse-width slot period is Tc.

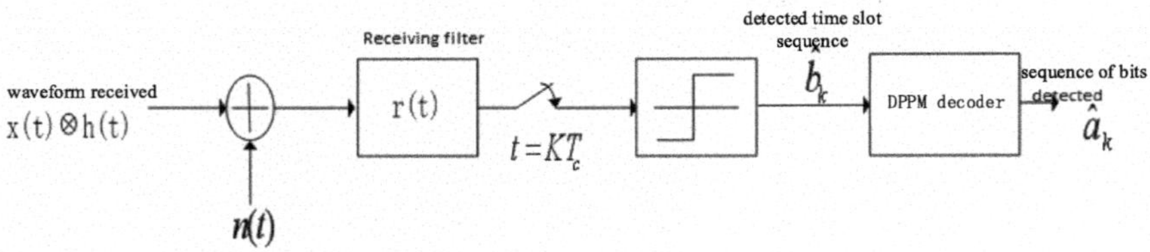

(b) Continuous-time block diagram of an unbalanced hard-decision DPPM receiver

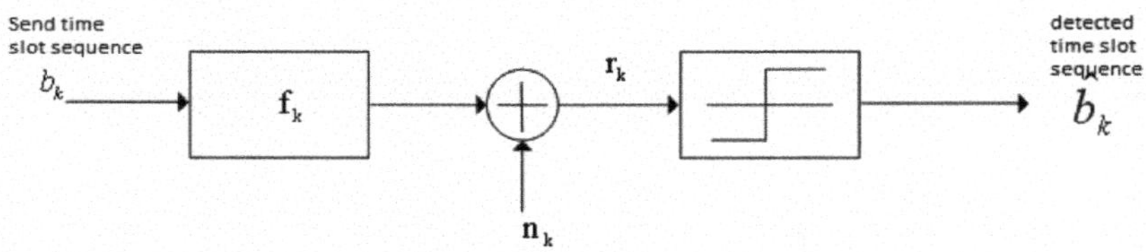

(c) Discrete-time equalization system of a time-slot sequence

Fig. 4.3 Conceptual diagrams of a DPPM system, **a** Time-structure diagram for a DPPM transmitter. The unit-amplitude pulse p(t) can be generated in the high-level slot in the transmission sequence, and the pulse-width slot period is Tc, **b** continuous-time block diagram of an unbalanced hard-decision DPPM receiver, **c** discrete-time equalization system of a time-slot sequence

[6]

$$P_F = 1 - (1 - P_0)^{N-(F/\log_2^L)}(1 - P_1)^{(F/\log_2^L)} \approx \left(N - \frac{F}{\log_2^L}\right)P_0 + \frac{F}{\log_2^L}P_1 \quad (4.3)$$

If $P_0 = P_1$则:

$$P_F \approx NQ\left(RP_t\sqrt{\frac{(L+1)\log_2^L}{8R_bN_0}}\right)F,$$

where N_0 is the spectral-energy density of bilateral noise and $((L-1)/4)(F/\log_2 L)^2$ is the average bit rate.

4.3 Maximum-Likelihood Detection of DPPM

4.3.1 *Maximum-Likelihood Detection in Continuous Channels*

Because the length of each frame of a DPPM signal is unknown before detection, when detecting the n-slot DPPM superframe encoded by the F-bit signal, we must compare all the superframes with the same length. Considering that each received slot sample represents an orthogonal dimension, maximum-likelihood detection is equivalent to minimum Euclidean-distance detection. When the detected superframe and the transmitted superframe are at different positions in a specific high-level slot, minimum-distance error detection is equivalent to position error detection, and the Hamming distance between the two superframes is two.

Compared with PPM, the minimum-distance error of DPPM can be arranged by moving the high-level time slot. As long as the final superframe does not exceed L low-level time slots, it can be placed anywhere in the superframe. The number of packets located in Hamming distance 2 is not fixed and the valuation is $((L-1)/4)(L/\log_2 L)^2$. This estimate shows that the average superframe length in the time slots is $\overline{N} = (F/\log_2 L)((L+1)/2)$, and about half of the high-level time slots are the above. When the superframe is short and the signal-to-noise ratio is very high, the superframe error is mainly the minimum-distance error. The maximum estimated superframe error rate is [6]

$$P_{F,DPPM} \cong \left(\frac{L-1}{4}\right)\left(\frac{F}{\log_2 L}\right)^2 Q\left(RP_t\sqrt{\frac{(L+1)\log_2 L}{4R_bN_0}}\right). \qquad (4.4)$$

4.3.2 *Maximum-Likelihood Detection in Discrete Channels*

In equalized discrete-time systems, maximum-likelihood detection (MLSD) and equalization techniques are often used in channels with intersymbol interference and Gaussian white noise [9], as shown in Fig. 4.4a. The continuous time system consists of a transmission filter $p(t)$, channel impulse response $h(t)$, matched filter $r(t) = p(-t) \otimes h(-t)$, and white-noise filter $w(z)$, where $w(z)$ is the inverse transformation of the maximum phase $h_k = p(t) \otimes h(t) \otimes r(t)|_{t=KT_c}$. The impulse response of the discrete-time channel is $g_k = P_c h_k \otimes W_k$, where $p(t)$, $h(t)$, $r(t)$, and g_k must be greater than zero because they all represent optical energy.

To calculate the superframe bit-error rate for MLSD detection, let $\overline{e}$ be the vector $\{-1, 0, 1\}$; the first and last terms of $\overline{e}$ should be non-zero. If the zero vector in $\overline{e}$ is less than that in f_k, $\overline{e}$ is defined as the error unit. As long as all time-slot values plus error units are zero, any non-overlapping error units can be detected through MLSD.

In a single-error event, each superframe contains only one error unit, and two error units exist in a double-error event. In addition, the total of all time-slot values plus any error units is not 0. A double-error event is not equal to the combination of two error events. In addition, S is all error units $\overline{e}$ that cause single errors, and D is all error units $\left(\overline{e}^1, \overline{e}^2\right)$ that cause double errors; therefore, the maximum error rate of the superframe is approximately [6]

$$P_{FDPPM} \approx \sum_{\in S} a(\overline{e}) Q\left(\frac{1}{2}\sqrt{\frac{\sum_k |e_k \otimes g_k|^2}{N_0}}\right) + \left(\sum_{(\overline{e}^1, \overline{e}^2)\in D} \frac{F}{2} a\left(\overline{e}^1, \overline{e}^2\right) Q\left(\frac{1}{2}\sqrt{\frac{\sum_k \left|e_k^1 \otimes g_k\right|^2 + \sum_k \left|e_k^2 \otimes g_k\right|^2}{N_0}}\right)\right). \tag{4.5}$$

The a priori probabilities of $a(\overline{e})$ and $a\left(\overline{e}^1, \overline{e}^2\right)$ are $\overline{e}$ and $\left(\overline{e}^1, \overline{e}^2\right)$, respectively, and the maximum error rates of OOK and PPM can be approximated [6]:

$$P_{FOOK} \le \sum_{\overline{e}} 2^{-H(\overline{e})} Q\left(\frac{1}{2}\sqrt{\frac{\sum_k |e_k \otimes g_k|^2}{N_0}}\right) \tag{4.6}$$

$$P_{FPPM} \le \sum_{\overline{e}} L^{-H(\overline{e})} Q\left(\frac{1}{2}\sqrt{\frac{\sum_k |e_k \otimes g_k|^2}{N_0}}\right). \tag{4.7}$$

$H(\overline{e})$ is the sum of non-zero elements in error event $\{\overline{e}\}$.

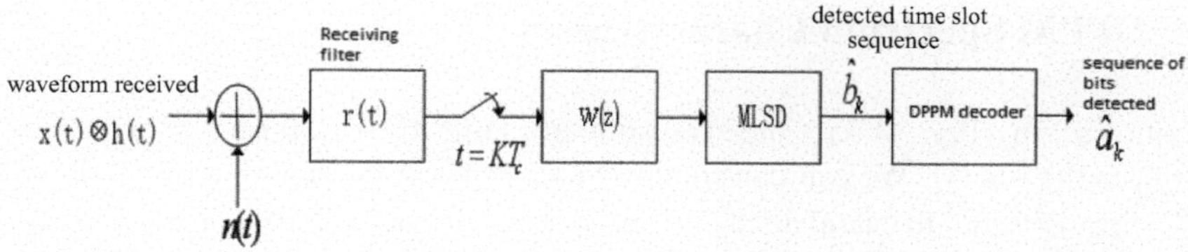

(a) Block diagram of a continuous-time white-noise matched-filter MLSD receiver

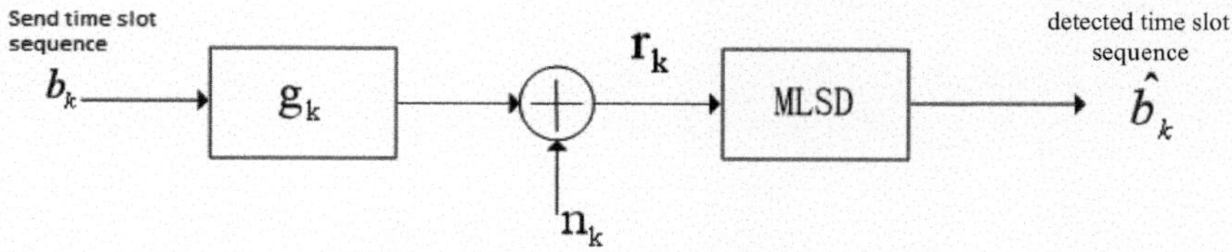

(b) Discrete-time equalization system of the received detection-slot sequence

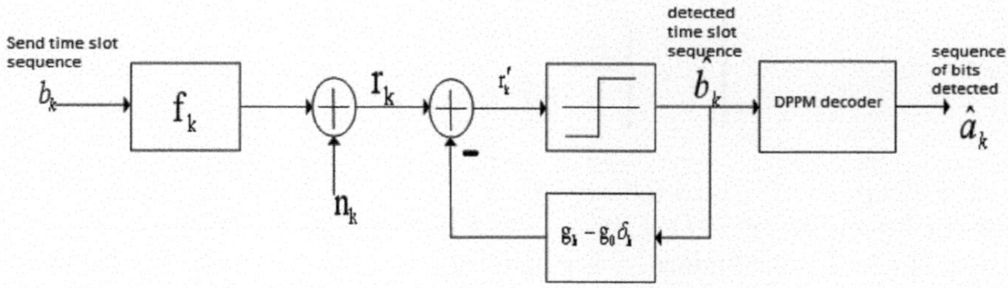

(c) Decision-feedback equalization of hard-decision decoding

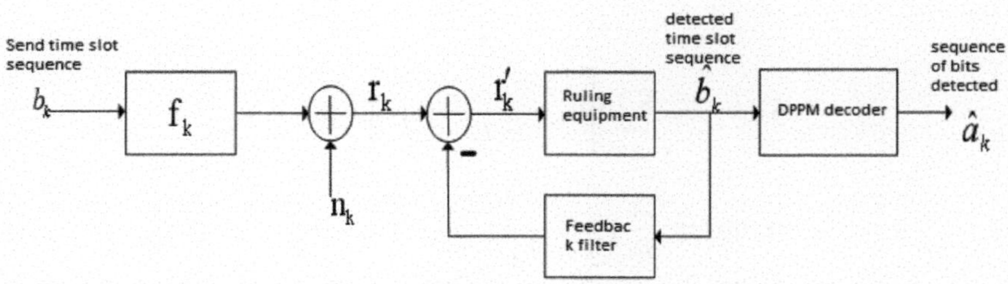

(d) Multislot-rate decision-feedback equalization

Fig. 4.4 Conceptual diagrams of a maximum-likelihood receiving system, **a** Block diagram of a continuous-time white-noise matched-filter MLSD receiver, **b** discrete-time equalization system of the received detection-slot sequence, **c** decision-feedback equalization of hard-decision decoding, (d) Multislot-rate decision-feedback equalization

4.4 DPPM Spectrum Characteristics

As shown in Fig. 4.3a, the transmitted DPPM waveform is $x(t) = \sum_{\mathrm{k}=-\propto}^{\propto} b_k \mathrm{P}(t - kT_{\mathrm{c}})$. We can consider $x(t)$ as a periodic process and use $S_x(\omega) = (1/T_{\mathrm{c}}) S_{\mathrm{b}}(\omega)|P(\omega)|^2$ to calculate the power-spectral density (PSD). For square wave $p(t)$, $|P(\omega)|^2 = T_{\mathrm{c}}^2 \mathrm{sinc}^2(\omega T_{\mathrm{c}}/2\pi)$ and $\mathrm{sinc(x)} \equiv \sin(\pi \mathrm{x})/\pi \mathrm{x}$. The spectral energy density $S_b(w)$ of the time-slot sequence can be obtained by the discrete Fourier transform of time-slot autocorrelation function r_k, which is defined as $r_{n-m} = E(b_n b_m) rnmbE$. The recursive relation of r_k is [6]

$$r_k = \frac{2}{L+1} \sum_{j=1}^{\min\{L,K-1\}} P_r\left(b_k = 1,\, b_{k-1} = \cdots = b_{k-j+1} = 0,\, b_{k-j} = 1 \middle| b_0 = 1\right). \tag{4.8}$$

It can be simplified to

$$\left\{ \begin{array}{c} r_0 = 2/(L+1) \\ r_k = \left\{ \begin{array}{ll} \frac{2}{L^k}(L+1)^{k-2} & 1 \le k \le L \\ \frac{1}{L}\sum\limits_{\mathrm{i}=1}^{\mathrm{L}} r_{k-\mathrm{i}} & k > L. \end{array} \right. \end{array} \right. \tag{4.9}$$

The algorithm of r_k is as follows: When k > L, $r_k = \sum_{i=1}^{L} c_i x_i^k$, where x_i is the basis of the equation $x^L = (1/L)\left(x^{L-1} + x^{L-2} + \cdots + 1\right)$, where r_k and $k = 0, 1, \ldots, L-1$. If $L - 1$ is determined, c_i is determined. It is not difficult to see that one x_i has only one value and the other x_i has only one absolute value.

We find that $S_b(\omega) = S_{\mathrm{c}}(\omega) + S_d(\omega)$, where $S_c(w)$ is a continuous component and $S_d(w)$ is a discrete component obtained from a nonzero value in the time-slot sequence; then, we obtain

$$S_{\mathrm{c}}(\omega) \cong \sum_{\mathrm{k}=-5L}^{5L} \left[r_k - \left(\frac{2}{L+1}\right)^2 \right] \exp(-jk\omega) \tag{4.10}$$

$$S_d(\omega) = \frac{2\pi}{T_{\mathrm{c}}} \left(\frac{2}{L+1}\right)^2 \sum_{\mathrm{k}=-\propto}^{\propto} \delta\left(\omega - \frac{2\pi \mathrm{k}}{T_{\mathrm{c}}}\right). \tag{4.11}$$

The spectrum diagrams of 8-DPPM, 16-DPPM, and 32-DPPM are shown in Fig. 4.5. Compared to PPM, the continuous part of the power-spectral density of DPPM is not close to zero in the low-frequency band.

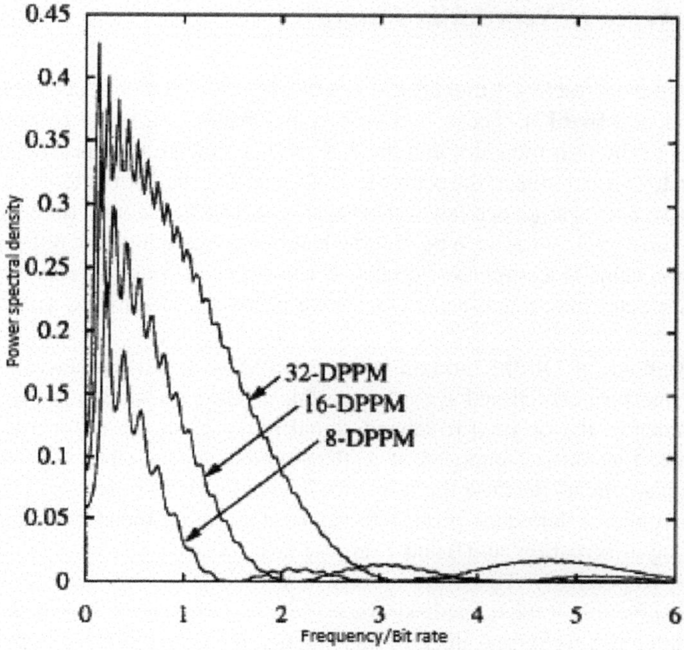

Fig. 4.5 DPPM spectra

4.5 Schemes for Realizing DPPM [10]

One characteristic of DPPM is that the time required to transmit the same amount of information is not fixed [6, 7, 11], and its average bit rate is $\overline{R}_b = 2/(L+1)T_c$, whereas the actual bit rate varies from $((L+1)/2L)\overline{R}_b$ to $((L+1)/2)\overline{R}_b$. If superframe information is transmitted for a longer time than the superframe, it could cause a transmission–buffer overflow and waste network resources; conversely, it could cause transmission-buffer underflow, reception-buffer overflow, or a temporary eye injury, owing to the high frequency of the optical signal [6].

In general systems, the time required to transmit the same amount of information is the same. Because of this characteristic of DPPM, the design of the system is more complex than that of PPM. There are two options: Having the transmitter detect the content of each information frame before transmission, and using large superframes to reduce the bit-error rate.

4.5.1 Scheme 1: Network in Frame

It can be seen from Table 4.1 that DPPM has two types of time-slot correspondence: positive and negative [7]. These two correspondence relations correspond to the same signal in the mth time slot and the $L - m + 1$ time slot, respectively. Taking hexadecimal as an example, if the source is 1110, it will be the 14th time slot according to the positive correspondence law; otherwise, it will be the second time slot. Thus, an average time of $(1 + L)/2$ can be obtained. When we calculate the pulse position, if the position value is less than or equal to 8, we follow the positive correspondence law; if it is greater than 8 and less than 16, we follow the negative correspondence law.

The advantage of DPPM modulation over PPM is that it can save bandwidth [7]. The frame formats defined according to this scheme are shown in Fig. 4.6. We divided frames of the original baseband signal into 32 time slots for a long time. The modulated signal is composed of a silent segment (protection segment) and a transmission segment. Because the transmission segment occupies up to eight time slots at this time, we define the protection segment as eight time slots. The first time slot is the flag information, and 0 and 1 represent the positive and inverse correspondence laws, respectively. To ensure a consistent time length for each information frame, if the position of the transmission section is less than eight, zero time slots are used to supplement eight time slots. In this manner, the time length of each frame of information after modulation is 16 time slots; that is, half that of the baseband signal.

Table 4.1 Mapping between source bits and transmission symbols

Source	4-PPM	4-DPPM	4-DPPM (back)
00	1000	1	0001
01	0100	01	001
10	0010	001	01
11	0001	0001	1

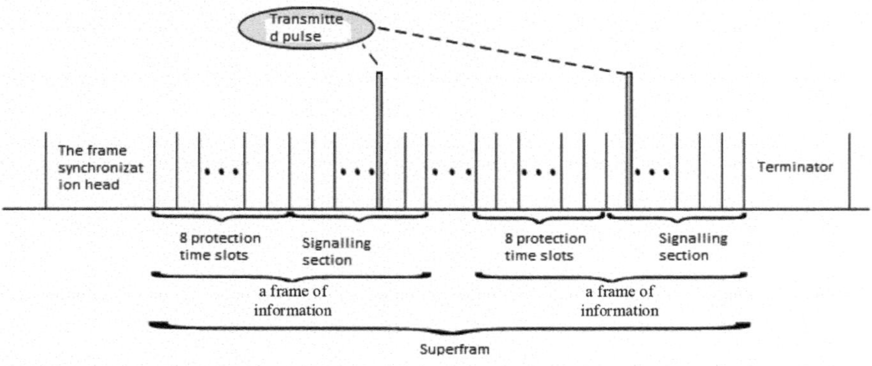

Fig. 4.6 System frame structure

As shown in Fig. 4.6, this scheme can save system bandwidth and improve energy utilization. In addition to reducing the time per frame, the other formats are consistent with PPM. The length of each frame is fixed and the position value cannot be determined by the distance between two adjacent pulses. Therefore, we should define a superframe structure similar to that of PPM, and then add a superframe synchronization header before each superframe to capture the frame-synchronization header at the demodulation end.

However, from other data, we know that DPPM does not require frame synchronization because the pulse position value can be determined by the distance between two pulses, so the signal can be demodulated. Therefore, the frame format defined in this manner is not the traditional DPPM format but can be defined as an improved form of PPM and DPPM.

However, to modulate according to this frame format, it is necessary to judge the source before modulation to select the appropriate modulation correspondence law to complete the modulation with a fixed length of time. This method only requires elastic memory in the modulation part; real-time demodulation can be completed in the demodulation part. This method can also support time-division multiplexing and save bandwidth more effectively; however, the interference between the transmitted multichannel signals is unacceptably large. A hardware implementation was carried out using a field-programmable gate array (FPGA), and the system flow is shown in Fig. 4.7.

4.5.2 *Scheme 2: Network Based on Superframe [12]*

(1) Design scheme of a DPPM modulation system

The modulation system transforms a signal from a bit sequence to a time-slot sequence, including the serial–parallel conversion of the signal, calculation of the pulse position, and transmission of the pulse information through the buffer and modulation module. Figure 4.8 shows a hardware block diagram of the system implementation.

We now take 16-DPPM as an example to analyze the modulation scheme. The original signal was a randomly generated binary sequence. It is defined that every four-bit ($\log_2^L = \log_2^{16} = 4$) binary signal of the baseband signal is a frame, so the amount of information contained in a frame of the modulated signal is still four. However, because the DPPM signal has no fixed average period per frame, the time required to transmit the same bit-information frame is uncertain; if a frame of information is transmitted for a longer time than the frame, it will cause the transmission buffer to overflow and waste network resources.

Alternatively, it will cause the transmission buffer to underflow, the reception buffer to overflow, or a temporary eye injury. Owing to the time-varying characteristics of DPPM, the core part of the modulation part is the buffer and modulator. One

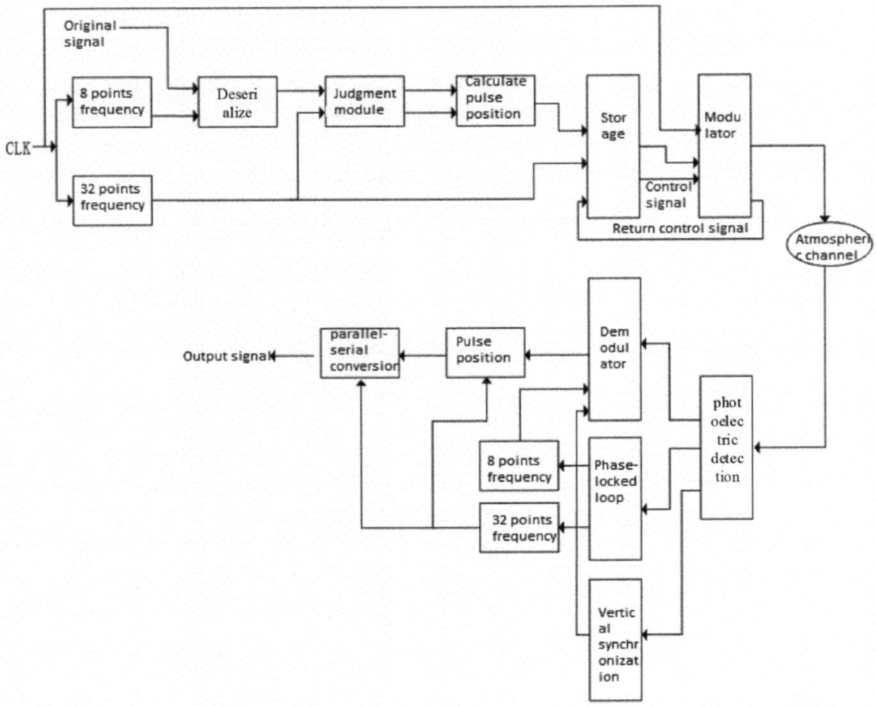

Fig. 4.7 System-hardware block diagram

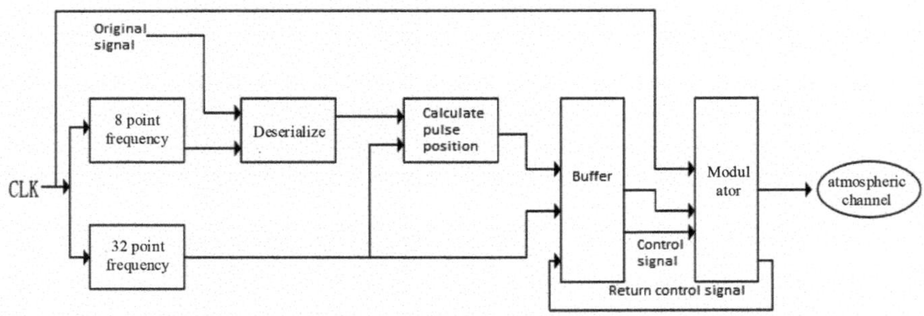

Fig. 4.8 Hardware block diagram of the modulation system

module controls the other, which avoids transmission- and reception-buffer overflow or underflow.

Figure 4.9 shows a flowchart of the buffer and modulator, where *i* is the address of the write buffer, that is, the address of the pulse position stored in the buffer; M is the capacity of the buffer; a_1, a_2, …, a_{m+1} is the output corresponding to the buffer; pulse is the transmitted pulse; enread is the read-enable signal, which is the signal of the buffer-control modulator; *j* is the address of the buffer read by the modulator, which is the signal of the modulator's anti-control buffer; and *cnt* is the time-slot value corresponding to the pulse position.

To improve the anti-interference ability of the transmitted pulse, we divided the frame information into protection and transmission time slots. The number of protection time slots was set to 16, and the number of transmission time slots was 1–16. In this manner, the modulated frame information is composed of 17–32 time slots. The information frame format is shown in Fig. 4.10a; therefore, *cnt* = pulse-position value + 16. The relationship between the read and pulse signals, according to the flow relationship in Fig. 4.9, is shown in Fig. 4.10b.

Here, the buffer capacity is set to 4×10, that is, 10 lines of a four-bit buffer. First, the pulse-position signal is stored in the buffer according to the baseband-signal frame rate. When the buffer is full, the enread signal is enabled to allow the modulator to read the pulse position. As the frame rate of the modulated signal information is

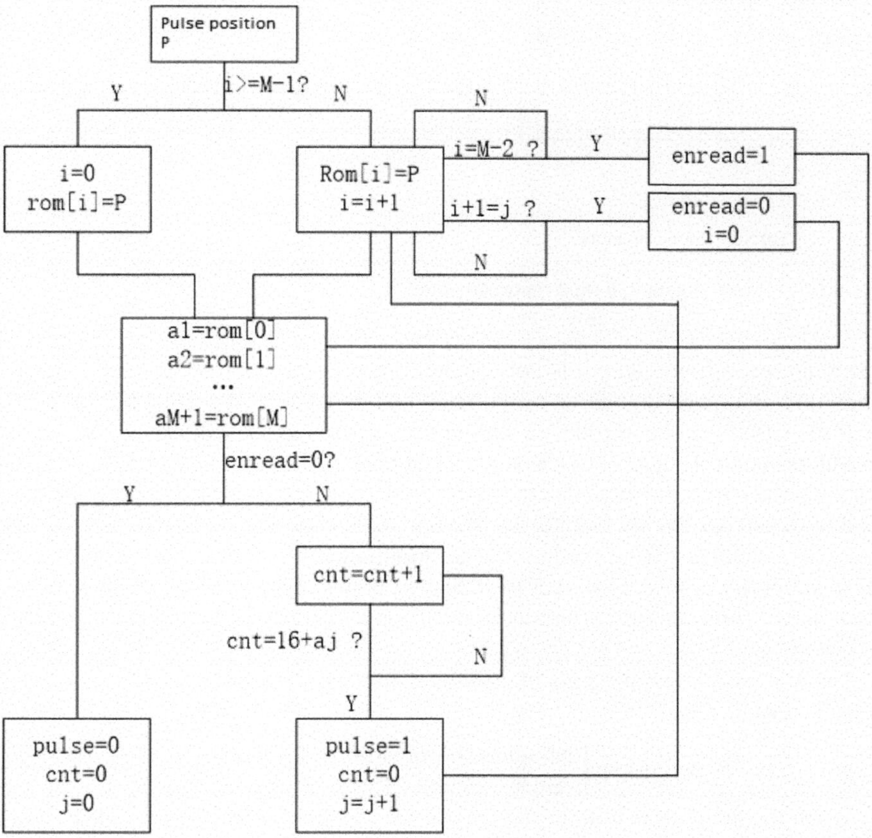

Fig. 4.9 Flowchart of buffer and modulator

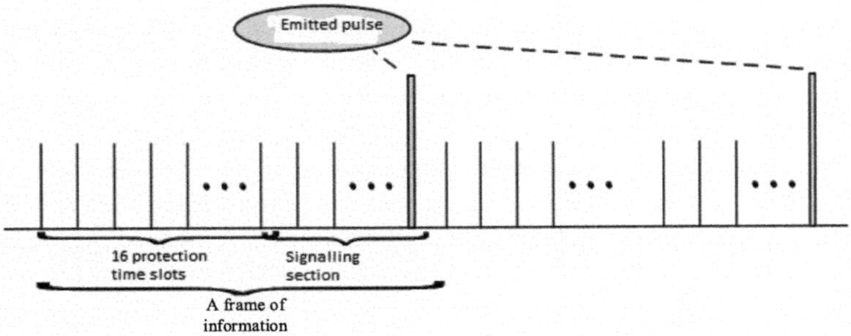

(a) Frame structure

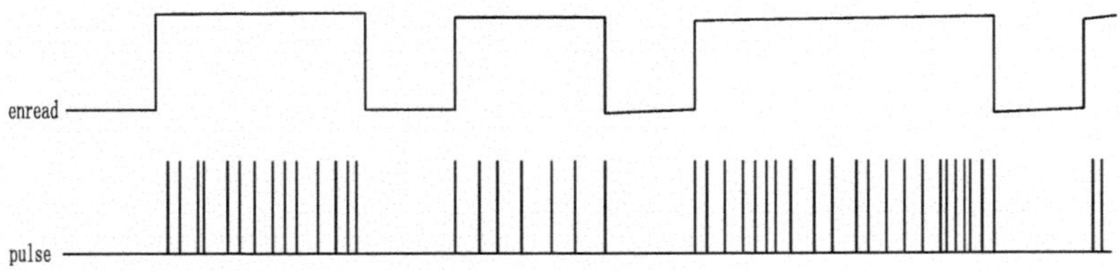

(b) Superframe structure

Fig. 4.10 **a** Frame structure, **b** superframe structure

uncertain, each frame length is correspondingly shorter than the baseband signal. Therefore, the speed of reading the buffer is faster than that of writing the buffer.

If the speed is not controlled, a sliding-code phenomenon will occur. Therefore, an address-control signal exists between the buffer and the modulator; that is, address i of the write buffer mentioned in the above paragraph and address j of the modulator read buffer. When the two addresses overlap, the enread signal is reset to 0, which stops the modulator from reading the information in the buffer, and address i of the write buffer is reset to 0; that is, it starts a new storage area. According to this relationship, information can be continuously modulated, and a sliding-code problem can be avoided.

The implementation program of the modulation buffer using the Verilog language is as follows:

- 4×10 buffer definition: $\begin{cases} \text{parameter wordwidth} = 4\text{,memsize} = 10; \\ \text{reg[wordwidth - 1:0] rom[memsize - 1:0];} \end{cases}$
- When the memory is full, the address of the memory buffer is reset and stored.
- Output the buffer.

The implementation program of the modulation module using the Verilog language is as follows:

- If enread is not enabled, the read address is the first address 0, and there is no pulse output; that is, the pulse signal is 0, and the slot-count signal *cnt* is counted again.
- If enread is enabled, the slot-count signal *cnt* is incremented every time the clock rises and holds.

The read address is incremented using a case statement, and the initial value is 0. According to the frame structure defined in Fig. 4.10b, when the slot-count signal reaches protection slot 16 plus the pulse-position value of the first bit of the memory, the pulse-signal pulse becomes 1 at the corresponding slot, the slot count signal is reset, and the read address is incremented to 1. Then, when the read address is 1, the previous calculation is repeated, the read address is increased to 2, etc. The read address increases to 9, and then the read address is reset to restart a new round of calculation.

(2) Design scheme of a DPPM demodulation system

The demodulation system extracts clock-synchronization information from the channel, extracts pulse-position information, converts the position information into *n* metadata groups according to the corresponding mapping relationship, conducts a parallel–serial conversion, and outputs binary information. Figure 4.11 shows a hardware block diagram of the system implementation.

As shown in Fig. 4.10b, when enread is enabled, a pulse is transmitted; otherwise, if there is no pulse, the form of the information received by the receiver is the same as the pulse. The time when enread is not enabled is fixed; that is, the time taken to fill the buffer. Each time enread is enabled, the transmitted pulse signal is composed of several frames. Because these frame signals are DPPM-modulated, the period of each frame is not fixed. If they are demodulated in this manner, the demodulated pulse-position value, that is, the length of the four-bit parallel data, is not fixed. Therefore,

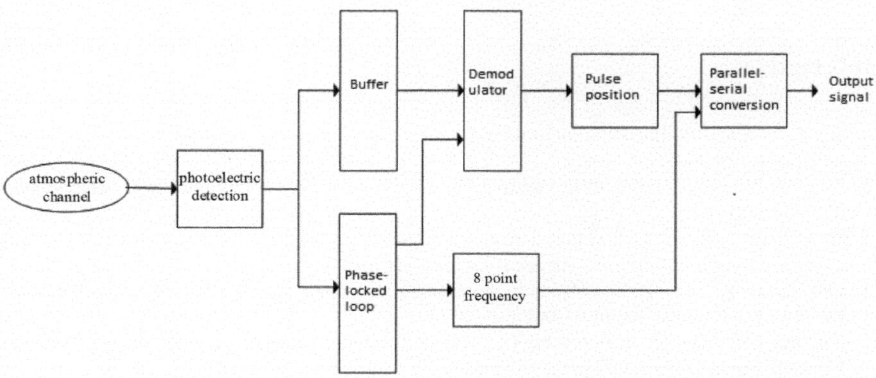

Fig. 4.11 Hardware block diagram of the demodulation system

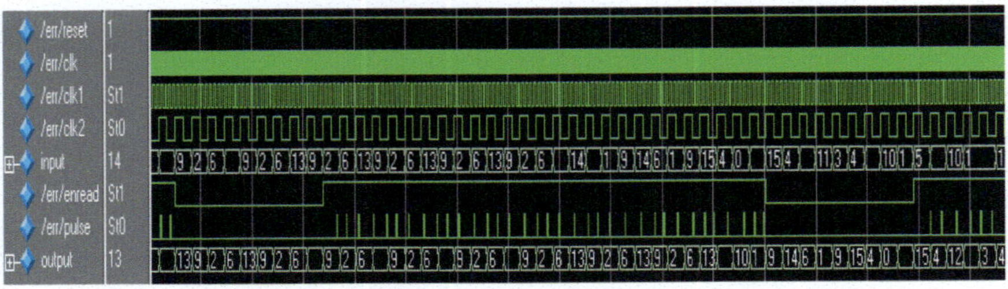

Fig. 4.12 Simulation results

there is no fixed clock to control the inverse transformation from the pulse-position value to the original signal, so it cannot be demodulated successfully.

However, according to the analysis of the modulation part, the time of each enread-enabled segment plus its nonenabled segment is simply the sum of the baseband time of the transmitted frame. In this manner, we can define each enread-enabled segment plus its non-enabled segment as a superframe, and each superframe is independent of the others. The non-enabled segment of enread can be used as the superframe-synchronization header, which is why DPPM does not need to design a frame-synchronization header, but only needs to use a phase-locked loop (PLL) for clock synchronization at the receiving end [13–20]. In this manner, we need only one superframe to demodulate one superframe, and the internal delay of the system is the time when the buffer at the modulation end is filled.

The system was implemented according to the scheme described above. It was written in the Verilog language [21, 22] and simulated using the ISE integrated tool. The obtained waveform is shown in Fig. 4.12. The reset signal is the system reset signal, CLK is the master clock, CLK1 is the clock after the frequency division of main clock 8, CLK2 is the clock after the frequency division of main clock 32, input is the original signal, output is the demodulated signal, and pulse is the transmitted pulse, that is, the received signal form.

References

1. Ke X, Xi X (2004) Introduction to wireless laser communication. Beijing University of Posts and Telecommunications Press
2. Wang J, Zhang M (2000) PPM modulation in optical wireless communication. Telecommun Technol 5:81–84
3. Pang Z, Pu D, Zou C (2022) Performance comparison of several modulation modes in optical communication. J Guilin Inst Electron Technol 22(10):1–4
4. Guo J, Tan Y, Ai Y (2006) Research on modulation technology in satellite optical communication. Opt Commun Technol (4):45–46
5. Zhou X, Chen W, Xiong Z, Huang T (1994) Performance research of direct detection optical PPM coding system under background noise limit. Telecommun Sci 10(12):28–31
6. Shui D, Kahb JM (1999) Differential pulse-position modulation for power-efficient optical communication. IEEE Trans Commn 47(8):1201–1209

7. Daniel Z (1988) Differential PPM has a higher throughput than PPM for the band-limited optical channel. IEEE Trans Inf Theory 34(5):1269–1273
8. Zhou L, Yin D (1998) Basic analysis of coding and modulation of multi pulse PPK. J Univ Electron Sci Technol 27(5):51–521
9. Pu D, Zou C (2002) Maximum likelihood detection of sampled PPM signal under white Gaussian noise. J Guilin Inst Electron Technol 22(6):5–8
10. Li Z (2003) Research on DPPM modulation and demodulation technology of atmospheric laser communication. Xi'an University of Technology
11. Tomoaki O, Iwao S, Shinsaku M (1994) Differential overlapping pulse position modulation in optical direct-detection channel. IEEE Trans Inf Theory 37(8):680–684
12. Li Z, Ke X, Liu J (2007) Research on DPPM modulation and demodulation technology in OW. Laser J 35(2):12–20
13. Ding D, Ke X (2005) Design and simulation of PPM modulation and demodulation system for atmospheric laser communication. Opt Commun Technol 15(1):50–52
14. He D, Wang F, Huang B, Huang T (1996) Laser underwater communication recognition decision frame synchronizer. J Huazhong Univ Technol 24(2):95–97
15. Wang F, Huang B, He D (1996) Trigger counting frame synchronizer for laser underwater communication. J Huazhong Univ Technol 24(2):92–94
16. Wang F, Huang B, He D (1995) Time slot synchronizer for laser underwater ppm Communication. J Huazhong Univ Technol 23(10):52–56
17. Davidson FM (1989) Slot Clock Recvery in optical PPM communication systems with avalanche photodiode photodetectors. IEEE Trans Commun 37(18):1164–1172
18. Vilnrotter VA, Rodemich ER (1986) A synchronization technique for optical PPM signals. Proc IEEE Military Com Conf (MILCOM'86), Montery, CA, 13(9):4521–4526
19. Ling G, Gagliardi RM (1986) Slot synchronization in optical PPM communications. IEEE Trans Commun 34(19):1202–1208
20. Otte R, de Jong LP, van Roermund AHM (1998) Slot synchronization by reducing the PPM pulsewidth in wireless optical systems. IEEE Trans Circuits Syst—II: Analog Dig Signal Process 45(18):901–903
21. Xia Y (2004) Verilog digital system design course. Beijing University of Aeronautics and Astronautics Press
22. Chang X (2003) Verilog HDL practice and application system design. Beijing University of Aeronautics and Astronautics Press

Chapter 5
TCM Code-Modulation Technology

This chapter analyzes the principle of trellis-coded modulation (TCM) in detail, focuses on the structure of a grid-coded modulator, summarizes the design method of the grid-coded modulator, analyzes the performance of the TCM system in terms of the error probability and coding gain of the grid-coded signal, and provides a hardware implementation scheme for TCM-coded modulation.

Trellis-coded modulation A method called trellis-coded modulation (TCM) has been developed since the 1980s. It combines coding and modulation and uses state memories and appropriate mappings to increase the distance between code-word sequences. TCM was proposed by Ungerboeck in 1982 [1]. This method does not reduce either the bandwidth utilization or the power utilization but trades the device complexity for the coding gain. This approach is certainly attractive at a time when integrated circuits are developing at a high rate, and the cost of transmission media is higher than the cost of end devices. and is the first consideration for communication costs. Nowadays, TCM technology has dominated in a large number of applications in frequency band and power limited channels simultaneously such as space, satellite, microwave, coaxial line and twisted pair communication [2–7].

5.1 OPPM Modulation Method

For the pulse-position modulation (PPM) transmission method [8, 9], when the length of the message frame is fixed, the transmission rate can be improved by shortening the width of the optical pulse; however, when the optical pulse width has been shortened to the limit, another method needs to be considered. If pulse positions are allowed to overlap, the transmission rate can continue to improve without shortening the optical pulse width, leading to overlapping pulse-position modulation (OPPM). OPPM is an improved form of PPM, which is an extension of the PPM-modulation signal set, and has a higher transmission efficiency than PPM [10–12].

X. Ke, *Coding Theory in Optical Wireless Communication Systems*, Optical Wireless Communication Theory and Technology, https://doi.org/10.1007/978-981-19-9837-9_5

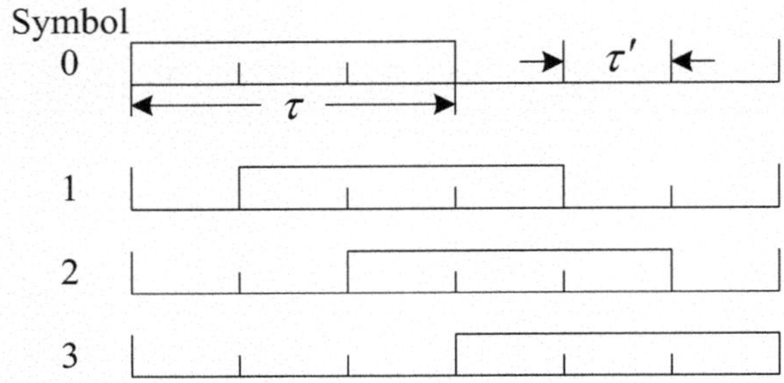

Fig. 5.1 Schematic diagram of OPP modulation

5.1.1 Basic Principle of OPPM Signal Generation

In the Q-element PP modulation method, a message frame of length T_L consists of Q time slots of length τ. In the OPP modulation method, each time slot is divided equally into N subintervals of length τ', N is called the overlap factor, and the information is transmitted using pulse positions of length τ with $t_k = (k - 1)\,\tau'$, $k = 1, 2, \ldots, J$ as the starting position. The relationship between J, Q, and N can be expressed as [10]

$$J = N(Q - 1) + 1. \tag{5.1}$$

Figure 5.1 presents a schematic diagram of OPP modulation with $Q = 2$, $N = 3$, and $J = 4$. The number of transmitted signals increases from Q to J. The pulse width of the OPPM signal remains the same as that of the PPM signal, provided that the pulses are allowed to overlap, and the Q-element PPM is converted into the J-element OPPM. After extending the modulated signal set, the OPPM signal will no longer remain orthogonal to the PPM signal, which means that the demodulation-error probability at the receiver side will increase, thus imposing higher requirements on the synchronization system.

5.1.2 OPPM Performance Analysis

(1) **OPPM modulated information-transmission rate**

We first examine the information-transmission rate of OPPM when the upper limit of the system bandwidth is B_L. Because the pulse width of the OPPM signal remains the same as that of the PPM signal and the system bandwidth is equal to the reciprocal of the pulse width, OPPM requires the same bandwidth as PPM. For PPM, the information-transmission rate r (in natural information units (*nats)/s*) can be expressed as

$$r \le B_L \frac{\ln(Q)}{Q} \le \frac{B_L}{2} \ln(2). \tag{5.2}$$

For OPPM,

$$r \leq B_L \frac{\ln[N(Q-1)+1]}{Q} \leq \frac{B_L}{2}\ln(N+1). \tag{5.3}$$

Both Eqs. (5.2) and (5.3) can be maximized when $Q = 2$, which means that for PPM and OPPM, when the information-transmission rate becomes the primary consideration, limiting $Q = 2$ can obtain the maximum information-transmission rate. When $N > 1$, it can be clearly seen that OPPM has advantages over PPM in terms of the information-transmission rate.

(2) **Demodulation-error probability of OPPM**

Assuming a Poisson distribution for the channel model, a pulse position of length τ with $t_k = (k - 1)\,\tau'$, $k = 1, 2, \ldots, J$ as the starting position is used to transmit information in OPPM. The photon count in each pulse position is denoted by N_k, $k = 1, 2, \ldots, J$, which obeys a Poisson distribution with parameter $(i\lambda_s\,\tau' + \lambda_n\,\tau)$, where λ_s is the signal strength, λ_n is the noise strength, and i is the overlap factor. The receiver determines the largest N_k as the location of the signal pulse, and if there are two equal N_k, it randomly selects one of them.

First, we consider the quantum-limited case, that is, $\lambda_n \ll \lambda_s$, with the limiting case of $\lambda_n = 0$. Binary vector $x = (x_1, x_2, \cdots, x_{NQ})$, where $x_i = \begin{cases} 1, n_i > 0 \\ 0, n_i \leq 0 \end{cases}$, n_i $(i = 1, 2, \ldots, NQ)$, denotes the photon count in the ith subinterval of length τ'. The channel model in the absence of background noise is transformed into a deletion channel, and the deletion probability in each subinterval of length τ' can be expressed as

$$\varepsilon = e^{-\lambda_s \tau'}. \tag{5.4}$$

This means that when the signal pulse appears in the ith subinterval of length τ', the probability that $x_i = 0$ is ε and the probability that $x_i = 1$ is $1 - \varepsilon$.

The OPPM schematic shows that the receiver has an equal probability of a correct decision when the first or Jth symbol is sent, which is denoted by P_1; the receiver has an equal probability of a correct decision when the remaining $J - 2$ symbols are sent, which is denoted by P_2. The demodulation-error probability P_e of OPPM can be expressed as

$$P_e = 1 - \frac{1}{J}[2P_1 + (J-2)P_2], \tag{5.5}$$

where

$$P_1 = (1-\varepsilon)\left[1 + \sum_{k=1}^{N-1} \frac{\varepsilon^k}{k+1}\right] + \frac{\varepsilon^N}{J} \tag{5.6}$$

$$P_2 = (1-\varepsilon)^2 + (1-\varepsilon)(2-\varepsilon)\sum_{k=1}^{N-1}\frac{\varepsilon^k}{k+1} + \frac{\varepsilon^N}{J} + \frac{\varepsilon^N}{N}(1-\varepsilon). \quad (5.7)$$

Assuming that symbol 1 is sent, then only subintervals 1, 2, …, N may appear as nonzero photon counts. The probability that a nonzero photon count occurs in subinterval 1 is $1-\varepsilon$, and if this event occurs, the probability of a correct verdict at the receiver is 1. The probability that a nonzero photon count does not occur in subinterval 1, but occurs in subinterval 2 is $\varepsilon\ (1-\varepsilon)$; if this event occurs, the probability that the receiver can make a correct verdict in subintervals 1 and 2 is 1/2. The probability that a nonzero photon count does not occur in subintervals 1 and 2, but occurs in subinterval 3 is $\varepsilon^2\ (1-\varepsilon)$; if this event occurs, the probability that the receiver can make a correct verdict in subintervals 1, 2, and 3 is 1/3.

Applying this reasoning to the other subintervals, while noting that if no nonzero photon counting occurs in all N subintervals, the receiver will make a random selection among the J transmit symbols, leads to the expression for P_1, and similarly for P_2.

We define the number of signal photons per information nat as μ; $\lambda_s\ \tau = \mu R\ln(J)$ and R is the code rate (the number of information symbols in the channel symbols; for an uncoded system $R = 1$). This leads to the following for an uncoded system:

$$\varepsilon = \mathrm{e}^{-\frac{\mu}{\mathrm{N}}\ln(J)}. \quad (5.8)$$

Substituting Eqs. (5.6) and (5.7) into Eq. (5.5) yields

$$P_e = 1 - \frac{1}{J}\left\{2\left[(1-\varepsilon)\left(1+\sum_{k=1}^{N-1}\frac{\varepsilon^k}{k+1}\right) + \frac{\varepsilon^N}{J}\right] + (J-2)\right.$$
$$\left.\left[(1-\varepsilon)^2 + (1-\varepsilon)(2-\varepsilon)\sum_{k=1}^{N-1}\frac{\varepsilon^k}{k+1} + \frac{\varepsilon^N}{J} + \frac{\varepsilon^N}{N}(1-\varepsilon)\right]\right\}. \quad (5.9)$$

Two specific cases are discussed below: when $Q = 2$ and $J = N(Q-1)+1 = N+1$, substituting into Eq. (5.9) yields

$$P(e) = 1 - \frac{1}{N+1}\left\{2\left[(1-\varepsilon)\left(1+\sum_{k=1}^{N-1}\frac{\varepsilon^k}{k+1}\right) + \frac{\varepsilon^N}{N+1}\right] + (N-1)\right.$$
$$\left.\left[(1-\varepsilon)^2 + (1-\varepsilon)(2-\varepsilon)\sum_{k=1}^{N-1}\frac{\varepsilon^k}{k+1} + \frac{\varepsilon^N}{N+1} + \frac{\varepsilon^N}{N}(1-\varepsilon)\right]\right\}, \quad (5.10)$$

where $\varepsilon = e^{-\frac{\mu}{N}\ln(J)} = e^{-\frac{\mu}{N}\ln(N+1)}$.

For a particular overlap factor N, the trend of P_e with μ can be observed by plotting the curve with the number of signal photons per information nat μ as the horizontal coordinate and the demodulation-error probability P_e of OPPM as the

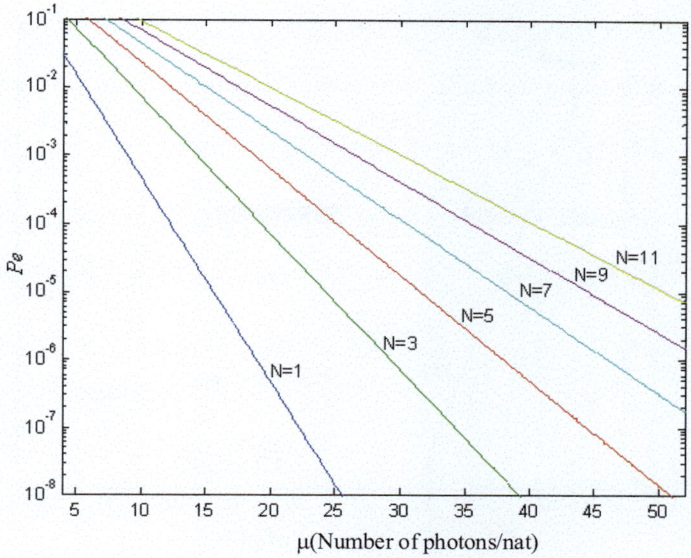

Fig. 5.2 Variation curve of demodulation-error probability P_e with μ for $Q = 2$, corresponding to different values of N

vertical coordinate. Figure 5.2 shows the $P_e - \mu$ curve corresponding to $N = 1$, 3, 5, 7, 9, and 11 for $Q = 2$. From Fig. 5.2, it can be observed that.

(i) As the overlap factor N increases, the probability of a demodulation error corresponding to the same μ value gradually increases;
(ii) For a fixed value of N, the demodulation-error probability gradually decreases as μ increases.

When $Q = 4$ and $J = N(Q-1)+1 = 3N+1$, substituting into Eq. (5.9) yields

$$P_e = 1 - \frac{1}{3N+1}\left\{2\left[(1-\varepsilon)\left(1+\sum_{k=1}^{N-1}\frac{\varepsilon^k}{k+1}\right)+\frac{\varepsilon^N}{3N+1}\right]+(3N-1)\right.$$
$$\left.\left[(1-\varepsilon)^2+(1-\varepsilon)(2-\varepsilon)\sum_{k=1}^{N-1}\frac{\varepsilon^k}{k+1}+\frac{\varepsilon^N}{3N+1}+\frac{\varepsilon^N}{N}(1-\varepsilon)\right]\right\}, \quad (5.11)$$

where $\varepsilon = e^{-\frac{\mu}{N}\ln(J)} = e^{-\frac{\mu}{N}\ln(3N+1)}$. The same method can be followed to obtain Fig. 5.3.

The same conclusion can be drawn from Fig. 5.3 as from Fig. 5.2, and it can also be observed that the demodulation-error probability decreases as Q increases from 2 to 4, corresponding to the same N and μ values. When $\lambda_n \gg \lambda_s$, the demodulation-error probability is the average of the demodulation-error probabilities at the receiving end of each symbol sent, which can be expressed as

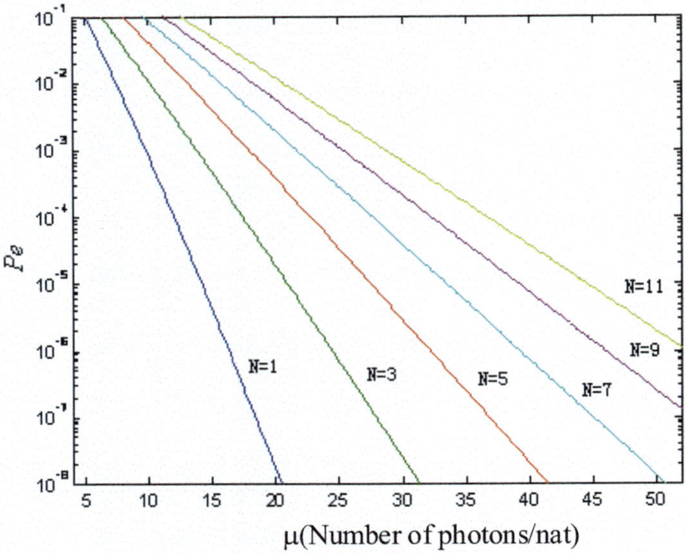

Fig. 5.3 Variation curves of demodulation-error probability P_e with μ corresponding to different N values for $Q = 4$

$$P_e = \frac{1}{J} \sum_{j=1}^{J} P(e|j), \tag{5.12}$$

where $P(e|j)$ denotes the probability that a demodulation error occurred at the receiver when symbol j was sent. Assuming that the photon count at the location of the transmitting light pulse is equal to the photon count at other locations, the following expression can be obtained:

$$P(e|j) \leq \Pr\left[\bigcup_{k \neq j} \{N_j \leq N_k\} | j \right]. \tag{5.13}$$

According to the concatenation bound of probability we can obtain

$$\Pr\left[\bigcup_{k \neq j} \{N_j \leq N_k\} | j \right] \leq \bigcup_{k \neq j} \Pr[N_j \leq N_k | j]. \tag{5.14}$$

The overlap between the light pulses of symbols j and k is $(N - |j - k|)$ and symbols j and k become orthogonal when $|j - k| > N$. The conditional probability in Eq. (5.14) can be expressed as

$$\Pr\left[N_j \le N_k | j\right] = \Pr\left[N_j' \le N_k'\right], \quad (5.15)$$

where N_j' and N_k' are two mutually independent random variables following the Poisson distribution, and the average photon counts are $\frac{(\lambda_s+\lambda_n)\tau}{N}d_{jk}$ and $\frac{\lambda_n\tau}{N}d_{jk}$, respectively, where d_{jk} is the distance between symbols j and k:

$$d_{jk} = \begin{cases} |j-k|, |j-k| \le N; \\ N, \ |j-k| \succ N \end{cases}. \quad (5.16)$$

From Eqs. (5.15) and (5.16), we obtain

$$\Pr\left[N_j \le N_k | j\right] = Q_1\left[\sqrt{\frac{2d_{jk}\lambda_n\tau}{N}}, \sqrt{\frac{2d_{jk}(\lambda_s+\lambda_n)\tau}{N}}\right], \quad (5.17)$$

where $Q_1(\alpha, \beta) = \exp\left(-\frac{\alpha^2+\beta^2}{2}\right)\sum_{k=0}^{\infty}\left(\frac{\beta}{\alpha}\right)^k I_k(\alpha\beta)$ is the Q function proposed by Marcum. Using the Chernoff bound in Eq. (5.15), it can be reduced to

$$\Pr\left[N_j \le N_k\right] \le \exp\left[-\frac{d_{jk}}{N}\left(\sqrt{\lambda_s\tau + \lambda_n\tau} - \sqrt{\lambda_n\tau}\right)^2\right]. \quad (5.18)$$

Combining Eqs. (5.12), (5.14), (5.17), and (5.18) yields

$$\begin{aligned} P_e &\le \frac{1}{J}\sum_{j=1}^{J}\sum_{k\ne j} Q_1\left[\sqrt{\frac{2d_{jk}\lambda_n\tau}{N}}, \sqrt{\frac{2d_{jk}(\lambda_s+\lambda_n)\tau}{N}}\right] \\ &\le \frac{1}{J}\sum_{j=1}^{J}\sum_{k\ne j}\exp\left[-\frac{d_{jk}}{N}\left(\sqrt{(\lambda_s+\lambda_n)\tau} - \sqrt{\lambda_n\tau}\right)^2\right]. \end{aligned} \quad (5.19)$$

Note that when $j \ne k$, $d_{jk} \in \{1, 2, \ldots, N\}$, denoted by $a(i)$, the number of times $d_{jk} = i$. $a(i)$ can be expressed as

$$a(i) = \begin{cases} 2(J-i), i = 1, 2, \cdots, (N-1); \\ (J-N)(J-N+1), i = N \end{cases}. \quad (5.20)$$

In this case, the demodulation-error probability of the uncoded system can be expressed as

$$P_e \le \frac{1}{J}\sum_{i=1}^{N} a(i) Q_1\left\{\sqrt{\frac{2i\lambda_n\tau}{N}}, \sqrt{\frac{2i[\lambda_n\tau + \mu\ln(J)]}{N}}\right\}$$

$$\leq \frac{1}{J}\sum_{i=1}^{N} a(i)e^{-i\gamma}, \tag{5.21}$$

where.$\gamma = \frac{1}{N}\left[\sqrt{\gamma_n\tau + \mu\ln(J)} - \sqrt{\lambda_n\tau}\right]^2$

Two specific scenarios are discussed below.

(i) When $Q = 2$ and $J = N(Q-1)+1 = N+1$, substituting into Eq. (5.21), we obtain

$$P_e \leq \frac{1}{N+1}\sum_{i=1}^{N} a(i)e^{-i\gamma}, \tag{5.22}$$

where $a(i) = \begin{cases} 2(\mathrm{N}+1-i), i = 1, 2, \cdots, (N-1); \\ 2, i = N \end{cases}$ and $\gamma = \frac{1}{N}[\sqrt{\lambda_n\tau + \mu\ln(N+1)} - \sqrt{\lambda_n\tau}]^2$; taking $\lambda_n\tau = 1, \gamma = \frac{1}{N}[\sqrt{1+\mu\ln(N+1)} - 1]^2$.

Taking $\lambda_n\tau = 1$, for a particular overlap factor N, the trend of P_e with μ can be observed by plotting the curve with the number of signal photons per information nat as the horizontal coordinate and the demodulation-error probability P_e of OPPM as the vertical coordinate. Figure 5.4 shows the $P_e - \mu$ curves corresponding to $N = 1, 3, 5, 7, 9$, and 11 for $Q = 2$ and $\lambda_n\tau = 1$.

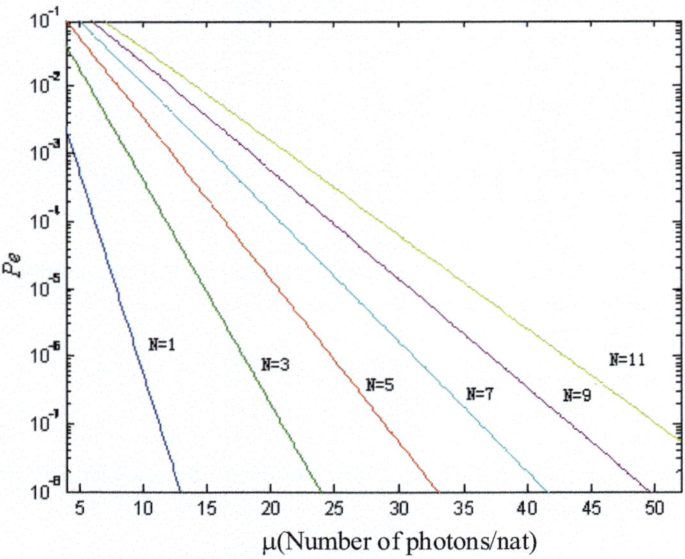

Fig. 5.4 Demodulation-error probability P_e versus μ for $Q = 2$ and $\lambda n\tau = 1$ corresponding to different values of N

The same conclusions can be drawn from Fig. 5.4 as in Fig. 5.2 above, except that the overall probability of the demodulation error for various N values increases compared to Fig. 5.3 because the test is performed at the background-noise limit.

(ii) When $Q = 4$ and $J = N(Q-1)+1 = 3N+1$, substituting into Eq. (5.21), we obtain

$$P_e \le \frac{1}{3N+1}\sum_{i=1}^{N} a(i)e^{-i\gamma}, \tag{5.23}$$

where $a(i) = \begin{cases} 2(3\text{N}+1-i), i = 1, 2, \cdots, (N-1); \\ 2(2N+1)(N+1), i = N \end{cases}$ and $\gamma = \frac{1}{N}[\sqrt{\lambda_n\tau + \mu\ln(3N+1)} - \sqrt{\lambda_n\tau}]^2$; taking $\lambda_n\tau = 1, \gamma = \frac{1}{N}[\sqrt{1+\mu\ln(3N+1)} - 1]^2$.

Following the same approach, Fig. 5.5 can be obtained. The same conclusions can be drawn from the figure as in Fig. 5.3 above. As the test is performed at the background-noise limit, the demodulation-error probability for all values of N increases in general, compared to Fig. 5.4.

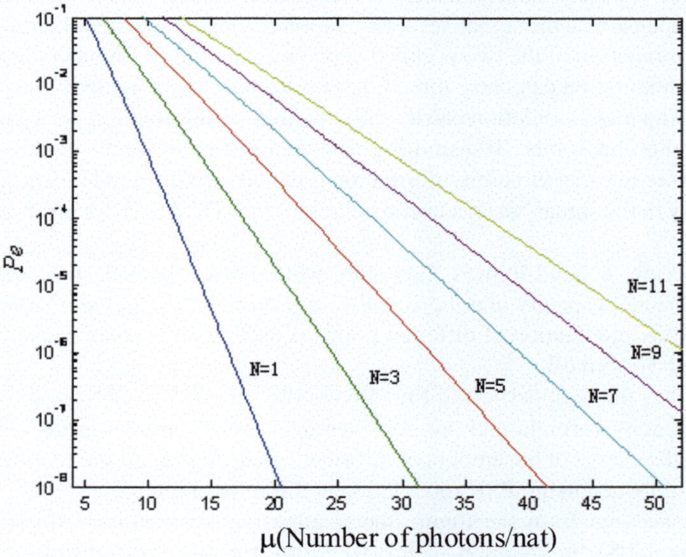

Fig. 5.5 Demodulation-error probability P_e versus μ for Q = 4 and λnτ = 1, corresponding to different values of N

5.2 TCM Principle Analysis

In the following, we analyze the theoretical basis of trellis-coded modulation (TCM), discuss the relationship between signal-set redundancy and coding gain, present the system model of TCM, and further analyze the structure and design method of the integrated grid-coded modulator in detail.

5.2.1 TCM Rationale

Grid-coded modulation is a “signal-space code” that combines coding and modulation to obtain error-correction capability by using the redundancy of signal sets [13]. For example, after encoding a two-bit message to generate a three-bit code set, a symbol’s ability to carry information can be increased by increasing the number of signal-set constellation points, such as changing 4-amplitude-shift-keying (ASK) modulation to 8-ASK modulation or 4-phase-shift-keying (PSK) modulation to 8-PSK modulation, to avoid widening the bandwidth while keeping the baud rate unchanged.

In the above process, the generation of redundant bits belongs to the coding category, the expansion of the signal-set constellation belongs to the modulation category, and the combination of the two is called coded modulation. Without coding, 8-ASK or 8-PSK modulation can carry three bits of information per symbol; the combination of coding and modulation carries only two bits of information per symbol after removing the check bits. Transmitting two-bit information with modulation that has the ability to carry three-bit information is called signal-set redundancy. It is this redundancy in the signal-set space (constellation) that TCM uses for error-correction capability.

Considering a band-limited Gaussian white-noise channel, the relationship between channel capacity and the signal-to-noise ratio (SNR) is shown in Fig. 5.6 when the PSK modulation of different points is used, with channel capacity in bits/symbol and SNR in dB.

The curves in the upper left of Fig. 5.6 are the curves derived from the Shannon channel-capacity formula and are considered as the theoretical limits. The lower curves are the curves of the amount of information carried versus the required signal-to-noise ratio when using PSK modulation at different points.

As can be seen from the figure, the required signal-to-noise ratio is 12.9 dB when using 4-PSK modulation with a 10^{-5} bit error rate and transmitting two-bit information per symbol; if 8-PSK modulation is used instead, with a 10^{-5} bit error rate and still transmitting two-bit information per symbol, the required signal-to-noise ratio is only 5.9 dB, which means that a 7-dB coding gain has been achieved. In the past, it was always taken for granted that 8-PSK should transmit three bits per symbol; however, TCM code breaks out of this traditional thinking and transmits two bits of information with 8-PSK, thus obtaining a considerable coding gain.

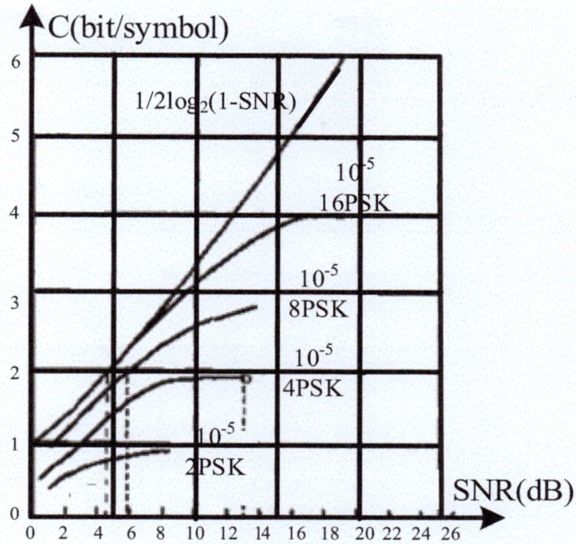

Fig. 5.6 Band-limited additive white Gaussian noise (AWGN) channel with PSK modulation. Channel capacity versus signal-to-noise ratio curve [1]

We can also use 16-PSK, 32-PSK, etc. to transmit two-bit information, and the required signal-to-noise ratio can be further reduced; however, it is not possible to exceed the 4.7-dB limit specified by Shannon's formula. This means that, regardless of the amount of effort, at most 1.2 dB of coding gain can be obtained with 8-PSK, instead of a 4-PSK transmission of two-bit information to obtain 7 dB of gain. Continuing to expand the signal set will make the equipment more complex and costly, for only a small benefit.

Therefore, TCM codes generally add only one redundant check digit, and the code rate R can be written as $m/m + 1$, which means that m-bit information is transmitted with a signal constellation of 2^{m+1} points per symbol. This 7-dB gain is the theoretical limit. Research on grid-coding modulation attempts to convert this potential gain into real gain; however, current engineering can achieve a maximum coding gain of no more than 6 dB.

5.2.2 *Analysis of the Structure and Design Method of a Grid-Coded Modulator*

While the channel encoder (decoder) and modulator (demodulator) in a general communication system are two independent modules, the TCM system is characterized by the fact that the encoder and modulator at the transmitter side are connected by the mapping module as a whole, and the demodulator and decoder at the receiver side are connected by the reflector module as a whole. The TCM system block diagram is shown in Fig. 5.7 [13].

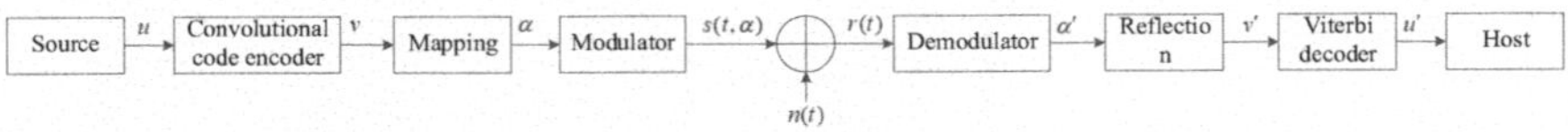

Fig. 5.7 TCM-system block diagram [13]

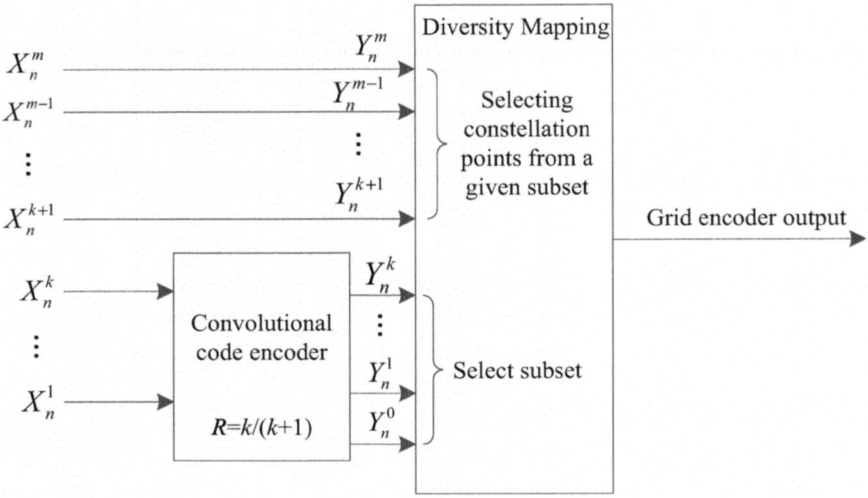

Fig. 5.8 General structure diagram of the integrated grid-coded modulator [14, 15]

The functions of the encoder and mapper can be implemented together in one module, which is called an integrated grid-coded modulator. The general structure of the integrated grid-coded modulator [14, 15] is shown in Fig. 5.8.

In Fig. 5.8, it can be seen that k bits of the m-bit binary message sequence from the source are fed to the convolutional code encoder with parameters ($k + 1$, k, L) for encoding. The k bits of information input to the convolutional code encoder are denoted by $X_n^1 \cdots X_n^k$, and the subscript n denotes the moment. The remaining m–k bits do not pass through the convolutional-code encoder, but go directly to the output. The $k + 1$ bit code set, $Y_n^0 \cdots Y_n^k$, output from the encoder is used to select a different subset of modulated signals, and the uncoded bits, $Y_n^{k+1} \cdots Y_n^m$, select different signal points in the given subset. The process of "selecting" a subset and "selecting" signal points in the subset is actually the process of "mapping."

The presence of uncoded bits in the integrated grid-coded modulator results in a grid graph where the path to transfer from one state to another is not unique. There may be several paths, called parallel-transfer paths. The existence of parallel-transfer paths is the most important feature of the grid diagram of a TCM system. When there are no parallel transfers, the free distance d_f^2 is the minimum value of the sequence distance between all paths that bifurcate from the zero state, those that return to the zero state, and the all-zero path. When parallel transfers exist, the parallel-transfer path of the all-zero path also affects the free distance. Figure 5.9

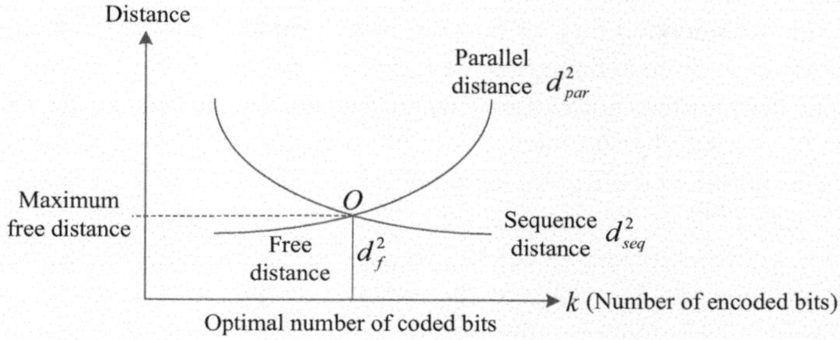

Fig. 5.9 Relationship between free distance, sequence distance and parallel distance

shows the relationship between the free distance, sequence distance, and parallel distance as [14]

$$d_f^2 = \min\left(d^2_{seq}, d^2_{par}\right). \tag{5.24}$$

As shown in Fig. 5.9, as the number of coded bits k increases, the sequence distance d^2_{seq} decreases monotonically, the parallel distance d^2_{par} increases monotonically, and the two curves intersect at point O. The free distance d_f^2 takes the smaller route between the sequence distance and parallel distance; the curve increases first and then decreases, and achieves its maximum value at point O. The number of coded bits corresponding to point O is the optimal number of coded bits.

The parallel distance to the left of point O is smaller than the sequence distance, so the free distance takes the parallel-distance value. To the right of point O, the sequence distance is smaller than the parallel distance, so the free distance takes the sequence-distance value; at point O, the two are equal. For the grid graph without parallel transfers, the parallel distance tends to infinity; however, because the sequence distance is smaller at this time, the free distance takes the sequence-distance value.

We know that the size of the free distance is directly related to the bit error rate (BER) of the TCM system; the larger the free distance, the lower the BER. The introduction of parallel transfers into the TCM system is beneficial for improving the system performance. The introduction of parallel transfers can achieve a "balance" between sequence distance and parallel distance, so that the free distance is maximized and the system performance can be optimized.

When designing a grid-coded modulator, the allocation of the ratio between coded and uncoded bits depends on the relationship between the sequence distance and parallel distance. When $d^2_{seq} > d^2_{par}$, d^2_{par} is the main factor that hinders the increase of d_f^2, so we should consider increasing the number of coded bits to increase d^2_{par}; when $d^2_{seq} < d^2_{par}$, d^2_{seq} is the main factor that hinders the increase of d_f^2, so we should consider decreasing the number of coded bits to increase d^2_{seq}. $d^2_{seq} = d^2_{par}$ is the best design.

By introducing uncoded bits, or parallel-transfer paths, into the integrated grid-coded modulator, the free distance can be increased and the system performance can be improved. It should be noted that different subsets of modulated signals can be selected in the integrated grid-coded modulator using the $k+1$ bit code set $Y_n^0 \cdots Y_n^k$ of the encoder output, and different signal points are selected in a given subset using the uncoded bits $Y_n^{k+1} \cdots Y_n^m$.

This process consists of steps, such as diversifying the modulated signals, mapping the $k+1$ bit code set $Y_n^0 \cdots Y_n^k$ of the encoder output into different subsets of modulated signals, and mapping the uncoded bits $Y_n^{k+1} \cdots Y_n^m$ into different signal points in the given subset. The sequence and parallel distances in the grid diagram are subsequently mapped to the Euclidean distance (geometric distance between signal points in a signal constellation) in the modulated signal set. Proper diversification and mapping of the modulated signals affect the performance of the TCM system. We consider 8-PSK modulation as an example to illustrate the method of diversifying modulated signals, and Fig. 5.10 shows the subset partitioning of the 8-PSK signal constellation.

Set A in Fig. 5.10 is called the level-0 subset and contains all eight signal points in the 8-PSK signal constellation. The Euclidean distances between two of these eight signal points are Δ_0, Δ_1, and Δ_2, which satisfy the relation $\Delta_0 < \Delta_1 < \Delta_2$. Sets B_0 and B_1 are called level-1 subsets, each containing four signal points, and the Euclidean distances between these four signal points, two by two, are Δ_1 and Δ_2. Sets C_0, C_1, C_2, and C_3 are called level-2 subsets, each containing two signal points, and the Euclidean distance between these two signal points is Δ_2.

A simple summary of the rule is that the minimum Euclidean distance in a subset increases with each subset. A more complete description of the subsetting principle is as follows [14].

(i) The Euclidean distances between the signal points in the signal constellation are analyzed from smallest to largest, denoted by Δ_0, Δ_1, $\Delta_2 \cdots$.
(ii) The signal constellation is partitioned into a number of subsets (typically two) with a minimum Euclidean distance of Δ_1, called first-order subsets.
(iii) The individual first-order subsets are partitioned again into subsets with a minimum Euclidean distance of Δ_2, called second-order subsets.
(iv) The second-order subsets are partitioned again until 2^{k+1} subsets are partitioned, each containing 2^{m-k} signal points, at which point the partitioning is completed.

Diversifying modulated signals according to the above principles ensures that there are 2^{k+1} subsets of modulated signals, corresponding to 2^{k+1} combinations of $k+1$ bit code sets output by the encoder, 2^{m-k} signal points corresponding to 2^{m-k} combinations of m–k uncoded bits, or 2^{m-k} signal points corresponding to 2^{m-k} possible parallel-transfer paths; the Euclidean distance between parallel-transfer paths is maximum. To ensure the mapping process of “selecting” the subset and “selecting” the signal points in the subset to maximize the free distance, corresponding mapping rules are required. In his 1982 paper, Ungerboeck proposed three mapping rules [8], which are described as follows:

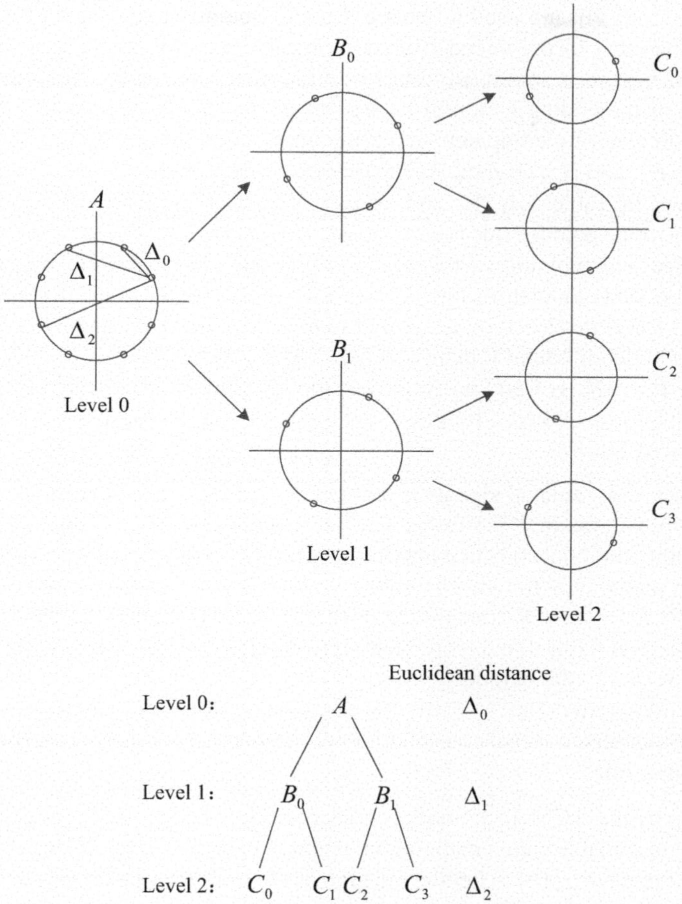

Fig. 5.10 Subset-segmentation diagram of an 8-PSK signal constellation

(i) All subsets with the same frequency are used.
(ii) For transfers departing from or merging into the same state of the grid graph, a subset with maximum Euclidean-distance separation should be specified.
(iii) Parallel transfers should specify the signal points with maximum Euclidean-distance separation.

Rule (i) reflects a good code. If the structure of the integrated grid-coding modulator is optimally designed to maximize the free distance and optimize the system performance, then the convolutional-code encoder is called a good code; it is also known as an optimal upper-bound (UB) convolutional code. (Because it is difficult to find them by analytical methods, most of them are currently found using a computer search.) Good codes should have a regular structure.

Rules (ii) and (iii) are used to ensure that the free distance of the TCM system is greater than that of the uncoded system, that is, to ensure that the performance of the TCM system is improved, relative to the uncoded system. The quantitative calculation of the coding gain will be studied shortly. Table 5.1 lists the (3, 2, *m*) UB codes obtained by computer search in combination with 8-PSK, and columns $H^0(D)$, $H^1(D)$, and $H^2(D)$ indicate the subcheck polynomials.

As shown in Table 5.1, the coding gain increases gradually as *m* increases; however, the price paid for this is an increase in the complexity of the convolutional-code encoder, which requires a balanced consideration in the design. The four-state 8-PSK grid-coded modulator shown in Fig. 5.11 was designed in the best form using the (3, 2, 2) UB code structure, according to the above diversity and mapping rules.

The grid-coded modulator mainly consists of a convolutional-code encoder with a code rate $R = k/k + 1$ and a diversity-mapping module. The design of the grid-coded modulator is centered on these two modules, and the design method can be summarized as follows.

(i) The set of modulated signals is divided according to the established diversity principle to obtain 2^{k+1} subsets, each of which contains 2^{m-k} signal points.
(ii) The convolutional code that maximizes the free distance d_f^2 for different modulation methods is found. The code rate of this convolutional-code encoder is usually $R = k/k + 1$, considering the need to introduce parallel-transfer paths, where several bits of the *k*-bit binary information sequence input to the encoder are directed to the output.
(iii) The code-word output from the convolutional-code encoder to the signal points in the modulated signal constellation was mapped, according to the established mapping rules.

The key to the TCM technique is constructing a good encoder structure and performing reasonable signal mapping to obtain the maximum Euclidean distance of the encoded sequence [16]. After the integrated grid-coding modulator, the signal can be sent according to the different modulation methods. The demodulation, reflection,

Table 5.1 (3, 2, m) UB codes combined with 8-PSK [1]

m	$H^0(D)$	$H^1(D)$	$H^2(D)$	相对4PSK的编码增益 (dB)
2	5	2	0	3.0
3	11	02	04	3.6
4	23	04	16	4.1
5	45	16	34	4.6
6	105	016	074	4.8
7	203	014	016	5.0
8	405	250	176	5.4
9	1001	0164	0260	5.7

Note From G. Ungerboeck, *Channel coding of multilevel/phase signals*, 1982, p. 63

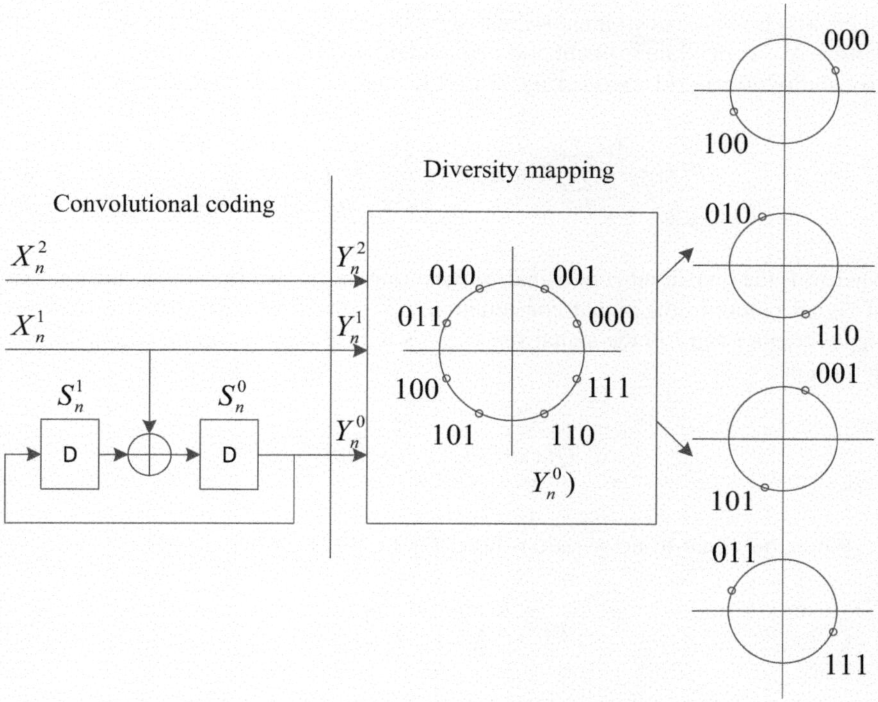

Fig. 5.11 Four-state 8-PSK grid-coded modulator

and Viterbi decoding performed at the receiving end are the inverse transformations of the corresponding modules at the sending end and are not described here.

5.3 Performance Analysis of the TCM System

5.3.1 Error-Probability Analysis of Grid-Coded Signals

The error probability of the grid-coded signal was analyzed according to the TCM system-model diagram shown in Fig. 5.8. The signal output from the modulator in Fig. 5.8 is $s(t, \alpha)$. Assuming that the signal passes through an AWGN channel, the signal received at the receiver is

$$r(t) = s(t, \alpha) + n(t), \tag{5.25}$$

where $n(t)$ is Gaussian white noise with a mean of 0 and a one-sided power-spectral density N_0. If maximum-likelihood demodulation is used at the receiver, the error probability of the grid-encoded signal is [13]

$$P(\varepsilon) = \frac{1}{S}\sum_{i=v}^{S} P(\varepsilon|s_i) \leq \frac{1}{S}\sum_{i=0}^{S-1}\sum_{\substack{j=0 \\ i\neq j}}^{S-1} Q\left[\sqrt{d_{ij}^2 \frac{E_b}{N_0}}\right], \tag{5.26}$$

where S is the total number of signal outputs from the transmitter, that is, the number of signal points in the signal constellation, d_{ij} is the Euclidean distance between signal points i and j in the signal space, $\frac{E_b}{N_0}$ is the signal-to-noise ratio, and $Q(x)$ is defined as

$$Q(x) = \frac{1}{\sqrt{2\pi}} \int_x^{\infty} e^{-\frac{t^2}{2}} dt. \tag{5.27}$$

When the signal-to-noise ratio is high, Eq. (5.26) is approximated as

$$P(\varepsilon) \approx CQ\left[\sqrt{d_f^2 \frac{E_b}{N_0}}\right], \tag{5.28}$$

where C is a constant independent of $\frac{E_b}{N_0}$ and d_f^2 is the free distance. The d_f^2 between two signal sequences, α and β, is defined as

$$d_f^2 = \min_{allU_\alpha, U_\beta} \frac{1}{2E_b} \int_0^{\infty} [s(t,\alpha) - s(t,\beta)]^2 dt, \tag{5.29}$$

where U_α and U_β are different information sequences input to the convolutional-code encoder.

$$\begin{aligned} U_\alpha &= (\cdots, u_{\alpha 0}, u_{\alpha 1}, u_{\alpha 2}, \cdots) \\ U_\beta &= (\cdots, u_{\beta 0}, u_{\beta 1}, u_{\beta 2}, \cdots) \end{aligned} \tag{5.30}$$

From Eq. (5.29), it can be seen that the system error probability depends on the free distance d_f^2 between the signal sequences. The role of coding is to increase d_f^2, thus improving the system performance.

5.3.2 Coding-Gain Analysis of a TCM System

We will use 8-PSK TCM as an example to quantitatively analyze the coding gain of this TCM system, designed with coding and modulation as a whole, with respect to the uncoded system, and then generalize the expression. Figure 5.12 shows the grid diagram of the 8-PSK TCM code, from which it can be seen that the subset sequence corresponding to the path with the smallest sequence distance between all paths bifurcating from the zero state, those returning to the zero state, and the all-zero path is (C_1, C_2, C_1). The subset sequence corresponding to the all-zero path is (C_0, C_0, C_0), and the Euclidean distance between the two is

$$\begin{aligned} d^2_{seq\min} &= dis^2[(C_0, C_0, C_0), (C_1, C_2, C_1)] \\ &= dis^2(C_0, C_1) + dis^2(C_0, C_2) + dis^2(C_0, C_1) \\ &= \Delta_1^2 + \Delta_0^2 + \Delta_1^2. \end{aligned} \tag{5.31}$$

Assuming that the 8-PSK signal-constellation diagram in Fig. 5.10 is a unit circle, it follows that

$$d^2_{seq\min} = \left(\sqrt{2}\right)^2 + [2\sin(\pi/8)]^2 + \left(\sqrt{2}\right)^2 = 4.586. \tag{5.32}$$

The parallel distance between an all-zero path and its parallel-transfer path is

$$d^2_{par} = \Delta_2^2 = 2^2 = 4. \tag{5.33}$$

According to the definition of the free distance in the TCM system, the following is obtained:

$$d_f^2 = \min\left(d^2_{seq}, d^2_{par}\right) = d^2_{par} = 4. \tag{5.34}$$

We define the coding gain as [14]

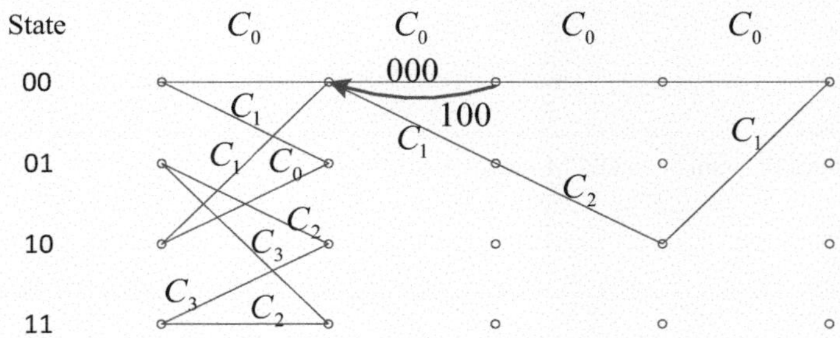

Fig. 5.12 Grid diagram of 8-PSK TCM code

$$\gamma = 10\log \frac{d_f^2 / E_c}{d_{un}^2 / E_{un}}, \tag{5.35}$$

where d_f^2 is the free distance in the TCM system and d_{un}^2 is the minimum distance between signal points when uncoded, which, for this example, refers to the minimum Euclidean distance in the 4-PSK signal constellation; E_c and E_{un} denote the average energy of the signal set in the coded and uncoded systems, respectively, and the average energy of the 8-PSK TCM system and the uncoded 4-PSK system in this example are equal. The coding gain is expressed as follows:

$$\gamma = 10\log \frac{d_f^2}{d_{un}^2} = 10\log \frac{4}{2} = 10\log 2 = 3.01\ \text{dB}. \tag{5.36}$$

The coding-gain formula indicates that, under the same bandwidth and transmit power conditions, if the BER is maintained, the signal-to-noise ratio required by the coding system is reduced, compared to that of the non-coding system by γ dB, which relaxes the channel requirements; that is, the coding system can obtain better performance compared to the non-coding system under the same channel conditions.

5.4 OPPM-Based TCM System Design

A block diagram of an OPPM-based TCM system is shown in Fig. 5.13. The convolutional-code encoder module, mapping module, and OPPM-modulator module at the transmitter side together form the grid-coded modulator, and the OPPM demodulator module, reflector module, and Viterbi decoder module at the receiver side together form the grid-coded modulated Viterbi decoder. We designed the TCM system with 8-OPPM as an example and divided the entire system into two modules: the grid-coded modulator and the grid-coded modulated Viterbi decoder.

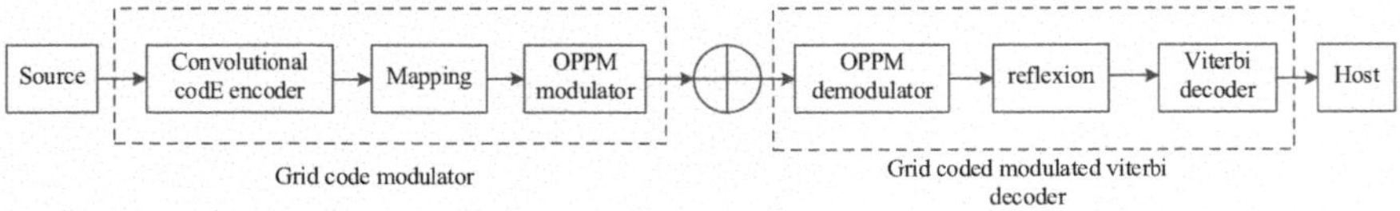

Fig. 5.13 Block diagram of an OPPM-based TCM system

5.4.1 Design of a Grid-Coded Modulator

The design of the grid-coded modulator follows the flow in the dashed box on the transmitter side of Fig. 5.13: It divides the modulation signal set, designs the convolutional-code encoder, and implements the mapping between the output code word of the convolutional-code encoder and the 8-OPPM signal.

First, the composition of the 8-OPPM signal set was analyzed. In a practical atmospheric laser-communication system, considering the minimum repetition period required by the laser to reconstruct the particle-number inversion and to avoid inter-frame symbol interference caused by channel dispersion, a certain protection time period must be set after the transmission time period. We set the protection period to be the same length as the transmission period and represent the actual 8-OPPM signal with the help of the $J = N(Q - 1) + 1 = 2(8 - 1) + 1 = 15$ metatheoretical OPPM signals of the first eight pulse positions derived from $Q = 8$ and $N = 2$. The resulting frame structure of the 8-OPPM signal and the correspondence between the 8-OPPM signal and the position of the optical pulse are shown in Fig. 5.14.

OPPM uses optical pulses of length τ, with $t_k = (k-1)\tau'$, $k = 1, 2, \cdots, J$ as the starting position to transmit information. For 8-OPPM, an optical pulse of length τ with 1, 2, 3, 4, 5, 6, and 7 times τ' is used as the starting position to convey three-bit binary information. It is expressed as (0, 1, 2, 3, 4, 5, 6, 7) for the 8-OPPM modulated

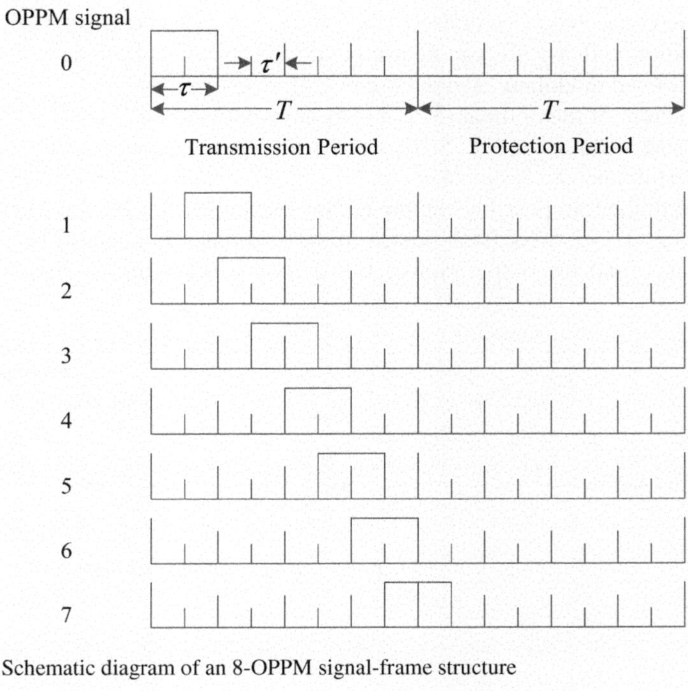

Fig. 5.14 Schematic diagram of an 8-OPPM signal-frame structure

signal set, corresponding to eight different optical-pulse positions. According to the previous diversity principle [14],

(i) The Euclidean distances between signal points in the signal constellation are analyzed, and expressed in descending order: $\Delta_0, \Delta_1, \Delta_2 \cdots$.
(ii) The signal constellation is partitioned into a number of subsets (typically two) with a minimum Euclidean distance of Δ_1, called first-order subsets.
(iii) Each first-order subset is partitioned into subsets with a minimum Euclidean distance of Δ_2, known as second-order subsets.
(iv) The second-order subsets are partitioned until 2^{k+1} subsets are partitioned, each containing 2^{m-k} signal points, at which point the partitioning is complete.

Here, the Euclidean distance between two OPPM signal points is first specified as the distance between the starting positions of the optical pulses. Because the distance between the starting positions of two adjacent optical-pulse signals is the length of a subinterval, the minimum-distance unit between two optical-pulse signals is a subinterval length. The above 8-OPPM modulated signal set is divided as shown in Fig. 5.15. The minimum Euclidean distance between the OPPM signal points in each subset of levels increases from one to two to four, as the subset proceeds.

The signal-diversity diagram has a form similar to the subset-partitioning diagram of the 8-PSK signal constellation shown in Fig. 5.10. They differ only in that the 8-PSK signal diversity is a two-dimensional signal constellation partitioned into subsets, whereas the 8-OPPM signal diversity is performed on a one-dimensional signal frame consisting of 14 subintervals. The (3, 2, m) UB codes listed in Table 5.1, in combination with 8-PSK, can be used to design the convolutional-code encoder in the grid-coded modulator, and we use $m = 2$ as an example.

The structure of the optimal (3, 2, 2) UB convolutional-code encoder combined with 8-OPPM is shown in Fig. 5.16, and the four-state lattice diagram [10] corresponding to this encoder is shown in Fig. 5.17. The four states from 0 to 3 in Fig. 5.17 correspond to the four possible combinations of states 00, 01, 10, and 11 in the two shift registers in Fig. 5.16. Each branch in the grid diagram contains two parallel-transfer paths, and the output convolutional-code word corresponding to a walk

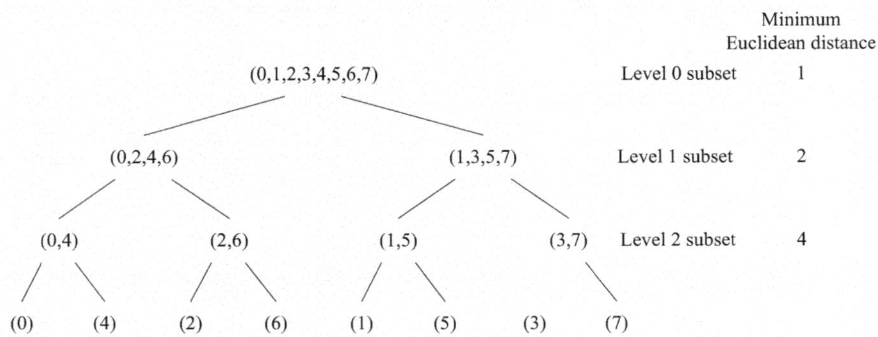

Fig. 5.15 8-OPPM signal-diversity diagram

along the two parallel-transfer paths is labeled above the branch as an octal number. T indicates the time elapsed for a state transfer to occur in the shift register.

According to the mapping rules described earlier [14], (i) all subsets are used with the same frequency; (ii) the subset with the maximum Euclidean-distance separation should be specified for transfers starting from or merging into the same state of the grid graph; and (iii) parallel transfers should specify the signal points with the maximum Euclidean-distance separation.

The output code word of the convolutional-code encoder can be mapped to the 8-OPPM signal. Figure 5.18 reflects the mapping between the convolutional code and the 8-OPPM signal, with the convolutional code corresponding to each OPPM signal marked below it. The three-bit convolutional code controls the laser to emit an optical pulse signal of length τ at the beginning of the corresponding subinterval, as shown in Fig. 5.14. The light-pulse signal is transmitted through an additive Gaussian white-noise channel to the receiver for demodulation.

Fig. 5.16 Structure of a (3, 2, 2) UB convolutional-code encoder

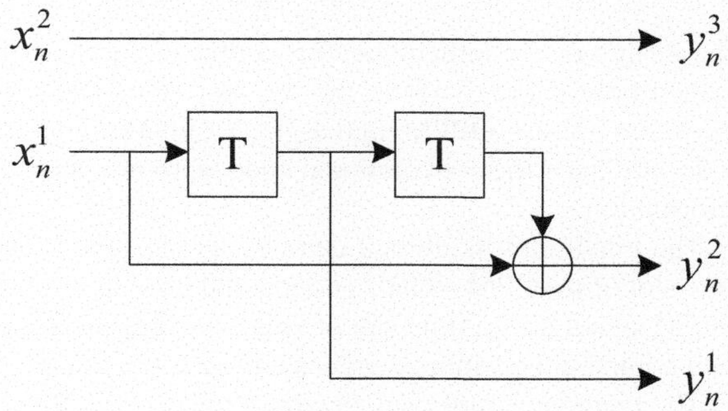

Fig. 5.17 (3, 2, 2) convolutional-code grid map

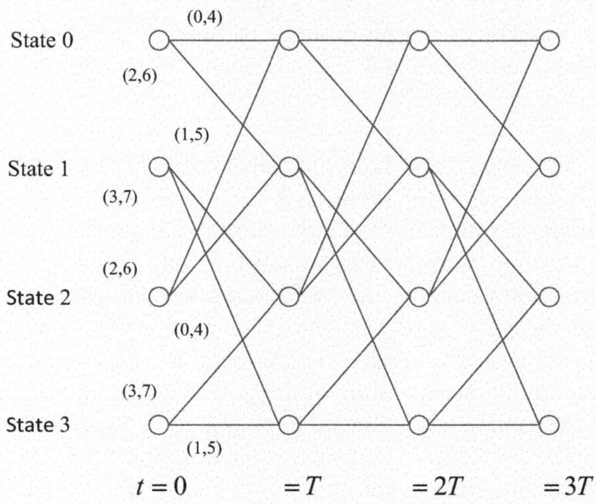

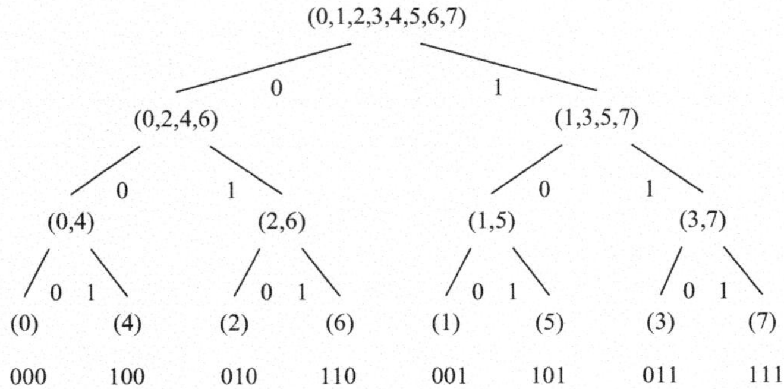

Fig. 5.18 8-OPPM signal-mapping diagram

5.4.2 Design of a Grid-Coded Modulated Viterbi Decoder

The design of the lattice-coded modulated Viterbi decoder follows the flow in the dashed box of Fig. 5.13 on the receiver side. The received optical-pulse signal is recovered as an 8-OPPM signal, the 8-OPPM signal is reflected to a convolutional code word, and the convolutional code word is Viterbi decoded to recover the binary information.

The maximum-likelihood demodulation of the PPM signal is performed at each time slot of the PPM signal. The method used in this study is to sample and sum the OPPM signal over a time slot of length τ, starting from the start of each subinterval, then compare the summed values and select the maximum value.

The method is improved, based on the original PPM-signal maximum-likelihood demodulation method, which can better solve the problem that one pulse received by the 8-OPPM demodulator occupies seven subintervals, that is, pulse overlap. Here, the number of sampling times includes m times in each subinterval of the OPPM signal frame; thus, mN times in one time slot, and a total of mNQ times in one signal frame. For 8-OPPM, 2 m samples were taken in a time slot, and 14 m samples were taken in a signal frame.

The sum of 2 m sampling values in a time slot of length τ, starting from the first sub-interval, is $\sum_1$, that is, $\sum_1 = y_1 + y_2 + \cdots + y_{2m}$, where y_i denotes the ith sampling value; similarly, the sum of 2 m sampling values in the time slot of length τ, starting from the second sub-interval, is $\sum_2$, and $\sum_3, \cdots, \sum_8$ are obtained in turn. Subsequently, the eight sums are compared to select the largest operation, and the pulse position with the largest sum is determined as the position of the transmitted optical pulse. In the specific implementation of the maximum-likelihood demodulation algorithm for the OPPM signal, if two sums are equal and a maximum is encountered, it is judged to be one of them with the same probability.

The 8-OPPM signal recovered from the 8-OPPM demodulator can be reflected to the three-bit convolutional code, according to the mapping rules. The process of inputting the binary information sequence to the convolutional-code encoder for encoding is the process of walking along a specific path of the grid map; each walking step will have a convolutional-code output corresponding to the input binary information.

For TCM systems, because there is a mapping between the convolutional codes and modulation signals, it can also be said that for each step, there is a modulation-signal output corresponding to the input binary information. The grid diagram can provide information about which state transfer is possible and the output signal (convolutional code or modulation signal) corresponding to that state transfer. Viterbi decoding involves calculating the branch and path measures of each state of the shifter at the current moment, based on known information, including the structure of the lattice diagram and the received convolutional codes, thus updating the retention path to obtain the decoded output.

Considering that, as the mapping from convolutional codes to OPPM signals proceeds, the sequence distance and parallel distance in the grid map will be subsequently mapped to the Euclidean distance in the modulated signal set, we use the Euclidean distance between the OPPM signal points as the distance measure, instead of using the Hamming distance as the distance measure, in the design of the Viterbi decoder, unlike previous Viterbi decoders. The existence of parallel transfers is a characteristic of TCM systems, and solving the problem of calculating the branch measure of a parallel-transfer path must also be considered when designing a Viterbi decoder. The above two points are the differences between the Viterbi decoder used in this system and a general-purpose Viterbi decoder, and need to be addressed in the design.

Following is a four-step implementation of the Viterbi decoding algorithm; solutions to both problems are provided in the implementation of the algorithm.

Step 1: Compute the branch measure $BM_n(i,j)$; that is, the Euclidean distance between the received signal R at moment n and the subset of signals corresponding to the jth possible state transfer that reaches state i.

The four states from top to bottom in Fig. 5.17 represent state 0 (0 0), state 1 (0 1), state 2 (1 0), and state 3 (1 1) of the shifter, respectively. Following is an example of calculating the branch measure for state 0. The calculation methods for the other states are similar. From the figure, it can be seen that the possible previous states of state 0 are states 0 or 2, and the signal subset corresponding to the transfer from state 0 to state 0 is (0, 4), containing two signal points $S_1 = 0$ and $S_2 = 4$.

Signal points 0 and 4 correspond to two parallel-transfer paths. The one with the minimum Euclidean distance between the received signal R and these two signal points at moment n (or the one with the maximum similarity λ) is defined as the Euclidean distance between the received signal R and the subset (0, 4), denoted by the branching measure $BM_n(0, 1)$, with the zero in parentheses denoting all possible previous states of the studied state 0 and 1 denoting the first previous state of state 0. As the second pre-state of state 0, the subset of signals corresponding to the transfer from state 2 to state 0 is (2, 6). The Euclidean distance between the received signal R and the minimum of these two signal points at moment n is defined as the Euclidean distance between the received signal R and subset (2, 6), denoted by branching measure $BM_n(0, 2)$.

For the four-state lattice diagram with parallel transfers shown in Fig. 5.17, there are two possible previous states that reach state i ($i = 0, 1, 2, 3$); hence, there are two branching measures $BM_n(i, j)$ to be computed for state i: $BM_n(i, 1)$ and $BM_n(i, 2)$. Because there are two parallel-transfer paths and two signal points in the subset of signals when the state transfer occurs, the Euclidean distance between the received signal R and the two signal points in the subset is calculated separately for each branch measure (expressed as a Euclidean distance), and the smaller one is selected as the branch measure. The branch measure [14] can be expressed as follows when there is a parallel transfer:

$$BM_n = \min_{i \in subsets} \left\{ d_f^2[R_n(t), S_i(t)] \right\} \tag{5.37}$$

or expressed in terms of similarity λ as

$$BM_n = \max_{i \in subsets} \{\lambda_n[R_n(t), S_i(t)]\}. \tag{5.38}$$

Step 2: Update the path measure $PM_n(i)$ for each state.

As time continues, the small segments of the branches are connected into paths, and the path measure is equal to the sum of the branch measures. For the four-state lattice diagram with parallel transfers shown in Fig. 5.17, there are two possible previous states for reaching state i ($i = 0, 1, 2, 3$): $p(i, 1) = i_{p1}$ and $p(i, 2) = i_{p2}$. The process of updating the path measure of state i involves calculating the sum of its two previous states' path measures $PM_{n-1}(i_{pj})$ ($j = 1, 2$) at moment $n - 1$ and the branch measure $BM_n(i, j)$ corresponding to this transfer, and using the smallest of them as the path measure at this moment.

$$PM_n(i) = \min_{j=1,2} \left\{ PM_{n-1}\left(i_{pj}\right) + BM_n(i, j) \right\} \tag{5.39}$$

We assume that the path measures for each state at moment 0 are $PM_0(0) = 0$, $PM_0(1) = \infty$, $PM_0(2) = \infty$, and $PM_0(3) = \infty$.

Step 3: Update the stay path $SUR_n(j)$ for each state.

The retention path is actually a sequence of L convolutional codes (L is the decoding delay of L code-word cycles generated by the Viterbi decoding algorithm; the role of the decoding delay will be explained below), which are stored as bits in a first-in-first-out shifter, corresponding to a certain transfer path.

$$SUR_n(i) = \{C_n(i), C_{n-1}(i), \cdots, C_{n-L+1}(i)\} \tag{5.40}$$

If the previous state, such that state i obtains the minimum $PM_n(i)$, is p, the retention paths of state i and state p at $n - 1$ moments, respectively, are

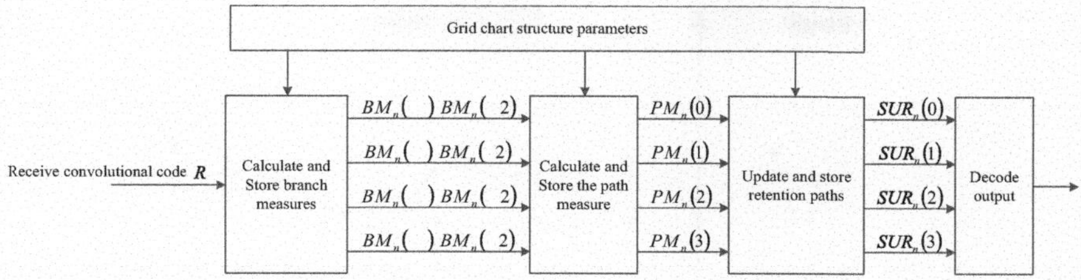

Fig. 5.19 Block diagram of Viterbi decoding

$$\begin{aligned} SUR_{n-1}(i) &= \{C_{n-1}(i), C_{n-2}(i), \cdots, C_{n-L}(i)\} \\ SUR_{n-1}(p) &= \{C_{n-1}(p), C_{n-2}(p), \cdots, C_{n-L}(p)\}. \end{aligned} \tag{5.41}$$

The convolutional code corresponding to the transfer from state p to state i is $C_n(i)$. Updating the retention path $SUR_n(i)$ of state i is the process of depositing the retention path of state p into the retention-path shifter of state i, then feeding $C_n(i)$ into the retention-path shifter of state i, and shifting out code word $C_{n\text{-}L}(p)$, which is at the front of the sequence; that is,

$$\begin{aligned} & SUR_n(i) \\ & = \{C_{n-1}(p), C_{n-2}(p), \cdots, C_{n-L}(p)\}\underline{\underline{\text{update to}}}\{C_n(i), C_{n-1}(i), \cdots, C_{n-L+1}(i)\}. \end{aligned} \tag{5.42}$$

Step 4: Decode the output.

In the previous step of updating the retention paths, a code word at the front of the sequence was "popped out" of the shifter for each state. The output of the state with the smallest path measure, $PM_n(i)$, is defined as the output of the grid-coded modulated Viterbi decoder at time n. A block diagram of the Viterbi decoder is shown in Fig. 5.19.

As can be seen from the decoding process, because the length of the retention path is equal to the length of the L convolutional codes, the retention-path memory is set to zero at initialization; hence, the first L code words of the decoding output are 000, and the convolutional codes received at n moments will be decoded at $n + L$ moments after a delay of L cycles. The Viterbi decoding algorithm uses this delay to improve the error-correction capability of the decoder.

The growth law of path measure PM over time [8] is shown in Fig. 5.20. There is only one correct path on the grid diagram, and although its path measures and those of other paths are continuously increasing, the reasons for the increase are different, and the statistical properties are not the same. The increase in PM of the correct path is caused by a code-word error, and the rate of increase depends on the error probability and its distribution characteristics. The increase in PM of other paths is caused by the path difference, and the PM increases continuously and rapidly with time.

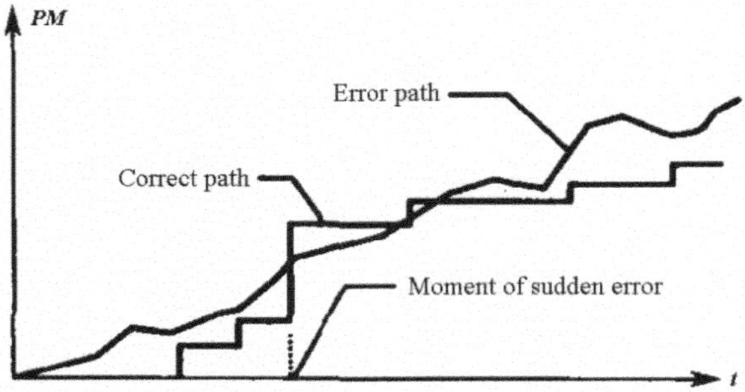

Fig. 5.20 Growth pattern of path measure PM [14]

When a burst error occurs in the channel, the PM of the correct path will suddenly increase and temporarily exceed the other paths. However, if the length of the burst error is within a certain limit and does not occur continuously, then, after a period of time, the PM of the correct path will return to the minimum, and time will smooth the peaks and valleys of the error.

The introduction of a delay time makes it possible to make judgments based on statistical characteristics over a period of time, instead of immediate, isolated, and code-by-code judgments, which reflect the principle of error-correction codes with noise equalization. A large decoding delay is good for improving the decoding accuracy; however, it is not good for real-time communication. The delay can vary, depending on the use of the channel, and the delay L is generally taken to be 5–10 times the constraint length m of the convolutional code (n, k, m) [14], such as when $L = 5\ m = 10$.

5.5 Computer Simulation and FPGA Design of TCM Coding

We performed a computer simulation of the TCM system with 8-OPPM using Simulink in the MATLAB 7.0 environment. Figure 5.21 shows the simulation results comparing the BER of the TCM system oriented to 8-OPPM and the uncoded 4-PPM system, with the vertical coordinate taken as the logarithm of base 10 and two-bit information transmitted per frame cycle T.

From Fig. 5.21, it can be seen that the bit error rate of the 8-OPPM-oriented TCM system drops to 0 when the signal-to-noise ratio is greater than 0.5 dB, and a coding gain of 1–3 dB can be achieved, relative to the uncoded 4-PPM. By calculating the cutoff rates of 4-PPM and 8-OPPM in references [17, 18], it is concluded that the theoretical coding gain is approximately 5–8 dB when transmitting two-bit information with 8-OPPM, which has the ability to transmit three-bit information per symbol.

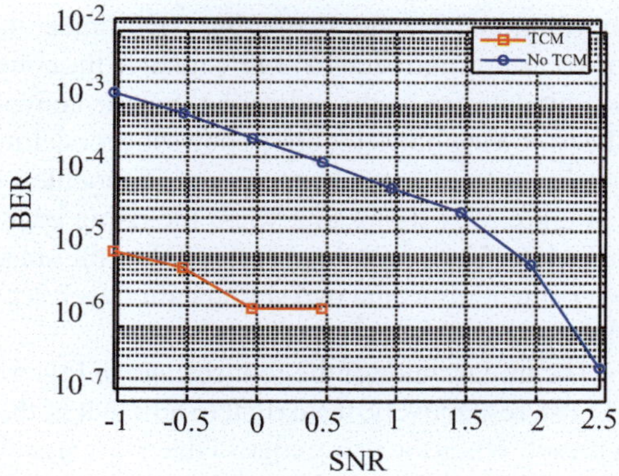

Fig. 5.21 Comparison of encoded 8-OPPM and unencoded 4-PPM bit error rates

5.5.1 Transmitter FPGA Design

Based on the block diagram of the TCM system, the field-programmable gate array (FPGA)-based functional-unit division of the transmitter side in the block diagram can be obtained as shown in Fig. 5.22, relying on the Verilog hardware-description language (HDL) to achieve a single-chip design result. The established model can functionally realize the system requirements on the one hand, and satisfy the implementability of Verilog HDL programming on the other.

The reference clock signal *clk* in Fig. 5.22 is divided by a frequency divider to generate two clock signals: frame clock *frameclk* and subinterval clock *subclk*, $f_{clk} = 64 f_{frameclk}$ and $f_{clk} = 4 f_{subclk}$ (It is assumed that the receiver samples four times in each subinterval of the 8-OPPM signal frame and each 8-OPPM signal frame contains 16 subintervals). When the rising edge of the *reset* signal appears, the source module reads one byte of the input eight-bit binary signal into memory, sends it to the

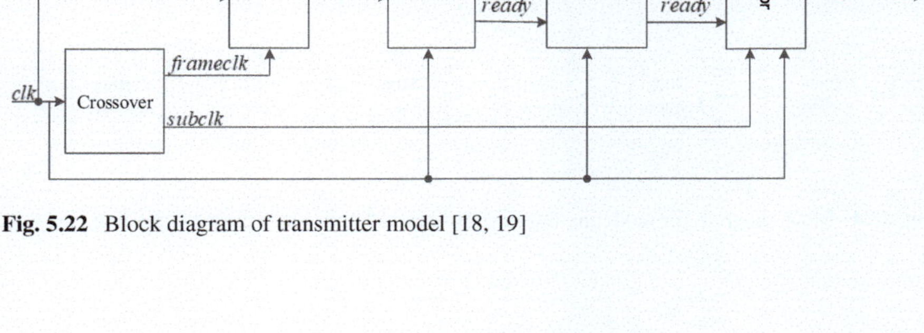

Fig. 5.22 Block diagram of transmitter model [18, 19]

convolutional encoder in two-byte steps under the action of the frame clock *frameclk* for encoding, and sends a *ready* signal to the convolutional-encoder module when the two-bit data are ready, indicating that the convolutional encoder can start encoding. The following block diagram shows the modeling of the convolutional-code encoder, synchronous code-sequence insertion module, and OPPM modulator.

Figure 5.23 shows that when the rising edge of the *ready* signal is detected, the input interface sends data *d*1 and *d*0 to the subsequent module for calculation. The *reset* signal resets the two shifters after each section of data is transmitted, returning them to state 0.

The initial states of the four shifters in Fig. 5.24 are 000 (0), 111 (7), 010 (2), and 011 (3), respectively, from right to left; that is, the synchronization code to be inserted is preset. When the rising edge of the *ready* signal is detected, the input interface sends the code word of a convolutional code composed of three binary numbers *c*2, *c*1, and *c*0 into the leftmost shifter, and the preset 000 (0) in the rightmost shifter is ejected to insert the synchronization-code sequence into the front of the data sequence.

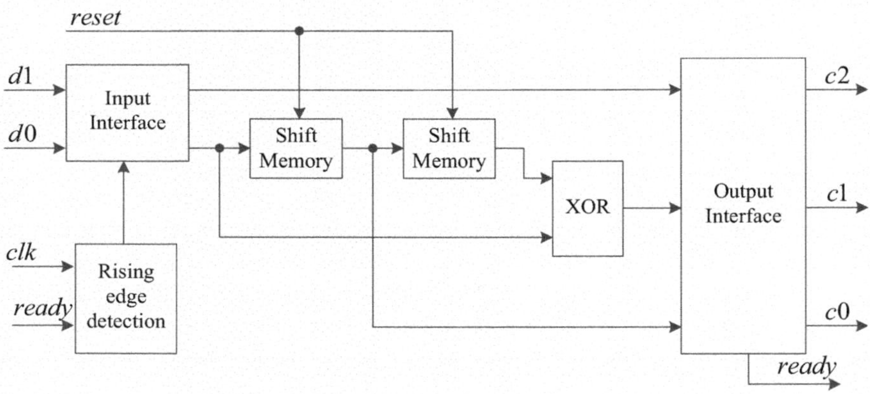

Fig. 5.23 Convolutional-code encoder modeling block diagram [18]

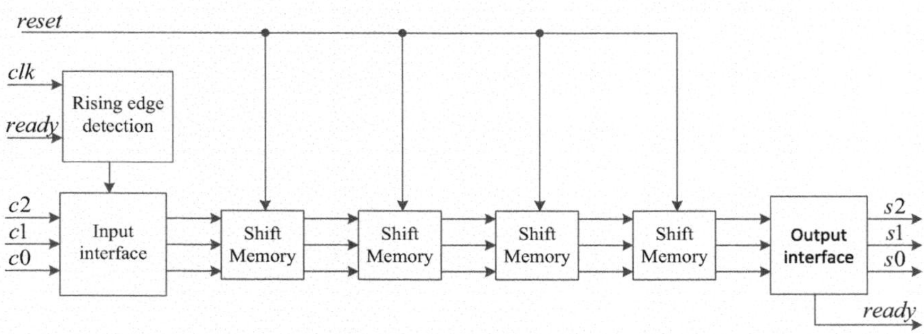

Fig. 5.24 Block diagram for modeling the synchronous code-sequence insertion module [18]

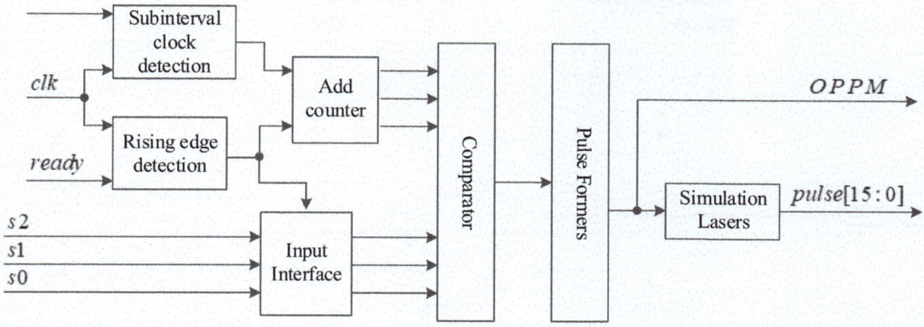

Fig. 5.25 Block diagram of OPPM modulator modeling [18]

Figure 5.25 shows that when the rising edge of the *ready* signal is detected, the counter is cleared, binary numbers are fed to the comparator from input interfaces *s*2, *s*1, and *s*0, and the counter is incremented and compared with the input three binary numbers under the action of *subclock*; when equal, a pulse signal is sent with the current count value as the starting position. The OPPM signal output from the pulse former is used as the modulation signal for the simulated laser connected to the actual laser. The role of the analog laser module is to send a pulse signal with a certain intensity (with [15:0]; that is, 16 binary digits to indicate its intensity) to the receiving end, which is used during the simulation phase when the actual laser is not connected.

The transmitter-modeling block diagram is a Verilog program written in modules under the ISE 6.2i software environment to generate Verilog modeling symbols for each module and connect each modeling symbol according to the transmitter-modeling block diagram, shown in Fig. 5.22, to draw the top-level circuit diagram of the transmitter, as shown in Fig. 5.26.

5.5.2 FPGA Design of the Receiver Side

Figure 5.27 shows a block diagram of the receiver-side modeling that includes the synchronous code-sequence identifier, synchronous signal-recovery module, OPPM demodulator, and Viterbi decoder.

Figure 5.28 shows that *pulse* [15:0], which represents the intensity of the light signal under the action of clock *clk*, is fed into shifter 1. Because a light pulse occupies the width of two subintervals, the receiver samples each subinterval four times. There are eight sampled values in the width of a light pulse, so the length of shifter 1 is set to eight bits to store the eight sampled values in the width of a light pulse. The 8 sampled values are summed and fed into the shift memory 2, which is used to store a total of $4 \times 16 \times 4 = 256$ sum values in four frames.

The location of the sum of the eight samples of the light pulses in the four synchronous frames in shifter 2 has been marked in Fig. 5.28, and the specific value

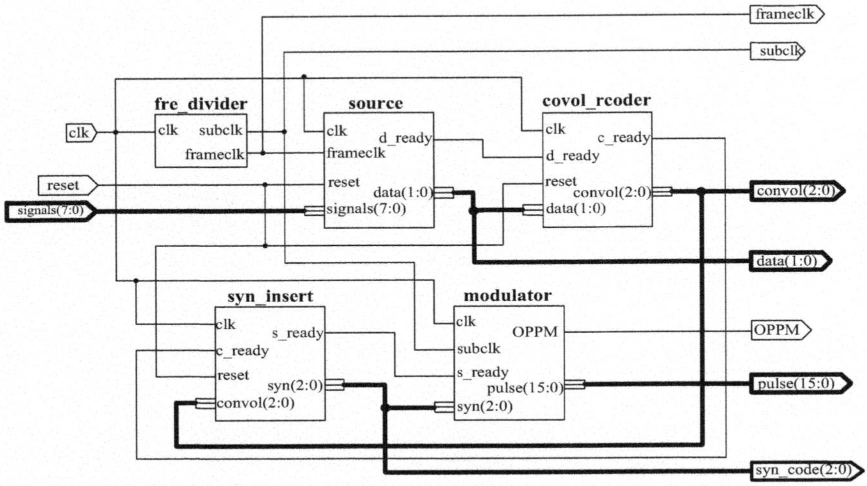

Fig. 5.26 Top-level circuit diagram of the transmitter [18]

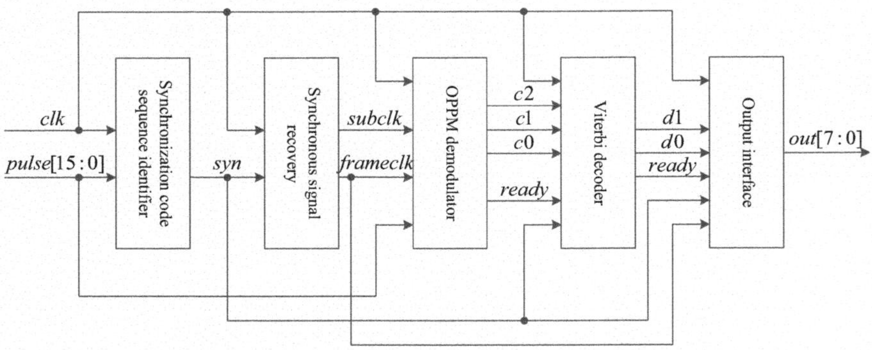

Fig. 5.27 Block diagram of the receiver-side modeling [18]

can be calculated by using the following formula: $S = 64n + 4(m + 2) - 1$, where $n = 0, 1, 2, 3$ indicates the sequence number of the synchronous s frames, $m = 0, 7, 2, 3$ indicates the starting position of the light pulses in the four synchronous frames, and S indicates the location of the eight samples of the light pulses in shift memory 2. The above four positions are summed in the adder, and the narrow pulse signal *syn* is output when the sum value is greater than the set threshold, indicating that a synchronous frame is detected.

In the fourth synchronization frame, the light pulses are located in subintervals 3 and 4; therefore, *syn* is located at the beginning of subinterval 5. Because the distance between *syn* and the beginning of the next frame is known, the frame clock *frameclk*

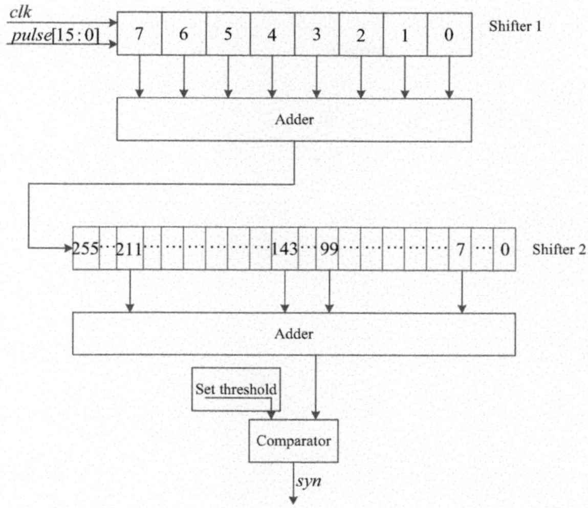

Fig. 5.28 Block diagram for modeling the synchronous code-sequence recognizer [18]

and subinterval clock *subclk* can be recovered using the counter under the action of clock *clk*, as shown in Fig. 5.29.

The structure of the shifter in Fig. 5.30 is the same as that of shifter 1 in Fig. 5.29. *pulse* [15:0], which represents the intensity of the optical signal, is fed into the shifter under the action of clock *clk*, and the summation operation is performed when the output of comparator 1 is high. The role of the frame clock *frameclk* is to clear the two counters to zero; following the subinterval clock *subclk,* the two counters start working at the same time, when the count value of counter 1 is equal to 3 (at this time, to complete the sampling of the first two subintervals). When comparator 1 outputs a high level to drive the adder, the feedback to counter 1 sets counter 1–2 (because there is an overlap of subintervals between the two adjacent light pulses).

The output of the adder is compared with the threshold value. When it is greater than the preset threshold value, the sum value is sent to the buffer for storage and the threshold value is adjusted to the current sum value in time. When frame clock

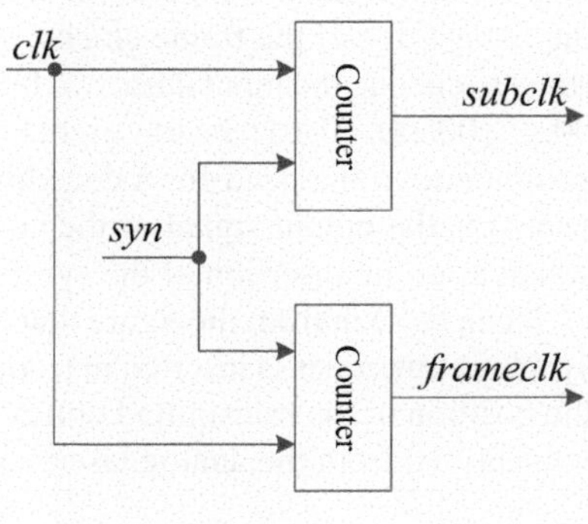

Fig. 5.29 Block diagram for modeling the synchronous signal-recovery module [18]

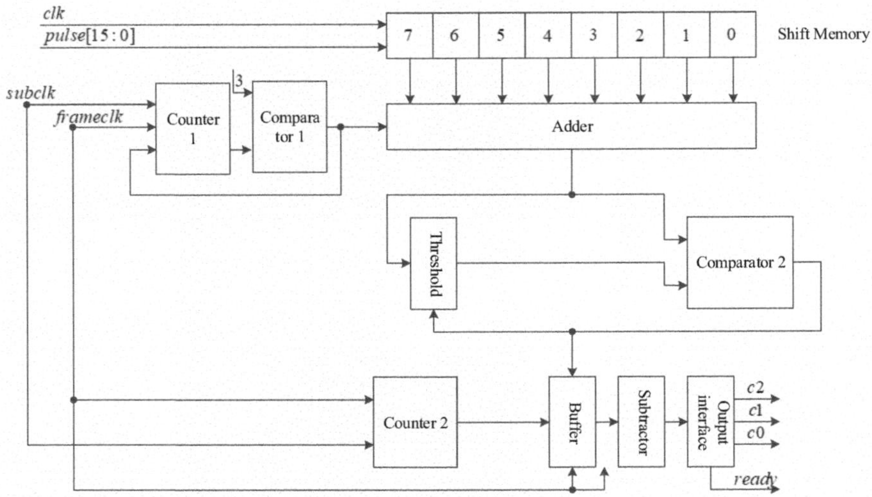

Fig. 5.30 Block diagram of the OPPM demodulator modeling [18]

frameclk rises, it means that the eight sum values in a frame are calculated. At this time, the data kept in the buffer must be the maximum, and the count value of counter 2 minus 3 can be used as the output of the recovered OPPM signal.

The four steps of the Viterbi decoding algorithm can be drawn as shown in Fig. 5.31 of the Viterbi-decoder modeling block diagram. It should be noted that while following clock *clk*, by detecting the rising edge of the *syn* signal, the matrix-iteration module, the path-input module, and the reset signal of the path-storage module are activated, and by detecting the rising edge of the *ready* signal, the matrix-iteration module, the path-input module, and the operating clock of the path-storage module are activated. The output interface corresponds to the input interface and outputs one byte of an eight-bit binary number. A top-level circuit diagram of the receiver is shown in Fig. 5.32.

Figure 5.33 shows the results of the functional simulation of the TCM system using the ModelSim SE 6.0 simulation software. After the rising edge of the *reset* signal, the rising edge of the frame clock *frameclk* is when the convolutional-code encoder starts to encode the input *data*, two bits by two bits. The input data in Fig. 5.33 are 00, 00, …, 00, 00, 10, 01; the low eight bits are valid data. The high-side complementary zero is due to the need for a decoding delay in the Viterbi decoder at the receiver; *convol* is the output signal of the convolutional-code encoder; and *syn_code* is the signal after the insertion of the synchronous code sequence.

It can be seen from the figure that the sequence consisting of 0, 7, 2, and 3 four-bit synchronous codes is inserted in front of the data sequence; the OPPM signal reflects the position of the transmitted optical pulse; and the *pulse* signal is the light pulse of intensity 10 from the analog laser.

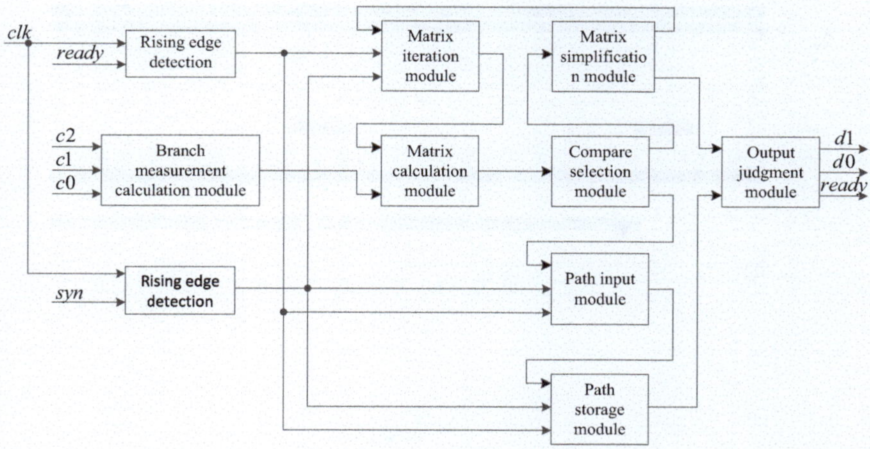

Fig. 5.31 Block diagram of the Viterbi decoder modeling [18]

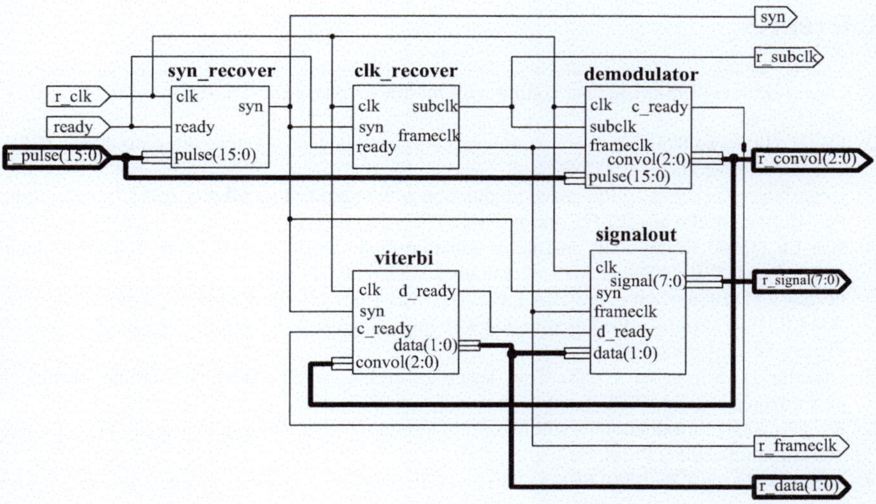

Fig. 5.32 Top-level circuit diagram of the receiver side [18]

At the receiving end, a narrow pulse signal *syn* is emitted when the synchronization code sequence is detected; subsequently, the subinterval clock and frame clock can be recovered; the convolutional code recovered by maximum-likelihood demodulation is reduced to the data signal output after Viterbi decoding. The simulation results show that the designed system can meet the expected functional requirements.

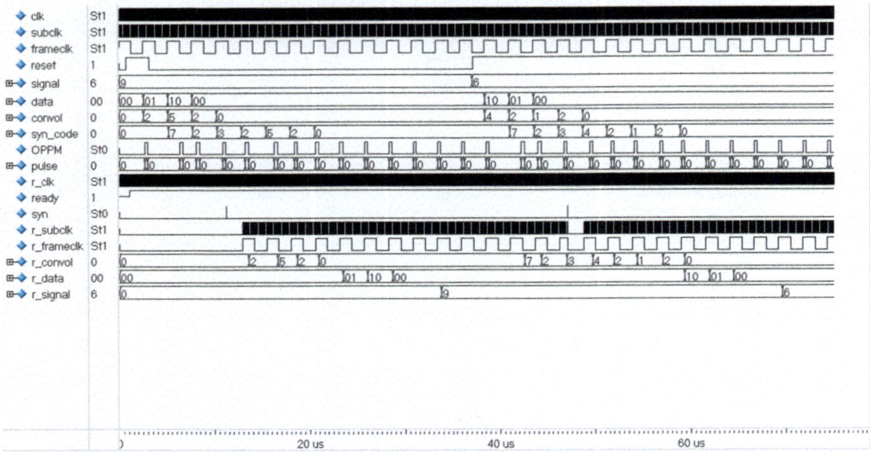

Fig. 5.33 Functional simulation results graph [18]

References

1. Ungerboeck G (1982) Channel coding with multilevel/phase signals. IEEE Trans Inf Theory 28(1):55–67
2. Ima H, Hirakawa S (1977) A new multilevel coding method using error-correcting codes. IEEE Trans Inf Theor IT-23(3):371–377
3. Ungerboeck G (1987) Trellis-coded modulation with redundant signal sets: part I, introduction; Part II, State of the art. IEEE Commun Mag 25(2):5–21
4. Wei LF (1984) Rotationally m-variant convolutional channel coding with expanded signal space, Part I. IEEE J Select Areas Commun SAC-2:659–671
5. Forney GD (1988) Cost codes-Part I, Part II. IEEE Trans Inf Theory 5:1123–1187
6. Wei LF (1987) Trellis-coded modulation with multidimensional constellations. IEEE Trans Inf Theory IT-33(4):483–501
7. Divsalar D, Simon M (1988) The design of trellis coded MPSK for fading channels: performance criterin. IEEE Trans Commun 36(9):1004–1012
8. Ke XZ (2004) Introduction to wireless laser communication. Beijing University of Posts and Telecommunications Press, Beijing, pp 8–9
9. Wang JY, Zhang ZX (2000) PPM in wireless infrared communication. Telecommun Eng 5:81–84
10. Georghiades CN, Crutcher JK (1991) Throughput efficiency considerations for optical OPPM. IEEE Globecom 91:430–433
11. Shalaby HMH (1995) Maximum achievable throughputs for uncoded OPPM and MPPM in optical direct-detection channels. J Lightwave Technol 13(11):2121–2128
12. Lin XG, Liu FQ (1997) Simulation method of eight-state lattice code encoding and decoding. J Harbin Eng Univer 18(5):58–64
13. Zhang ZC (2003) Principle and application of error correction coding. Electronic Industry Press, Beijing, pp 202–222
14. Xin X, Li YS (2002) Computer simulation and performance analysis of TCM modems. Radio Commun Technol 28(5):14–16
15. Zou CY (2000) The efficiency of optical PPM transmission. J Guilin Univer Electron Technol 20(4):53–55

16. Guo SX, Wu MJ, Fei Y, Huang JG (2006) Simulation research of the serial concatenated TCM coding method based on underwater communication channel. Wireless Commun Technol 3:7–10
17. Li YY, Da XY, Wang Y (2004) Application of TCM on satellite high speed data transmission system. Space Electron Technol 3:50–53
18. Huang L (2007) Study on the application of TCM technology in atmospheric laser communication system. Xi'an University of Technology, Xi'an
19. Ding DQ (2005) Design of PPM modulation and demodulation system for atmospheric laser communication. Xi'an university of technology, Xi'an

Chapter 6
Duplex Pulse Interval Modulation

The pulse-position modulation (PPM) method is simple, has the highest power-utilization rate, has a small average-transmission power and error time-slot ratio, and a strong anti-jamming ability; however, it has poor bandwidth efficiency and requires symbol synchronization. Digital pulse-interval modulation (DPIM) has a complex structure, shorter symbol length, higher bandwidth efficiency, and higher power efficiency than on–off keying (OOK) without symbol synchronization.

Dual-header pulse-interval modulation (DHPIM) is a highly efficient modulation method that shortens the symbol length and improves the bandwidth efficiency and transmission capacity, compared with DPIM. It does not require symbol synchronization and significantly simplifies the complexity of the system. It can exchange a higher modulation order for higher power utilization. It is an efficient modulation method that has a significant advantage in optical wireless communication systems.

As part of a modulation-mode analysis of optical wireless communication systems, the symbolic structures of OOK, PPM, multipulse position modulation (MPPM), DPIM, and DHPIM are compared, and their average symbol lengths, average transmission power, bandwidth, and transmission rates are analyzed.

6.1 Modulation Mode Analysis of Optical Wireless Communication

The symbolic structures of OOK, PPM, MPPM, DPIM and DHPIM are compared, and their average symbol length, average transmission power, bandwidth comparison and transmission rate are analyzed.

X. Ke, *Coding Theory in Optical Wireless Communication Systems*, Optical Wireless Communication Theory and Technology, https://doi.org/10.1007/978-981-19-9837-9_6

6.1.1 Dual-Head Pulse Interval Modulation

Dual-header pulse-interval modulation (DHPIM) is an improved form of pulse-interval modulation. The communication time is divided into equal-width periods. Each time period is called a time slot. DHPIM is similar to PIM, and the modulation is complex. The number of time slots contained in each symbol changes. It uses two different starting pulses, each of which has two possible widths, followed by one or two protection slots, and then a number of slots with the modulation information, which depends on the decimal value of the modulation data.

Using a symbol modulated as the interval between two-headed pulses, the number of information gaps in the information frame, the order of modulation, and the decimal number correspond to the modulation symbol; the symbol consists of a header slot and protection slot.

The header consists of a time slot with a pulse width and a protection time slot [1]. The specific symbol structure is shown in Fig. 6.1.

The frame structure for DHPIM is shown in Fig. 6.1. It begins with a rectangular pulse at frame *n*, which starts at $t = T_n$ and lasts for τ, $\tau = (1 + h_n)\alpha T_s/2$, where $h_n \in \{0, 1\}$, *n* represents the *n*th frame, α is time slot. The DHPIM pulse can be represented by the following mathematical equation [2]:

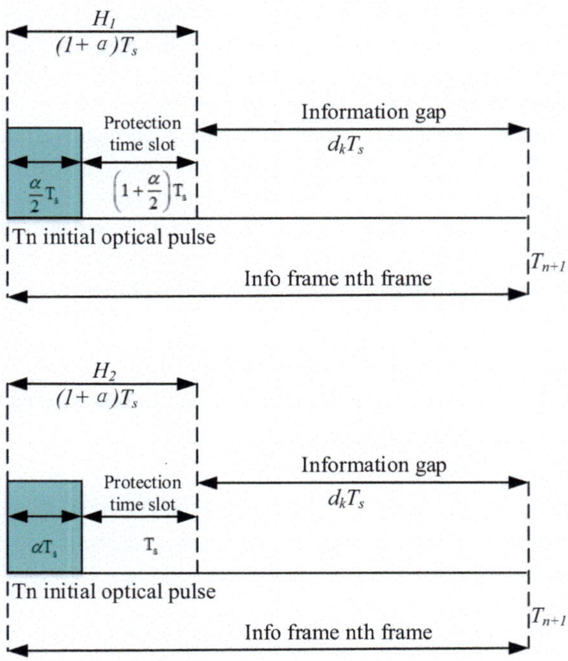

Fig. 6.1 DHPIM frame structure

$$x(t) = A \sum_{n=0}^{\infty} \left\{ rect\left[\frac{2(t - T_n)}{\alpha T_s} - \frac{1}{2} \right] + h_n rect\left[\frac{2(t - T_n)}{\alpha T_s} - \frac{3}{2} \right] \right\}, \tag{6.1}$$

where A is the pulse-amplitude value, and *rect* is the rectangular-function expression, which is mathematically represented by the following equation:

$$rect(u) = \begin{cases} 1; \ -0.5 < u < 0.5 \\ 0; \ \text{otherwise} \end{cases}. \tag{6.2}$$

The start time of the nth frame is defined as $T_n = T_0 + T_s\left[n(\alpha + 1) + \sum_{k=0}^{n-1} d_k\right]$, where T_0 is the start time of the first pulse at $n = 0$ and $d_k \in \{0, 1, \ldots (2^{M-1} - 1)\}$ denotes the decimal value of the information time-slot representation of the kth frame.

$$d_k = \begin{cases} k, & k < 2^{M-1} \\ 2^M - 1 - k, & k \geq 2^{M-1} \end{cases}. \tag{6.3}$$

Assuming T_s is the duration of each time slot in the modulation symbol, Fig. 6.2a shows the structure of the DHPIM symbols at $\alpha = 2$, where the modulation-pulse time slots of each symbol are integer timeslots; Fig. 6.2b shows the structure of the DHPIM symbols at $\alpha = 1$. The length of the DHPIM modulation symbols is the shortest at $\alpha = 1$; however, the modulation pulse appears as half of a time-slot pulse, which is more complicated to implement in hardware. Therefore, this paper mainly discusses DHPIM modulation at $\alpha = 2$, where the modulation waveform appears as a whole time-slot pulse.

DHPIM modulation uses different pulse widths as start pulses, and its pulse headers can be used as symbol starts and clock references, thus giving the DHPIM a built-in symbol-synchronization function. DHPIM not only eliminates the number of redundant time slots, reduces the average symbol length, and increases the data throughput, but also reduces the transmit bandwidth while further increasing the transmission rate. While DHPIM does not require symbol synchronization, it is at the cost of requiring a higher average-transmission power than PPM and DPIM [3].

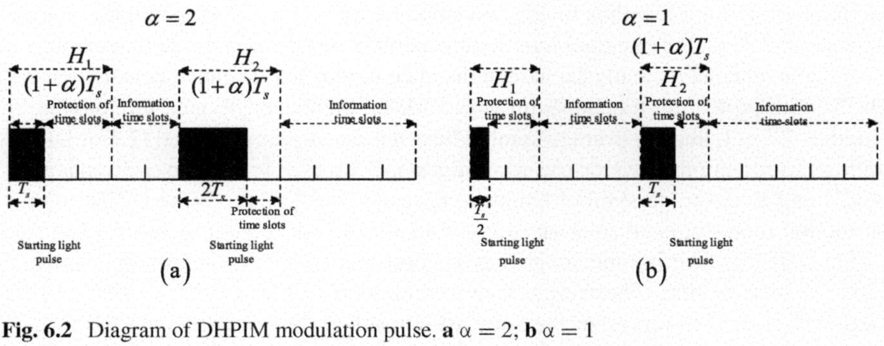

Fig. 6.2 Diagram of DHPIM modulation pulse. **a** $\alpha = 2$; **b** $\alpha = 1$

6.1.2 Performance Comparison of Two-Head Pulse Interval Modulation

Intensity modulation/direct detection systems are typically used in optical wireless communication. OOK is the most common and easiest modulation mode. Pulse-position modulation (PPM) is used to enhance the anti-interference ability of the transmission channel and improve the data-transmission rate. Digital pulse-interval modulation (DPIM) and dual-headed pulse-interval modulation (DHPIM) were also generated. Based on analyses of the OOK, PPM, DPIM, and DHPIM modulation structures, the overall performances of these four modulation modes were simulated and compared.

(1) **Average symbol length**

Assuming that the bit rate sent by the source is R_b bits/s, the modulation bit number per symbol is M bits/symbol, the transmission bit rates of OOK and PPM are the same, and the slot width of OOK modulation is T_b, then $T_b = 1/R_b$ and T_s is the slot width for PPM modulation. From $2^M T_s = MT_b$, the information slot under the same bit rate is $T_s = MT_b/2^M$, and the number of slots per PPM symbol is $L = 2^M$. Only one of the time slots in each PPM frame has a light pulse, whereas the other time slots have no pulse [4]. The number of time slots per symbol of MPPM is L, and multiple light pulses can exist in the time slots per frame.

DPIM eliminates the "0" slot after the "1" pulse in the L slots in the PPM; therefore, it reduces the number of time slots required for DPIM to transmit the same bit source. Because DHPIM modulation uses two different pulses as its starting pulses, this modulation further shortens the average length of the symbol. Different modulation modes exhibit different performance characteristics, depending on the symbol structure.

Taking modulation order M = 4 as an example, the modulation waveforms of OOK, PPM, (8, 2)MPPM, DPIM, and DHPIM are given, as shown in Fig. 6.3.

Table 6.1 lists the corresponding relationship between the modulation modes and information bits at the transmitter, in order of modulation.

The number of time slots contained in the modulation symbol is the symbol length of the modulation mode. If the modulation order is M bits/symbol, and the PPM modulation-symbol length is fixed, the symbol length of the PPM modulation mode is $L_{PPM} = 2^M$, and the symbol length of the DPIM modulation mode is variable.

For convenience of analysis, we use the average symbol length to express that for DPIM with protection time slots, in which the minimum length is 2, the maximum length is $2^M + 1$, and the average symbol length is $L_{DPIM1} = (2^M + 3)/2$. For DPIM without a protection time slot, the minimum symbol length is 1, the maximum length is 2^M, and the average symbol length is $L_{DPIM0} = (2^M + 1)/2$ (the DPIM with a protection time slot is taken as an example in a later analysis in the paper) [5].

For DHPIM, the minimum length has $\alpha+1$ time slots, and the maximum length has $(2^{M-1} + \alpha)$ time slots. The average symbol length is $L_{DHPIM} = (2^{M-1} + 2\alpha + 1)/2$. DHPIM1 is used to represent the simulation results when $\alpha = 1$, and DHPIM2 is

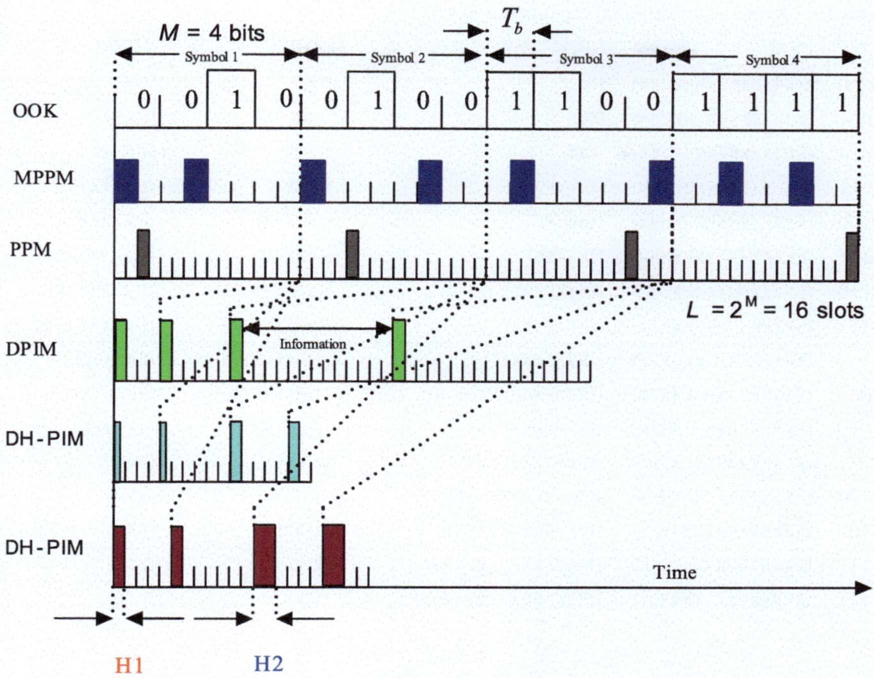

Fig. 6.3 Pulse waveforms with several modulation modes

used to represent the simulation results when $\alpha = 2$. The average symbol lengths for different modulation modes were compared, as shown in Fig. 6.4.

As shown in Fig. 6.4, for all modulation modes, the number of time slots required for encoding increases as the modulation order increases. For the same modulation order M, the average symbol length of PPM is the longest, followed by DPIM, and DHPIM is the shortest. In DHPIM modulation, $\alpha = 2$ is typically used to make the modulation results appear as full-time-slot pulses.

(2) **Analysis of the average transmit-power requirement**

Owing to the eye-security and portability requirements of mobile communication devices, the transmission power of optical wireless communication should be as low as possible, which requires the maximum power utilization or power efficiency [6]. Under the same peak-power conditions, the average transmission-power requirements of different modulation modes were analyzed and compared.

Because these modulation modes can be considered to emit a "0, 1" pulse sequence, no power is required when sending the "0" pulse; when sending the "1" pulse, peak power is required and the probability of inputting "0" or "1" bits is equal [7]. Then, $p_{ave,OOK} = p_t/2$ and $p_{ave,PPM} = p_t/2^M$, where $p_{ave,OOK}$ and $p_{ave,PPM}$ are the average transmission power of OOK and PPM, respectively. For DPIM with a protection time slot, the minimum length is 2, the maximum length is $2^M + 1$, and the average

Table 6.1 Symbols for different modulation modes

4 bit	PPM	DPIM (Protected time slot)	DHPIM (a = 2)	DHPIM (a = 1)
0000	1000000000000000	10	100	(10)0
0001	0100000000000000	100	1000	(10)00
0010	0010000000000000	1000	10000	(10)000
0011	0001000000000000	10000	100000	(10)0000
0100	0000100000000000	100000	1000000	(10)00000
0101	0000010000000000	1000000	10000000	(10)000000
0110	0000001000000000	10000000	100000000	(10)0000000
0111	0000000100000000	100000000	1000000000	(10)00000000
1000	0000000010000000	1000000000	1100000000	100000000
1001	0000000001000000	10000000000	110000000	10000000
1010	0000000000100000	100000000000	11000000	1000000
1011	0000000000010000	1000000000000	1100000	100000
1100	0000000000001000	10000000000000	110000	10000
1101	0000000000000100	100000000000000	11000	1000
1110	0000000000000010	1000000000000000	1100	100
1111	0000000000000001	10000000000000000	110	10

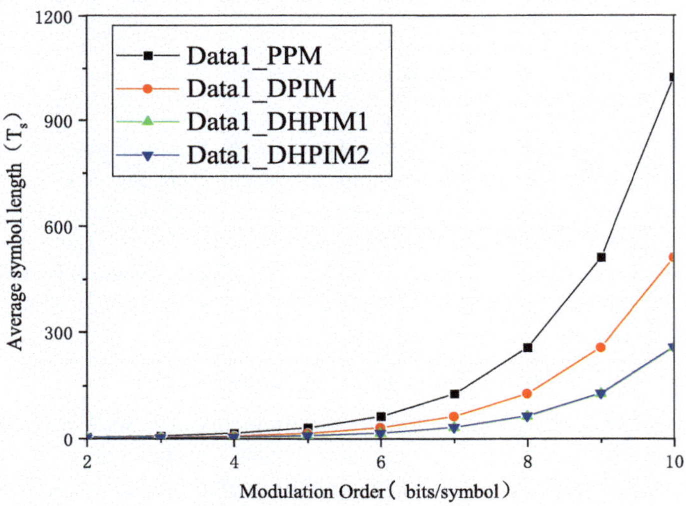

Fig. 6.4 Average symbol lengths for different modulation modes

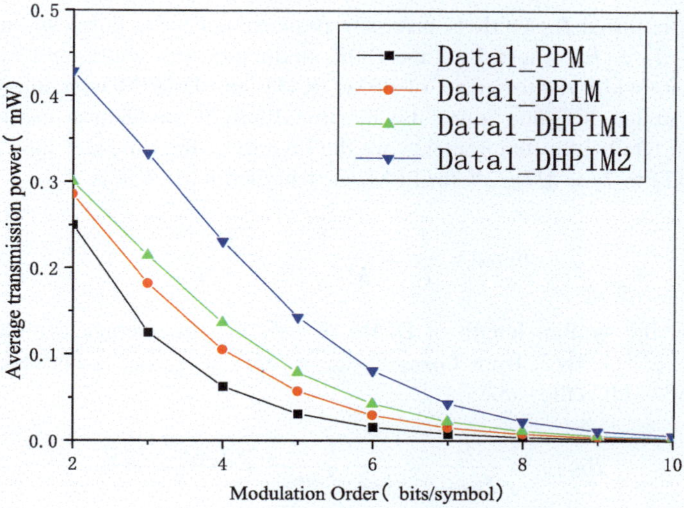

Fig. 6.5 Average transmission power for different modulation modes

symbol length is $L_{DPIM} = (2^M + 3)/2$. Each symbol of DPIM has a "1" pulse, so the average transmission power of DPIM is $p_{ave,DPIM} = (2p_t)/(2^M + 3)$.

For DHPIM, the minimum length in DHPIM is the $\alpha + 1$ time slot. The maximum length is $(2^{M-1} + \alpha)$ time slots, the average symbol length is $L_{DHPIM} = (2^{M-1} + 2\alpha + 1)/2$, and each symbol has an average of $(0.5\alpha + \alpha)/2 = 3\alpha/4$ "1" pulses. Then, the average transmission power of DHPIM is $p_{ave,DHPIM} = 3\alpha p_t/(2^M + 4\alpha + 2)$.

DHPIM1 represents the simulation results obtained when $\alpha = 1$ and DHPIM2 represents those obtained when $\alpha = 2$. Figure 6.5 shows the simulation results when $p_t = 1$ mW is set as the peak power.

As shown in Fig. 6.5, for a fixed modulation order M, because DHPIM has a shorter symbol length and a higher duty cycle than the corresponding PPM, its average required power is higher, followed by that of DPIM. Therefore, the average power utilization of DHPIM is lower than that of PPM; however, for the same average bit-transfer rate and the same bandwidth, DHPIM can use a higher modulation order. Its average power utilization [8] was higher than that of PPM.

(3) **Comparison of the bandwidth requirements**

To obtain a higher signal-to-noise ratio in optical wireless communication, a large-area photodetector with high sensitivity is typically used. Its high capacitance limits the bandwidth of the system, and the smaller the bandwidth requirement of an optical wireless communication system, the better. Therefore, the bandwidth efficiency of the modulation method is an important evaluation criterion.

We assume that the transmitter transmits a signal at a fixed source bit rate of R_b bits/s. If calculated at the first zero of the power spectrum, $B_{OOK} \approx R_b$ and the bandwidth required for OOK is inversely proportional to its pulse width: that is, $B_{OOK} = 1/T_b = R_b$, where T_b is the OOK modulation slot width, and $T_b = 1/R_b$ [9]. The bandwidth requirements of PPM, DPIM, and DHPIM were analyzed and compared under the same source bit rate for different modulation methods. The modulation bandwidth requirements are the inverse of the time slot interval: from $2^M T_s = MT_b$ to $T_s = MT_b/2^M$ for PPM, the bandwidth of PPM is

$$B_{PPM} = \frac{1}{T_s} = \frac{1}{MT_b/2^M} = \frac{R_b 2^M}{M}. \tag{6.4}$$

Because the symbol length of DPIM is variable, the average symbol length, $L_{DPIM1} = (2^M + 3)/2$, from $L_{DPIM1} T_{sdpim} = MT_b$ to T_{sdpim}, is derived, and the DPIM bandwidth is the same:

$$B_{DPIM} = \frac{1}{T_{sdpim}} = \frac{1}{MT_b \Big/ \frac{2^M+3}{2}} = \frac{R_b(2^M + 3)}{2M}. \tag{6.5}$$

Similarly, the average symbol length of DHPIM is $L_{DHPIM} = (2^{M-1} + 2\alpha + 1)/2$. The time-slot width of DHPIM varies from $L_{DHPIM} T_{sdhpim} = MT_b$ to T_{sdhpim}, and the time-slot duration is $T_{sdhpim} = (2M)/((2^{M-1} + 2\alpha + 1)R_b)$. Because DHPIM starts with two pulse slots of different widths, $T_{\min}$ is the minimum pulse duration of DHPIM and $T_{\min} = 0.5\alpha T_{sdhpim} = (\alpha M)/((2^{M-1} + 2\alpha + 1)R_b)$. The DHPIM bandwidth can be calculated as follows:

$$B_{DHPIM} = \frac{1}{T_{\min}} = \frac{R_b(2^{M-1} + 2\alpha + 1)}{\alpha M}. \tag{6.6}$$

The simulation results when $\alpha = 1$ are represented by DHPIM1, and the simulation results when $\alpha = 2$ are represented by DHPIM2. The bandwidth requirements were compared with those of the OOK normalization process, as shown in Fig. 6.6.

As shown in Fig. 6.6, the bandwidth requirement of DHPIM modulation with $\alpha = 1$ is similar to that of DPIM modulation, but less than that of PPM modulation. In addition, the minimum pulse time slot of DHPIM increases with the increasing of α. so the bandwidth requirement of DHPIM modulation with $\alpha = 2$ is less than that of the other modulation modes.

With an increase in the modulation order M, the average symbol length of DHPIM is smaller than that of the other modulation methods, and the bandwidth requirement of PPM is smaller. Therefore, DHPIM has a higher data-transmission efficiency when the bandwidth requirement is the same. DHPIM modulation has the highest bandwidth efficiency, and PPM has the lowest bandwidth efficiency.

(4) **Transmission rates**

One parameter for comparing different modulation modes is the unit transmission rate, which is the number of bits transmitted per Hertz/s. γ is used to represent

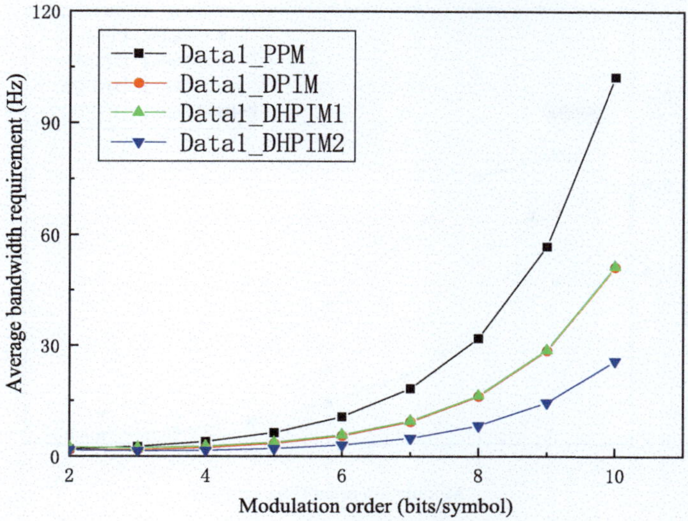

Fig. 6.6 Bandwidth requirements for different modulation modes

the transmission rate, $\gamma = R/B\left(\text{bits s}^{-1}\,\text{Hz}^{-1}\right)$, where R is the transmission rate $\left(\text{bit s}^{-1}\right)$ and B is the signal bandwidth (Hz). In optical communication, the laser usually works in a pulse state with a duration of τ_{ld} and a corresponding bandwidth of $B_{ld} = 1/\tau_{ld}$ (Hz). When the duty cycle is $\tau_p = 1$, $B = 1/T_s$ (Hz) (where T_s is the slot width). If R_b is the rate of information transmission, then the transmission rate is $Y = R_b/B$.

With OOK as a reference, we set the code-unit duty cycle to 1 (which is used in the following analysis), the bit rate to R_b, and the bandwidth to R_b. For PPM modulation with the same M bit rate, assuming equal probability of sending "0" and "1," as indicated by the last section of bandwidth requirements, the PPM bandwidth $B_{PPM} = \left(R_b 2^M\right)/M$, DPIM bandwidth $B_{DPIM} = \left(R_b(2^M + 3)\right)/2M$, and DHPIM bandwidth $B_{DHPIM} = \left(R_b(2^{M-1} + 2\alpha + 1)\right)/(\alpha M)$. Then, the transmission rates of different modulation modes can be obtained from the formula $Y = R_b/B$.

PPM transmission rate:

$$Y_{PPM} = \frac{R_b}{B_{PPM}} = R_b \Bigg/ \frac{R_b 2^M}{M} = \frac{M}{2^M}. \tag{6.7}$$

DPIM transmission rate:

$$Y_{DPIM} = \frac{R_b}{B_{DPIM}} = R_b \Bigg/ \frac{R_b(2^M + 3)}{2M} = \frac{2M}{2^M + 3}. \tag{6.8}$$

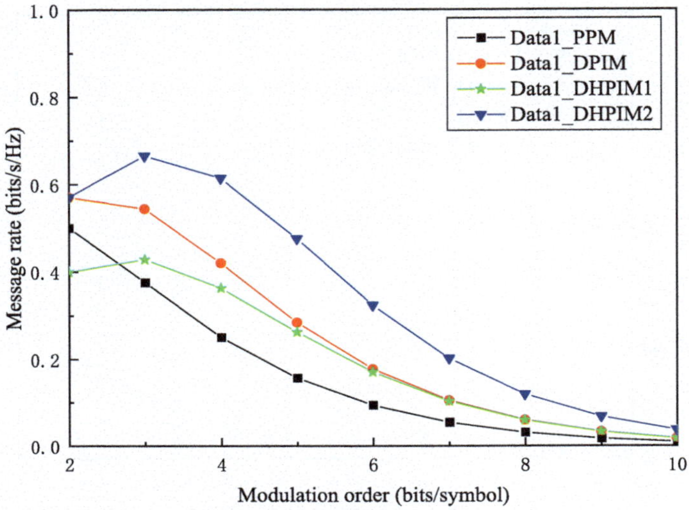

Fig. 6.7 Comparison of signal-transmission rates of different modulation modes

DHPIM transmission rate:

$$Y_{DHPIM} = \frac{R_b}{B_{DHPIM}} = R_b \Big/ \frac{R_b(2^{M-1} + 2\alpha + 1)}{\alpha M} = \frac{\alpha M}{2^{M-1} + 2\alpha + 1}. \tag{6.9}$$

The simulation results when $\alpha = 1$ are represented by DHPIM1 and those for $\alpha = 2$ are represented by DHPIM2. The transmission rates for the different modulation modes (normalized by OOK) are shown in Fig. 6.7.

As shown in Fig. 6.7, when the modulation order $M > 2$, the $\alpha = 2$ transmission rate in DHPIM is highest, followed by DPIM, $\alpha = 1$ in DHPIM, and PPM. When $M > 6$ bits, the transmission rates of $\alpha = 1$ and $\alpha = 2$ are equal, but both are better than those of DPIM and PPM.

(5) **Power-spectral density analysis**

The power spectrum calculation of OOK is a non-return-to-zero (NRZ) signal because the OOK signal can be expressed as

$$S_{OOK}(t) = \sum_{n=0}^{\infty} a_n g(t - nT_s), \tag{6.10}$$

where a_{n} takes “1” or “0” with the presence or absence of a pulse, respectively, and T_s represents the time-slot width. $g(t)$ is a rectangular pulse function, and its mathematical expression formula is as follows:

$$g(t) = \begin{cases} 1; & -T_s/2 < t < T_s/2 \\ 0; & \text{otherwise} \end{cases}. \tag{6.11}$$

The Fourier transform of a rectangular pulse function, that is, its corresponding spectrum is

$$\begin{aligned} G(f) &= \int_{-\frac{T_s}{2}}^{\frac{T_s}{2}} e^{-j2\pi ft} dt = \frac{1}{j2\pi}\left(e^{j\pi fT_s} - e^{-j\pi fT_s}\right) \\ &= T_s\left(\frac{\sin(\pi fT)}{\pi fT_s}\right). \end{aligned} \tag{6.12}$$

Then, the power spectrum of $S_{OOK}(t)$ is

$$P(f) = T_s P(1 - P)|G(f)|^2, \tag{6.13}$$

where P and $(1 - P)$ are the probabilities of taking "1" or "0" of a sequence, respectively. By substituting Eq. (6.12) into Eq. (6.13), the power spectrum of the OOK signal can be obtained as follows:

$$P(f) = \frac{T_s}{4}\left|\frac{\sin \pi fT_s}{\pi fT_s}\right|^2. \tag{6.14}$$

For the power-spectrum calculation of a DPIM signal, the DPIM pulse sequence can be expressed as

$$x(t) = \sum_{n=-\infty}^{\infty} a_n g(t - nT_s), \tag{6.15}$$

where $g(t)$ is a rectangular pulse waveform, T_s represents the time-slot width, and a_n is a random variable that indicates whether there is a pulse in the nth time slot. $x(t)$ is a periodic stationary process; its power-spectral density can be expressed as

$$P(f) = \frac{1}{T_s}|G(f)|^2 S_{slot}(f), \tag{6.16}$$

where $G(f) = T_s\left(\frac{\sin \pi fT_s}{\pi fT_s}\right)$, $G(f)$ is the Fourier transform of $g(t)$, and $S_{slot}(f)$ is the power-spectral density of the time-slot sequence. It can be expressed as a discrete Fourier transform corresponding to the R_k autocorrelation function of the time-slot sequence.

For DPIM with one protection slot, the final-slot autocorrelation function is given by Eq. (6.17) [10].

$$R_k = \begin{cases} L_{ave}^{-1} & k = 0 \\ 0 & k = 1 \\ \left(\frac{L_{ave}^{-1} L^{-1}}{\sqrt{1+4L^{-1}}}\right)\left[\left(\frac{1+\sqrt{1+4L^{-1}}}{2}\right)^{k-1} - \left(\frac{1-\sqrt{1+4L^{-1}}}{2}\right)^{k-1}\right] & 2 \le k \le L+1 \\ \frac{1}{L}\sum_{i=1}^{L} R_{k-1-i} & k \ge L+1 \end{cases}, \quad (6.17)$$

where $L = 2^M$, L_{ave} is the average number of time slots per DPIM symbol, and $L_{DPIM} = 2^M + 3/2$. The more k increases, the more similar R_k is to L_{ave}, which can be obtained from Eq. (6.17). When $k > 5L$, R_k can be approximated as L_{ave}^{-2}.

Because the mean value of the time-slot sequence is nonzero, the spectrum may contain δ functions. Nevertheless, rectangular pulses that occupy the entire time slot should be considered. $|G(f)|^2$ simplifies the remaining δ component of the δ function, and the continuous part in $S_{slot}(f)$ can be represented as $S_c(f)$; therefore, the discrete Fourier transformation of the time-slot sequence autocorrelation function R_k can be represented as

$$\begin{aligned} S_c(f) &= \sum_{k=-\infty}^{\infty} (R_k - L_{ave}^{-2}) e^{j2\pi kfT_s} \\ &\cong \sum_{k=-5L}^{5L} (R_k - L_{ave}^{-2}) e^{j2\pi kfT_s}. \end{aligned} \quad (6.18)$$

The power-spectral density formula for the DPIM signal, obtained by substituting Eq. (6.18) into Eq. (6.16), is [11]

$$\begin{aligned} P(f) &= \frac{1}{T_s}|G(f)|^2 \sum_{k=-5L}^{5L} (R_k - L_{ave}^{-2}) e^{j2\pi kfT_s} \\ &= \frac{1}{T_s}|G(f)|^2 \sum_{k=-5L}^{5L} \left(R_k - \left(\frac{L+3}{2}\right)^{-2}\right) e^{j2\pi kfT_s}. \end{aligned} \quad (6.19)$$

The DHPIM pulse can be expressed by the following mathematical formula:

$$x(t) = A\sum_{n=0}^{\infty}\left\{rect\left[\frac{2(t-T_n)}{\alpha T_s} - \frac{1}{2}\right] + h_n rect\left[\frac{2(t-T_n)}{\alpha T_s} - \frac{3}{2}\right]\right\}. \quad (6.20)$$

It starts with a rectangular pulse at frame n. Its starting time $t = T_n$ lasts for τ; $\tau = (1+h_n)\alpha T_s/2$, where $h_n \in \{0, 1\}$, n stands for frame n, A is the pulse amplitude value, and $rect$ is the rectangular function expression. When $h_n = 0$, Eq. (6.23) has only one slot pulse and the duration of the pulse is $\tau = \alpha T_s/2$.

The start time of frame n is defined as $T_n = T_0 + T_s\left[n(\alpha+1) + \sum_{k=0}^{n-1} d_k\right]$, where T_0 is the start time of the first pulse of $n = 0$ and $d_k \in \{0, 1, \ldots (2^{M-1} - 1)\}$ is the decimal value represented by the information time slot of frame k.

The Fourier-transform expression of the N-sign truncation function is [12]

$$X_N(\omega) = A\sum_{n=0}^{N-1}\int_{-\infty}^{\infty}\left\{rect\left[\frac{2(t-T_n)}{\alpha T_s} - \frac{1}{2}\right] + h_n rect\left[\frac{2(t-T_n)}{\alpha T_s} - \frac{3}{2}\right]\right\}\cdot e^{-j\omega t}dt \tag{6.21}$$

$$X_N(\omega) = A\sum_{n=0}^{N-1}\left\{\int_{T_n}^{T_n+\frac{\alpha T_n}{2}} e^{-j\omega t}dt + h_n\int_{T_n+\frac{\alpha T_n}{2}}^{T_n+\alpha T_n} e^{-j\omega t}dt\right\}, \tag{6.22}$$

where $\omega = 2\pi f$ is the angular frequency and f is the frequency.

The power-spectral density of the signal can be obtained by averaging the symbol length N and then taking the limit:

$$P(\omega) = \lim_{N\to\infty}\frac{E\left[X_N(\omega)\cdot X_N^*(\omega)\right]}{E[T_N - T_0]}, \tag{6.23}$$

where $E[x]$ is the expected value of x, $X_N^*(\omega)$ is the conjugate of $X_N(\omega)$, and $T_N - T_0$ is expected to be

$$T_N - T_0 E[T_N - T_0] = NT_s\left[1 + \alpha + \frac{2^{M-1}-1}{2}\right] \tag{6.24}$$

$$\begin{aligned} &E\left[X_N(\omega)\cdot X_N^*(\omega)\right] \\ &= \left(\frac{A}{\omega}\right)^2\left(\frac{e^{(-j\omega T_0)}}{j}\cdot\frac{e^{(j\omega T_0)}}{-j}\right)\times\left[\left(1-e^{-j\omega T_s\alpha/2}\right)\left(1-e^{+j\omega T_s\alpha/2}\right)\right]\cdot S_N(\omega) \\ &= 4A^2\frac{\sin^2(\alpha\omega T_s/4)}{\omega^2}\cdot S_N(\omega), \end{aligned} \tag{6.25}$$

where

$$\begin{aligned} &S_N(\omega) \\ &= E\left[\sum_{n=0}^{N-1}\sum_{q=0}^{N-1}\left(\left(1+h_n e^{-j\omega T_s\alpha/2}\right)\left(1+h_q e^{j\omega T_s\alpha/2}\right)\right.\right. \\ &\left.\left.\times e^{-j\omega T_s(\alpha+1)(n-q)} e^{-j\omega T_s\left[\sum_{k=0}^{n-1} d_k - \sum_{k=0}^{q-1} d_k\right]}\right)\right]. \end{aligned} \tag{6.26}$$

Equation (6.27) can be divided into three parts: (1) $S_{N1}(\omega)$ is the expression when $q < n$; (2) $S_{N2}(\omega)$ is when $q = n$; and (3) $S_{N3}(\omega)$ is the expression when $q > n$; then, they are summed to obtain the $S_N(\omega)$ expression:

$$S_N(\omega) = S_{N1}(\omega) + S_{N2}(\omega) + S_{N3}(\omega). \tag{6.27}$$

For Eq. (6.27), when $q < n$, the following expression applies:

$$S_{N1}(\omega) = \left[\frac{9}{4} - 2\sin^2\left(\frac{\alpha\omega T_s}{4}\right)\right] \times \frac{G}{(1-G)^2}\left[N(1-G) - (1-G^N)\right], \tag{6.28}$$

where G is given by the following equation:

$$G = \left(\frac{1 - e^{-j\omega T_s 2^{M-1}}}{1 - e^{-j\omega T_s}} \cdot \frac{e^{-j\omega T_s(\alpha+1)}}{2^{M-1}}\right). \tag{6.29}$$

For Eq. (6.27), when $q = n$, the following expression applies:

$$S_{N2}(\omega) = \frac{N}{2}\left[5 - 4\sin^2\left(\frac{\alpha\omega T_s}{4}\right)\right]. \tag{6.30}$$

For Eq. (6.27), the expression that applies when $q > n$ is

$$S_{N3}(\omega) = S_{N1}^*(\omega). \tag{6.31}$$

Substituting $S_{N1}(\omega)$, $S_{N2}(\omega)$, and $S_{N3}(\omega)$ into Eq. (6.28) is expressed as follows [12]:

$$S_N(\omega) = \frac{N}{2}\left[5 - 4\sin^2\left(\frac{\alpha\omega T_s}{4}\right)\right] + \left[\frac{9}{2} - 4\sin^2\left(\frac{\alpha\omega T_s}{4}\right)\right]$$
$$\mathrm{Re}\left\{\frac{G}{(1-G)^2}\left[N(1-G) - (1-G^N)\right]\right\}. \tag{6.32}$$

The power-spectral density of the truncated signal can be obtained from Eqs. (6.25) and (6.26) applied to Eq. (6.24):

$$P(\omega) = \frac{4A^2\sin^2\left(\frac{\alpha\omega T_s}{4}\right)}{\omega^2 T_s\left[1 + \alpha + \frac{2^{M-1}-1}{2}\right]} \cdot \lim_{N\to\infty}\left\{\frac{S_N(w)}{N}\right\}. \tag{6.33}$$

To simplify Eq. (6.34), all possible values of $S_N(\omega)$ in Eq. (6.33) must be considered, which depends mainly on the value of G. Therefore, there are two situations that need to be considered [12]:

First case: When $|G| < 1$:

$|G| < 1$, $e^{-j\omega T_s} \neq 1$, and $\omega \neq 2\pi K/T_s$, where K is a positive integer. Here, $\lim_{N\to\infty} G^N = 0$; thus, by Eq. (6.33) the following also applies:

$$\lim_{N\to\infty}\left[\frac{S_N(w)}{N}\right] = \frac{1}{2}\left\{\left[5 - 4\sin^2\left(\frac{\alpha\omega T_s}{4}\right)\right]\right. \\ \left. + \left[9 - 8\sin^2\left(\frac{\alpha\omega T_s}{4}\right)\right]\mathrm{Re}\left(\frac{G}{1-G}\right)\right\}. \quad (6.34)$$

By introducing Eq. (6.35) into Eq. (6.34), we obtain

$$P(\omega) = \frac{2A^2\sin^2\left(\frac{\alpha\omega T_s}{4}\right)}{\omega^2 T_s\left[1 + \alpha + \frac{2^{M-1}-1}{2}\right]} \cdot \left[5 - 4\sin^2\left(\frac{\alpha\omega T_s}{4}\right)\right] \\ + \left[9 - 8\sin^2\left(\frac{\alpha\omega T_s}{4}\right)\right]\mathrm{Re}\left(\frac{G}{1-G}\right). \quad (6.35)$$

Equation (6.36) gives the power-spectral density of a DHPIM signal when it is finite. Here, $|G| \neq 1$ and $\omega \neq 0$.

Second case: When $|G| = 1$.

$G = 1$, $e^{-j\omega T_s} = 1$, and $\omega = 2\pi K/T_s$, where K is a positive integer. Here, the value of Eq. (6.33) is indeterminate. For Eq. (6.36), according to L'Hopital's rule, when $G \to 1$:

$$\lim_{N\to\infty}\frac{G}{(1-G)^2}\left[N(1-G) - (1-G^N)\right] = \frac{N(N-1)}{2}, \quad (6.36)$$

so, from Eq. (6.29):

$$S_N(w) = \frac{N}{4}\left\{\left[10 - 8\sin^2\left(\frac{\alpha\omega T_s}{4}\right)\right] + (N-1)\left[9 - 8\sin^2\left(\frac{\alpha\omega T_s}{4}\right)\right]\right\}. \quad (6.37)$$

Equation (6.37) is introduced into Eq. (6.34):

$$P(\omega) = \frac{A^2\sin^2\left(\frac{\alpha\omega T_s}{4}\right)}{\omega^2 T_s\left[1 + \alpha + \frac{2^{M-1}-1}{2}\right]} \\ \cdot \lim_{N\to\infty}\left\{\left[10 - 8\sin^2\left(\frac{\alpha\omega T_s}{4}\right)\right] + (N-1)\left[9 - 8\sin^2\left(\frac{\alpha\omega T_s}{4}\right)\right]\right\}. \quad (6.38)$$

Depending on the difference in K values, Eq. (6.38) tends to be zero or ∞, so the following conclusions can be drawn:

(i) When $K = 0$, $\omega = 0$, Eq. (6.35) tends to infinity, according to L'Hopital's rule:

$$P(0) = \frac{A^2 \lim\limits_{x\to\infty}\{9N+1\}}{T_s\left[1+\alpha+\frac{2^{M-1}-1}{2}\right]} \cdot \lim_{\omega\to 0}\left\{\frac{\sin^2\left(\frac{\alpha\omega T_s}{4}\right)}{\omega^2}\right\} \tag{6.39}$$

$$P(0) = \frac{\alpha^2 A^2 T_s}{16\left[1+\alpha+\frac{2^{M-1}-1}{2}\right]} \cdot \lim_{N\to\infty}\{9N+1\} \to \infty. \tag{6.40}$$

(ii) $K = 2\upsilon/\alpha$, υ is an integer, $\upsilon \neq 0$, and $\omega = 2\upsilon(2\pi/\alpha T_s)$. Equation (6.30) is reduced to 0, so it depends on the space–time slot in all spectra, and the space–time slot depends on α.

(iii) For all other frequency values, $\omega = 2\pi K/T_s$, $P(\omega) \to \infty$, the potential exists for different time-slot compositions and their harmonics $\omega = 2\pi(K/T_s)$. Depending on the value of α, the time-slot composition coincides with its harmonic part and the space–time slot part of *sinc*. The formula is given in the results section.

In summary:

$$P(\omega) = \begin{cases} \frac{2A^2\sin^2\left(\frac{\alpha\omega T_s}{4}\right)\left\{\left[5-4\sin^2\left(\frac{\alpha\omega T_s}{4}\right)\right]+\left[9-8\sin^2\left(\frac{\alpha\omega T_s}{4}\right)\right]\mathrm{Re}\left(\frac{G}{1-G}\right)\right\}}{\omega^2 T_s\left[1+\alpha+\frac{2^{M-1}-1}{2}\right]} & \omega \neq \frac{2\pi K}{T_s} \\ 0 & \omega = \frac{2\pi K}{T_s},\ K = \frac{2\upsilon}{\alpha}\cdot \\ \infty & \omega = \frac{2\pi K}{T_s},\ K \neq \frac{2\upsilon}{\alpha} \end{cases} \tag{6.41}$$

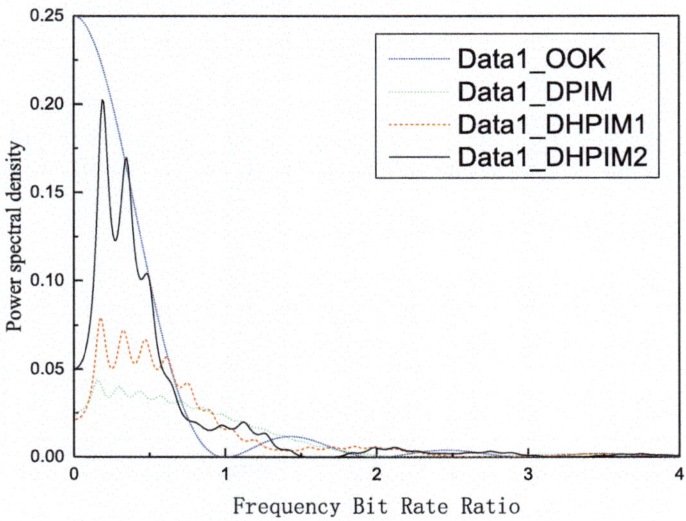

Fig. 6.8 Power-spectral density for different modulation modes

Based on the power-spectrum expressions of the various modulation modes obtained above, the power spectrum discussed above was simulated using the MATLAB software. The horizontal coordinate represents the ratio of the frequency to the information-transfer rate and the vertical coordinate represents the power-spectrum density. The simulation results indicate that the low-frequency part of two-headed pulse-interval modulation is not zero. Receiver systems typically use high-pass filters to filter out background-light noise (Fig. 6.8).

6.2 Channel Capacity and Error Performance Analysis

In this section, we analyze the channel capacity and error performance of pulse-position modulation (PPM), digital pulse-interval modulation (DPIM), and dual-headed pulse-interval modulation (DHPIM).

(1) **Channel capacity**

The channel capacity is the maximum average information rate that can be transmitted in a channel. Channels are divided into continuous and discrete channels; therefore, different methods are used to describe channel capacity. Because the two-header pulse-interval modulation discussed in this paper is a digital modulation method, only the discrete channel capacity is discussed.

There are two representations of discrete channel capacity: the maximum average amount of information that each symbol can transmit, and the maximum average amount of information that can be transmitted in a unit time. The two representations are essentially identical and can be converted into one another. The channel capacity can be expressed using the maximum information-transmission rate. The size of the channel capacity represents the information-transmission capability of the system per unit time. This is an important performance indicator of optical wireless communication systems [13].

For the OOK modulation mode, the modulation order is equal to the number of input information bits s and the modulation order is M; hence, the number of OOK slots is L_{OOK} and $L_{OOK} = M$. Assuming the input bit rate is R_b bits/s and the modulation rate is R_b bits/s according to the OOK modulation mode, the channel capacity of OOK per pulse slot $T_b = 1/R_b$ is

$$C_{\mathrm{OOK}} = \frac{M}{L_{OOK}T_s} = \frac{M}{MT_s} = R_b \text{ bit/s}. \tag{6.42}$$

Assuming that the PPM-modulated bit rate is R_b bits/s and each symbol of the PPM corresponds to M binary information bits s, then the pulse time slot $T_s = MT_b/2^M$ and the channel capacity of the PPM is

$$C_{PPM} = \frac{M}{L_{PPM}T_s} = \frac{M}{2^M MT_b/2^M} = R_b \text{ bit/s}, \tag{6.43}$$

where $L_{PPM} = 2^M$ denotes the PPM symbol length.

PPM and OOK have the same channel capacity [14], and they transmit the same average amount of information over the same time. Assuming that each time-slot duration of DPIM and DHPIM is the same as the PPM time-slot length, DPIM and DHPIM have higher channel capacities than other methods because their average symbol lengths are smaller, and more information can be transmitted than by PPM in a given time.

Under the same time-slot width, the symbol length of DPIM changes, and the number of time slots is shorter than that of PPM when sending the same bits; therefore, it has a higher channel capacity than PPM. If the average length of a DPIM symbol is $L_{DPIM} = (2^M + 3)/2$, the channel capacity of DPIM can be calculated as

$$C_{DPIM} = \frac{M}{T_s L_{DPIM}} = \frac{2R_b 2^M}{(2^M + 3)}. \tag{6.44}$$

Similarly, if the average length of a DHPIM symbol is $L_{DHPIM} = (2^{M-1} + 2\alpha + 1)/2$, then the channel capacity of DHPIM is

$$C_{DHPIM} = \frac{M}{T_s L_{DHPIM}} = \frac{2R_b 2^M}{(2^{M-1} + 2\alpha + 1)}. \tag{6.45}$$

Figure 6.9 shows different normalized modulation modes, relative to the PPM results in channel capacity.

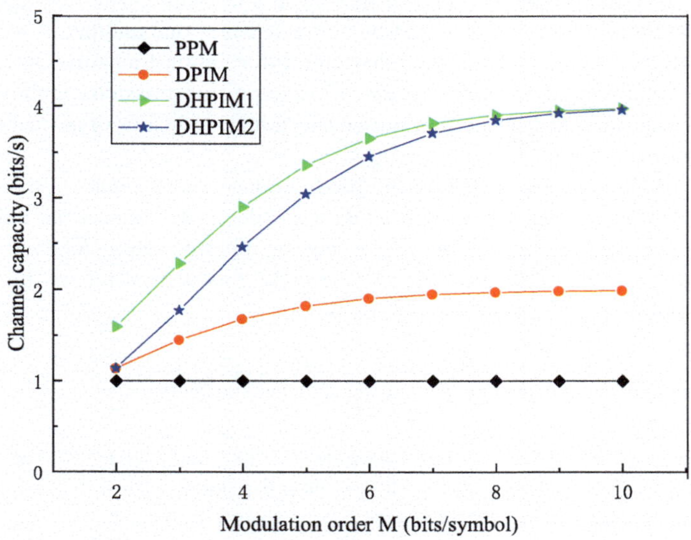

Fig. 6.9 Channel capacity for different modulation modes

From Fig. 6.9, it can be seen that as the modulation order increases, the channel capacity increases; DHPIM has the largest, DPIM has the second largest, and PPM has the smallest. DHPIM1 represents the channel capacity at $\alpha = 1$ and DHPIM2 represents the channel capacity at $\alpha = 2$. It can be seen DHPIM modulation has a great advantage over the other modulation modes in terms of channel capacity, can transmit more information in a given time, and is suitable for use in high-speed data transmission systems.

6.3 Implementation Principle and Simulation of DHPIM Modem

The software simulation was completed using MATLAB. DHPIM channel coding is discussed according to the characteristics of the DHPIM modulation mode to reduce the signal transmission-error rate and improve the reliability of the communication system. The channel coding of a forward error-correction mode was mainly analyzed, and the basic theory of Reed–Solomon (*RS*)error-correction coding and encoding was studied.

6.3.1 DHPIM Modem Software Simulation

(1) **Soft and hard judgments**

In intensity modulation/direct detection, when the receiver decoder decodes the received signal, the most important step is to judge the signal. In an optical wireless communication system, the optical pulse is affected by various factors, such as the optical-pulse waveform widening and the pulse amplitude changing, which affects the judgment of the receiver. This effect not only leads to errors in the code elements in a code group, but also causes interference in the signals between the code groups. This must be overcome or minimized.

Two types of judgments are currently available: hard and soft. Hard-judgment decoding refers to the direct decision of each sample based on a given decision-threshold value. An output greater than the threshold is a high level and the opposite is a low level. When using hard judgments, the decoder simply quantifies the sampled signal as "high level" or "low level" through threshold detection. The soft-judgment method does not quantify the sample values, but compares the size of each sample value. The maximum sample value is judged as "1," and the remaining time slots are judged as "0". Using soft judgments can reduce the error rate.

In PPM modulation, frame-synchronization and slot-synchronization clocks are required to determine the start and end of a code group. Soft-judgment decoding is easy to implement. A PPM signal can be demodulated by simply comparing the position of the largest pulse amplitude in a frame time slot, because PPM modulation transmits information in the position of the frame time slot in which the pulse is located. PPM can also be implemented using hard-judgment decoding. Under the same error rate, the light-power requirement for hard judgment is higher than that for soft judgment [15].

The coding-code length of a DHPIM modulation mode is uncertain. In the modulation mode discussed in this paper, the code-symbol length ranges from three to 10, which means a decoding error in any one time slot will result in a wrong judgment for the code-group time slot and also the adjacent time slot, which may have a disastrous impact on communication. DHPIM does not require symbol synchronization; it only needs to detect a "1" at the receiver to start the code group. Because there is a "1" connection in the DHPIM sequence, each modulation symbol starts with a pulse; therefore, soft judgment cannot be used in DHPIM demodulation.

In this study, when implementing the DHPIM software simulation, we used hard judgment; when a sample is greater than the threshold, it is judged to have a pulse, and when it is less than the threshold, it is judged to have no pulse output. The selection of the threshold determines the error-code performance of the system; therefore, we should analyze the performance of the system according to its requirements and determine an appropriate threshold to minimize the error-code performance of the system. This makes the communication system more reliable.

(2) **DHPIM system function**

In general, in PPM systems, the time required to transmit the same amount of information is the same. PPM symbols have fixed lengths, so the input-information time is guaranteed to equal the modulation time of each symbol. DHPIM symbols are not fixed in length. The modulation time of each symbol is related to the information that is modulated. This is because the modulation time of the same amount of information in DHPIM is not fixed, so the system design is more complex than that for fixed-length encoding, such as PPM.

For modulation order 4, the source information, modulation symbols, and corresponding decimal digits of DHPIM at $\alpha = 2$ are listed in Table 6.2. The lengths of the DHPIM symbols are variable and consist of different start pulses and gap time slots.

The modulation process converts information from a bit sequence to a time-slot sequence. The symbolic structure of DHPIM consists of pulse protection time slots with different widths, followed by the number of information slots related to the modulation information. Because the length of the DHPIM symbol is not fixed, and the time required to transmit the same amount of information is uncertain, the modulator rate is also not fixed. A data-cache method was used to solve the problem of an unstable modulation rate.

The data to be modulated are stored in the cache, and all the data in the cache are read at the beginning of the modulation. After the data in the buffer are modulated by DHPIM, a stop bit is added to demodulate correctly at the receiving end. When the detection indicates the end of the modulation, the next frame of data can be read so that the cycle completes all the data modulations.

Table 6.2 Mapping of source bits to modulation symbols

4 bit	DHPIM (a = 2)	Decimal value	4 bit	DHPIM (a = 2)	Decimal value
0000	100	0	1000	1100000000	8
0001	1000	1	1001	110000000	9
0010	10000	2	1010	11000000	10
0011	100000	3	1011	1100000	11
0100	1000000	4	1100	110000	12
0101	10000000	5	1101	11000	13
0110	100000000	6	1110	1100	14
0111	1000000000	7	1111	110	15

For any serial data, the modulation data are the decimal values corresponding to the serial–parallel conversion, and the modulation symbol is the last modulation output pulse slot. It can be observed that the length of each modulation symbol changes, starting with a different pulse width, followed by a protection and information slot, and having a pulse slot as the stop bit, after the data modulation in the cache is completed. Figure 6.10 shows the modulation data and modulation symbol diagrams.

Because the length of a DHPIM modulation symbol is not certain and the beginning of the modulation symbol begins with a pulse, the receiver does not need symbol synchronization during demodulation. It only needs time-slot synchronization; this makes the demodulation implementation more convenient and simplifies the complexity of the system implementation.

During demodulation, when it is determined that a time slot is at a high level, the system indicates the beginning of the modulation information, determines the pulse duration of the start time slot, counts the number of empty time slots between the start pulse and the next symbol pulse, and obtains the source data according to the corresponding law of DHPIM. When the last time slot of the superframe is detected, all the data of the superframe are demodulated and all the data are saved in the cache. With the next time slot, the system starts to read the symbol of the next superframe and enters the demodulation of the next superframe.

(3) **DHPIM simulation modeling**

According to the characteristics of the DHPIM modulation and demodulation modes, a software-simulation block diagram was designed and implemented. Figure 6.11 shows the overall block diagram of the DHPIM simulation. The DHPIM simulation

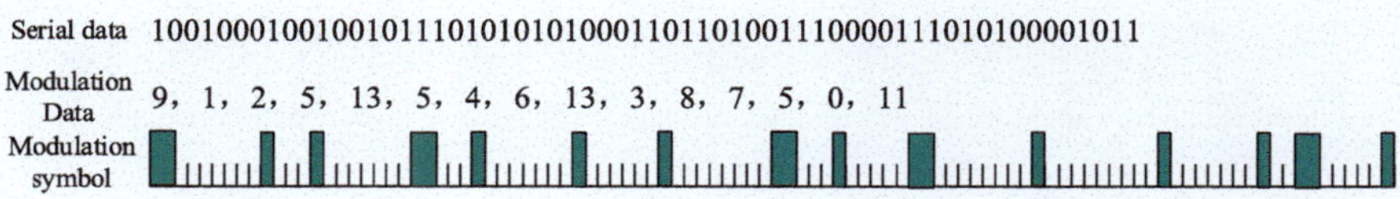

Fig. 6.10 Relationship between the modulation data and modulation symbols

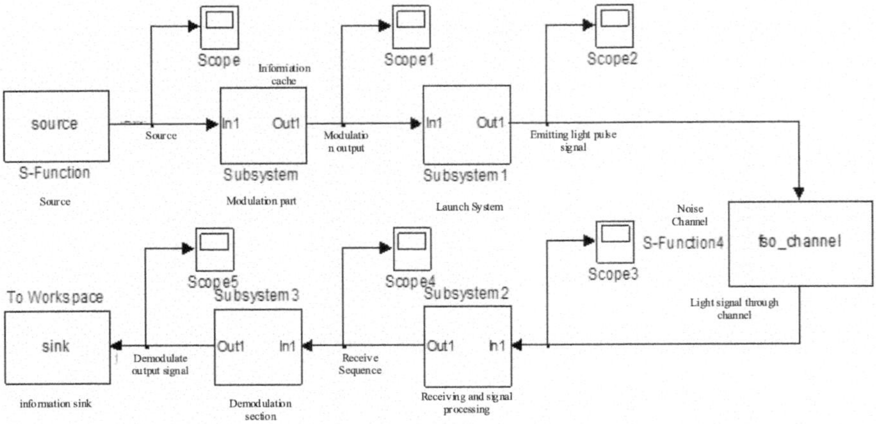

Fig. 6.11 DHPIM system-modeling diagram

realized by Simulink in MATLAB software conducted a software simulation with a modulation order of 4. It includes source, receiver, and signal-processing modules.

(4) **Implementation of the module subsystem**

In this study, the DHPIM model system was modeled using SIMULINK. The order of modulation $M = 4$ was chosen; that is, the modulation information is four bits/symbol. Therefore, the number of DHPIM modulation symbols is 3–10 slot lengths. Fifteen four-bit data points were selected as a set of data to modulate the 15 data points. The modulated signal was then converted into an optical signal using the transmission system. The optical signal was disturbed by noise through a wireless optical channel.

The optical signal was converted into a receiving sequence by the receiving and signal-processing modules. The signal-processing part uses the hard-judgment method. The receiving sequence is then decomposed by the demodulation part to save the data superframe. The data are sent to a mailbox when the demodulation is complete.

The source module completes the random binary sequence generated by the S function. The simulation results for the random binary sequence sent by the source are shown in Fig. 6.12.

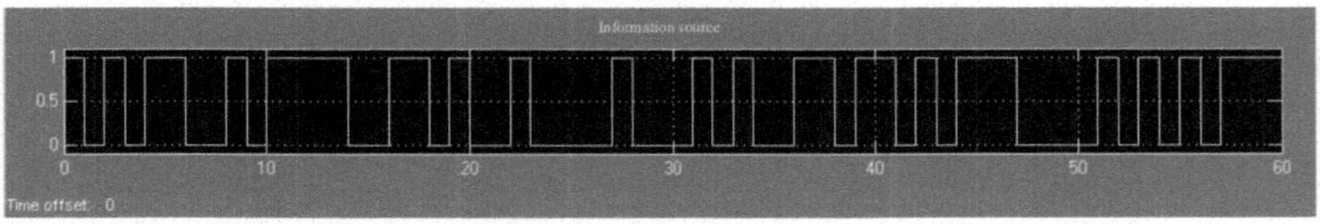

Fig. 6.12 Source output simulation

The modulation part completes the serial–parallel conversion of the binary sequence, converts the binary to decimal, and caches the modulation data. For each modulation data group, the pulse width of the head slot is determined by DHPIM according to the corresponding decimal value of the data.

Different pulse widths and the number of protection and information slots are generated by counting, according to the modulation data. This counting accomplishes the modulation function for all cached data, converts the binary sequence to the output of the time-slot sequence, and thus completes the signal modulation. The simulation results for the partial modulation output are shown in Fig. 6.13.

Because of the different lengths of the DHPIM modulation symbols, there is no fixed average period for each frame signal modulated by the DHPIM, which causes the time required to transmit the same number of bits of information frames to be uncertain; that is, instant modulation cannot be completed. Therefore, using hyperframe modulation, this simulation uses 15 frames of data as a hyperframe data group, and the modulation system receives and caches the 15 frames of data to be modulated. When the 15th data point is received, the modulation starts.

Therefore, the modulation of the source in Figs. 6.11 and 6.12 is not instant, but has a certain delay. Assuming that the duration of each time slot is seconds, this delay time can be calculated by multiplying the number of modulated data frames by the data bits in each frame. The delay time can be calculated as $t_{delay}=15 \times 4t_s = 60t_s$ s.

The function of the transmission system is to add the modulated output-pulse time-slot signal to the laser and complete the conversion of the electrical signal to an optical signal, which is changed by the information signal to carry information with the emitted light signal. The intensity of the emitted light is changed by the change in the modulated signal. The laser power used in the simulation is 10 mW. The simulation results for the transmission system are shown in Fig. 6.14.

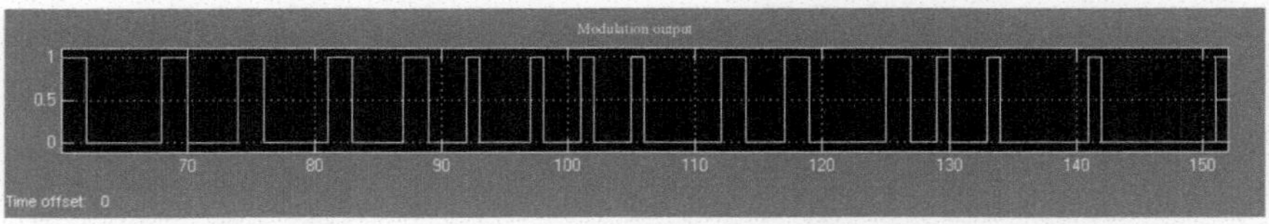

Fig. 6.13 Simulation of a partial modulation output

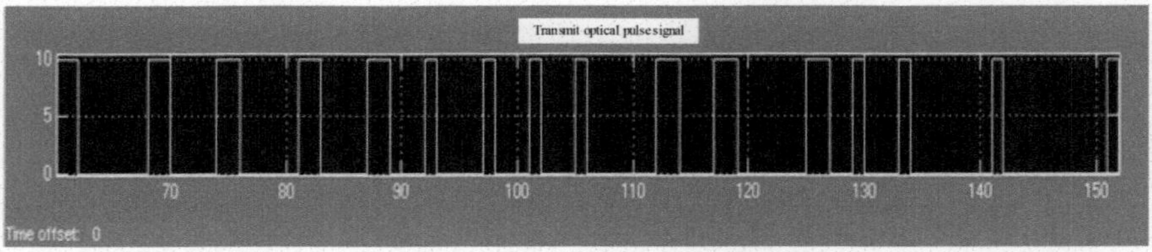

Fig. 6.14 Output simulation of the transmission system

In this simulation, the channel was a wireless optical channel. Because of the influence of noise, the optical signal is affected to a certain extent. The noise used in the simulation was Gaussian additive noise. Figure 6.15 shows the optical-signal simulation diagram with a signal-to-noise ratio of 10 dB through a wireless optical channel.

The receiving and signal-processing modules determine the amplitude of the received optical signal by setting an appropriate threshold value. Because DHPIM modulation starts with a pulse, a hard-judgment simulation is adopted. When the pulse is greater than the set threshold value, the receiving sequence is 1; otherwise, it is 0.

Figure 6.16 shows the simulation results of the receiving and signal-processing module. It can be seen that the module converts the optical signal into a sequence of electrical signals, and then sends it to the demodulation part to complete the signal-demodulation function. If we compare the signals before and after modulation (Figs. 6.13 and 6.16), they are the same, and the processed sequence is the modulated signal sequence.

The demodulation module demodulates the received sequence signal. First, the system determines whether the starting signal was a pulse signal. When a pulse signal is detected, it enters the detection state of the effective signal. During demodulation, the received starting pulse time slot should be judged according to the judgment, then the number of protection time slots and information time slots should be calculated, and this information should be mapped to the original source.

Because the modulation part transmits 15 data elements as a superframe after modulation, the receiver demodulates the 15 frames of data, stores the demodulated data, completes the parallel–serial conversion of the demodulated data, and finally

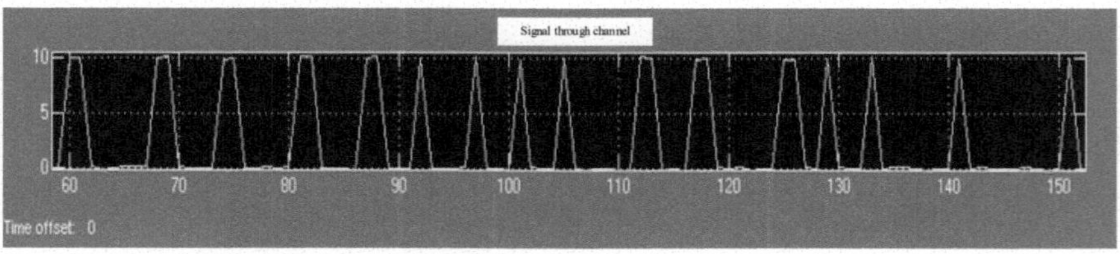

Fig. 6.15 Simulation diagram of a laser signal through an atmospheric channel

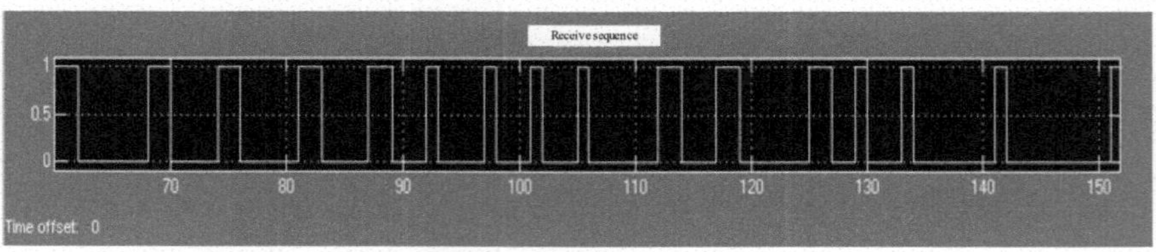

Fig. 6.16 Simulation diagram of the receiving and signal-processing module

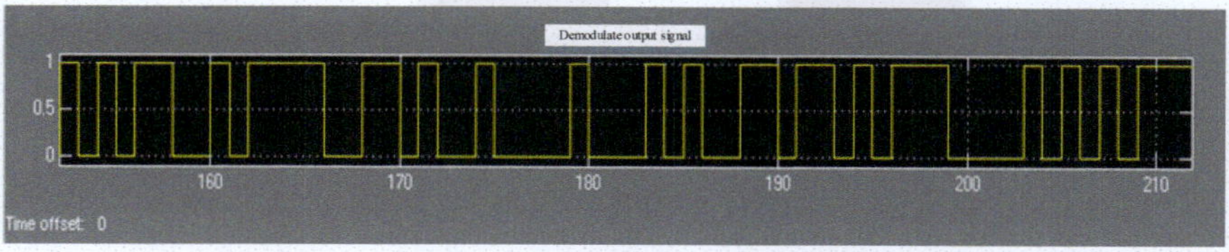

Fig. 6.17 Simulation diagram of the output signal of the demodulation

outputs binary information. Figure 6.17 shows the simulation results of demodulation. The sink saves the final demodulated binary information. Comparing Figs. 6.12 and 6.17, the binary code of the demodulated output signal is consistent with the binary code generated by the source, which verifies the theoretical correctness of the DHPIM modulation and demodulation system.

The design, based on a programmable logic device, is simple, and limited logic resources can be used to complete the modulation and demodulation design.

The DHPIM modulation and demodulation system was implemented using a schematic diagram and the Verilog hardware description language (HDL). An Altera Cyclone II-series field-programmable gate array (FPGA, ep2c5f256c8) was selected for FPGA simulation verification. The Altera Quartus II 7.0 software package was selected for synthesis, implementation, and post-wiring simulation, and generating FPGA download files.

References

1. Ghassemlooy Z, Popoola WO, Rajbhandari S (2007) A synopsis of modulation techniques for wireless infrared communication. ICTON mediterranean winter conference, ICTON-MW, pp 3–4
2. Nawras MA, Ghassemlooy Z (2001) Performance of dual header-pulse interval modulation (DH-PIM) for optical wireless communication systems. SPIE Proc Opt Wireless Commun III 37(19):4214–4218
3. Yebo C (2005) Performance analysis of dual header pulse code interval modulation based on dispersion channel. Photon Technol 3(27):166–170
4. Ghassemlooy Z, Rajbhandari S (2007) Performance of diffused indoor optical wireless links employing neural and adaptive linear equalizers. In: Information, communications and signal processing, 2007 6th international conference on, 2007, pp 1–6
5. Ghassemlooy Z, Hayes AR (2000) Digital pulse interval modulation for IR communication systems-a review. Int J Commun Syst 13(32):525–527
6. Cheng G, Wang H, Sun X et al (2008) A new modulation method for optical wireless communication. China Laser 35(12):1916
7. Kai P, Ping P (2009) Application of dhpim and LMS channel equalization in FSO. J Huazhong Univer Sci Technol 37(2):10–18
8. Kai Z (2004) Research on dual amplitude pulse interval modulation in optical wireless communication and its application in optical etc. Tsinghua University, Beijing
9. Meng L, Zhongyu W (2009) Research on joint coding and modulation of DPIM and turbo in optical wireless communication. Opt Commun 64(33):8–15

10. Hayes AR, Ghassemlooy Z, Seed NL (2000) Baseline-wander effects on systems employing digital pulse-interval modulation. IEEE Proc 147(4):295–300
11. Zhang J (2000) Modulation analysis for outdoors applications of optial wireless. In: Communication technology proceedings, WCC-ICCT2000, pp 1483–1487
12. Aldibbiat NM, Ghassemlooy Z, McLaughlin R (2001) Spectral characteristics of dual header pulse interval modulation (DH-PIM). IEEE Proc Online 17(31):280–286
13. Ma H (2003) Research on optical communication demodulation technology in mobile atmospheric laser communication. University of Defense Science and technology, Changsha, pp 22
14. Deng T (2007) Research on key technologies of free space optical communication system. Huazhong University of science and technology, Hubei, pp 64
15. Chen Z, Xu Z, Wang J (2007) Research on soft decision performance of IDPPM modulation in optical wireless communication. Nanjing: School Commun Eng, PLA Univer Technol 52(46):708–710

Chapter 7
Sixty-Four-QAM Modulation

Quadrature-amplitude modulation (QAM) is a type of joint amplitude and phase keying. It simultaneously modulates the amplitude and phase of a signal as two independent parameters. This modulation method has the advantages of a high spectrum efficiency, bandwidth scalability, and good anti-interference performance. It has been widely used in wireless-access and other fields.

Its application to optical wireless communication is conducive to breaking through the bottlenecks that restrict optical-communication development, and improving the reliability and information-transmission rate of optical communication. QAM can be divided into 16-QAM, 32-QAM, 64-QAM, 128-QAM, and 256-QAM, depending on the different levels, phases, and amplitudes. Among them, 64-QAM can simultaneously satisfy the anti-interference-ability and spectrum-utilization requirements.

7.1 Quadrature-Modulation Technology

7.1.1 64-QAM System Fundamentals

(1) **Modulation**

The generic expression for orthogonal amplitude modulation [1] is

$$S(t) = A_m \cos \omega_c t + B_m \sin \omega_c t, \tag{7.1}$$

where $0 \leq t \leq T_b$ and T_b is the code-element length. The modulated signal $S(t)$ consists of two mutually orthogonal carriers summed together. $\{A_m\}$ and $\{B_m\}$ are variables that can take multiple discrete values; hence, this modulation is called orthogonal amplitude modulation, where $m = 1, 2, \ldots, M$, and M is the number of

X. Ke, *Coding Theory in Optical Wireless Communication Systems*, Optical Wireless Communication Theory and Technology, https://doi.org/10.1007/978-981-19-9837-9_7

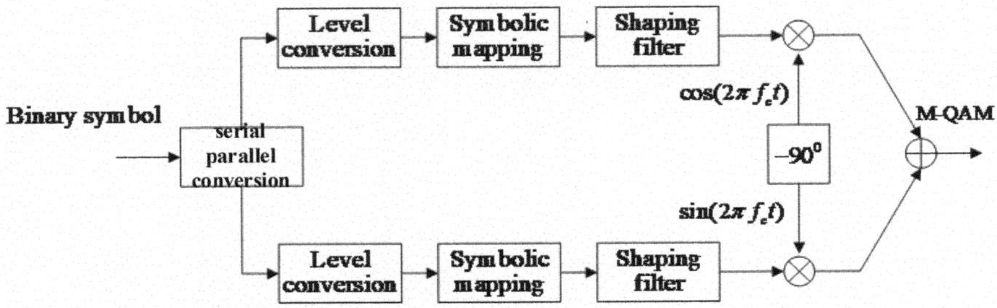

Fig. 7.1 Block diagram of an M-QAM modulator

levels of A_m and B_m. Amplitudes A and B in QAM can be expressed as

$$\begin{cases} A_m = d_m A \\ B_m = c_m A \end{cases}, \tag{7.2}$$

where A is the fixed amplitude and (d_m, c_m) is determined by the input signal, which is the coordinate point of the tuned signal in signal space.

Figure 7.1 shows the block diagram of an M-QAM modulator. First, the binary data b_n passes through a serial/parallel converter, which divides the binary information, at a rate of R_b bits per second, into two sequences consisting of odd and even positions.

With the M-QAM system, the symbol-conversion table of the upper branch converts each group of $k/2$ even bits into the transmitted symbol a_n^I, where $M = 2^k$. Similarly, each group of $k/2$ odd bits entering the lower branch is coded as the transmitted symbol a_n^Q. a_n^I and a_n^Q are random variables in the set $\{\pm 1, \pm 3, \cdots, \pm(\sqrt{M} - 1)\}$ that take the value of $\sqrt{M}$ decimals.

For example, with 64-QAM, the character set of a_n^I and a_n^Q is $\{-7, -5, -3, -1, 1, 3, 5, 7\}$. The pulse-shaping output in the upper branch is an eight-decimal polar pulse sequence $I(t) = \sum_{n=-\infty}^{\infty} a_n^I v(t - nT)$, modulated by $I(t)$ for the same carrier $A_c \cos(2\pi f_c t)$. Similarly, the eight-decimal polar pulse sequence $Q(t) = \sum_{n=-\infty}^{\infty} a_n^Q v(t - nT)$ generated by the pulse-shaping filter in the lower branch modulates the quadrature carrier $A_c \sin(2\pi f_c t)$.

The M-QAM signal can be expressed as [2]

$$x(t) = \sqrt{\frac{2E_s}{T}}[I(t)\cos(2\pi f_c t) - Q(t)\sin(2\pi f_c t)] \tag{7.3}$$

$$\begin{cases} I(t) = C_0 \sum\limits_{n=-\infty}^{\infty} a_n^I v(t - nT) \\ Q(t) = C_0 \sum\limits_{n=-\infty}^{\infty} a_n^Q v(t - nT) \end{cases}, \tag{7.4}$$

where $\sqrt{E_s/T}$ denotes the usual pulse per unit of energy. The purpose of introducing constant C_0 is to set the average energy of the signal equal to E_s.

The average energy E_s is given by

$$E_s = \sum_{m=1}^{\sqrt{M}} \sum_{n=1}^{\sqrt{M}} E\{a_m^I, a_n^Q\} \int_0^T |s_{mn}(t)|^2 dt, \tag{7.5}$$

where $s_{mn}(t)$ is the transmitted waveform corresponding to constellation point (a_m^I, a_n^Q):

$$s_{mn}(t) = \sqrt{\frac{2E_s}{T}} C_0 [a_m^I v(t) \cos(2\pi f_c t) - a_m^Q v(t) \sin(2\pi f_c t)]. \tag{7.6}$$

Substituting Eq. (7.6) into Eq. (7.5), we obtain

$$\begin{aligned} E_s &= \frac{2E_s C_0^2}{M} \sum_{m=1}^{\sqrt{M}} \sum_{n=1}^{\sqrt{M}} [(a_m^I)^2 + (a_n^Q)^2] \frac{1}{2} \int_0^T \left| \frac{v(t)}{\sqrt{T}} \right|^2 dt \\ &= \frac{E_s C_0^2}{M} \sum_{m=1}^{\sqrt{M}} \sum_{n=1}^{\sqrt{M}} [(a_m^I)^2 + (a_n^Q)^2]. \end{aligned} \tag{7.7}$$

From Eq. (7.7), we obtain

$$C_0^2 = \frac{M}{\sum_{m=1}^{\sqrt{M}} \sum_{n=1}^{\sqrt{M}} [(a_m^I)^2 + (a_n^Q)^2]}. \tag{7.8}$$

For the M-QAM constellation diagram, according to Eq. (7.7), we obtain

$$C_0 = \sqrt{\frac{3}{2(M-1)}}. \tag{7.9}$$

If the following basis function is chosen [3],

$$\begin{cases} \phi_1(t) = \sqrt{\frac{2}{T}} \cos(2\pi f_c t) \\ \phi_2(t) = \sqrt{\frac{2}{T}} \sin(2\pi f_c t) \end{cases}, \tag{7.10}$$

it is possible to represent all M-QAM signals in the nth symbol interval in the A-plane in vector form, as follows:

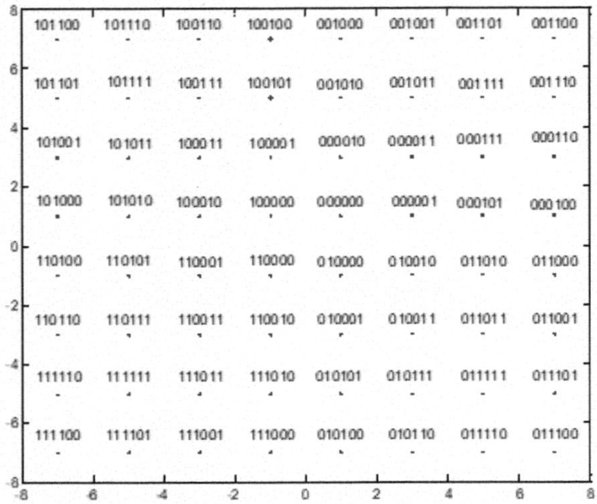

Fig. 7.2 64-QAM constellation-diagram details

$$s = (a_n^I C_0 \sqrt{E_s}, a_n^Q C_0 \sqrt{E_s}). \tag{7.11}$$

According to Eq. (7.11), we can obtain

$$d_{\min} = 2C_0 \sqrt{E_s}. \tag{7.12}$$

Substituting Eq. (7.9) into Eq. (7.12), we can obtain

$$d_{\min} = 2\sqrt{\frac{3E_s}{2(M-1)}}. \tag{7.13}$$

The 64-QAM constellation-diagram details are shown in Fig. 7.2.

(2) **Demodulation**

The basic principle is that the 64-QAM modulated signals are multiplied by two orthogonal carriers to generate two data channels: in-phase branch I and orthogonal branch Q. The I and Q signals are filtered using low-pass filters to remove the high-frequency components. The judgment module judges the filtered two-way signals to obtain the two-way code element information, and then converts the eight-level data to two-level. Finally, the two-way signals are combined into one-way by the serial/parallel conversion module to recover the signal (Fig. 7.3).

At the receiving end of the system, the tuned signal is multiplied by two mutually orthogonal carriers to recover the baseband signal, where the signal is expressed as

$$\begin{aligned} I(t) &= Y(t)\cos\omega_0 t \\ &= (A_m \cos\omega_0 t + B_m \sin\omega_0 t)\cos\omega_0 t \end{aligned}$$

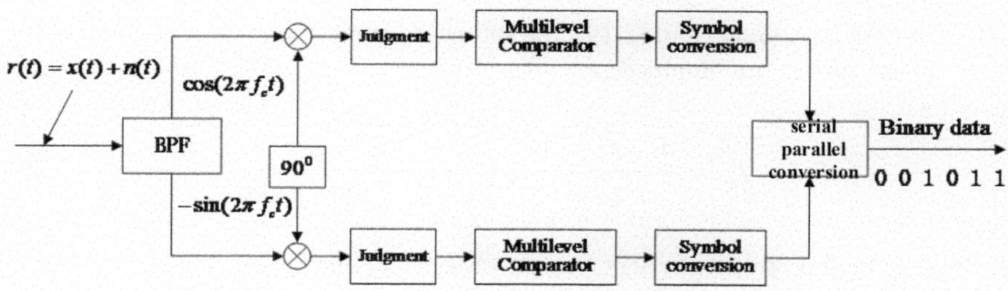

Fig. 7.3 64-QAM demodulation-schematic block diagram

$$= \frac{1}{2}A_m + \frac{1}{2}A_m \cos 2\omega_0 t + \frac{1}{2}B_m \sin 2\omega_0 t. \tag{7.14}$$

The expression for the quadrature-component signal is

$$\begin{aligned} Q(t) &= Y(t) \sin \omega_0 t \\ &= (A_m \cos \omega_0 t + B_m \sin \omega_0 t) \sin \omega_0 t \\ &= \frac{1}{2}B_m - \frac{1}{2}B_m \cos 2\omega_0 t + \frac{1}{2}A_m \sin 2\omega_0 t. \end{aligned} \tag{7.15}$$

(3) **QAM error performance**

In an M-QAM system, the probability of the correct reception of symbols [4] is

$$\begin{aligned} P_c &= \text{Probability of correct reception of } I \times \text{Probability of correct reception of Q} \\ &= (1 - P_0) \times (1 - P_0), \end{aligned} \tag{7.16}$$

where P_0 is equal to the probability of incorrect verdicts for the I paths and Q paths.

According to Eq. (7.13), the symbolic error-probability P_e [5] in the M-QAM system can be obtained as

$$P_e = 1 - P_c = 1 - (1 - P_0)^2 = 2P_0 - P_0^2 \approx 2P_0, \tag{7.17}$$

where $P_0 \ll 1$. Because the symbol-error probability P_0 [6] in a $\sqrt{M}$-ary pulse-amplitude modulation (PAM) system is

$$P_0 = \frac{2(\sqrt{M} - 1)}{\sqrt{M}} Q\left(\frac{d_{\min}}{\sqrt{2N_0}}\right), \tag{7.18}$$

substituting Eq. (7.12) into Eq. (7.18) we obtain

$$P_0 = \frac{2(\sqrt{M} - 1)}{\sqrt{M}} Q\left(\sqrt{\frac{3E_s}{(M - 1)N_0}}\right). \tag{7.19}$$

Combining Eqs. (7.17) and (7.19), the following symbolic-error probabilities for the M-QAM system are obtained:

$$P_e = \frac{4(\sqrt{M}-1)}{\sqrt{M}} Q\left(\sqrt{\frac{3E_s}{(M-1)N_0}}\right). \tag{7.20}$$

Because the bit error rate (BER) is related to the symbol error rate as

$$BER = \frac{P_e}{\log_2 M}, \tag{7.21}$$

according to Eqs. (7.20) and (7.21), by substituting $E_s = kE_b = \log_2 M \times E_b$, we can obtain the bit error rate of the M-QAM system as

$$BER = \frac{P_e}{\log_2 M} = \frac{4(\sqrt{M}-1)}{(\log_2 M)\sqrt{M}} Q\left(\sqrt{\frac{3E_b(\log_2 M)}{(M-1)N_0}}\right). \tag{7.22}$$

Therefore, the BER of the 64-QAM system is

$$\begin{aligned} BER_{64-QAM} &= \frac{P_e}{\log_2 64} = \frac{4(\sqrt{64}-1)}{(\log_2 64)\sqrt{64}} Q\left(\sqrt{\frac{3E_b(\log_2 64)}{(64-1)N_0}}\right) \\ &= \frac{7}{12} Q\left(\sqrt{\frac{2E_b}{7N_0}}\right). \end{aligned} \tag{7.23}$$

7.1.2 Protection Interval

In wireless communication systems, the multipath effect leads to intersymbol crosstalk and affects the signal-transmission quality. After the signal reaches the receiving end through multiple paths, the back-end of the previous symbol interferes with the front-end of the next symbol. This interference is called inter-symbol interference (ISI). ISI is likely to lead to sampling-decision errors, which seriously affect the transmission quality of digital signals.

The effect of intercoding crosstalk can be eliminated by adding a protection interval between the code elements. As shown in Fig. 7.4, after adding the protection interval, the code elements of Path 1 and Path 2 do not interfere with each other, which means that inter-coding crosstalk is eliminated. Note that the protection interval must be larger than the maximum multipath delay to completely eliminate the crosstalk caused by the multipath effect.

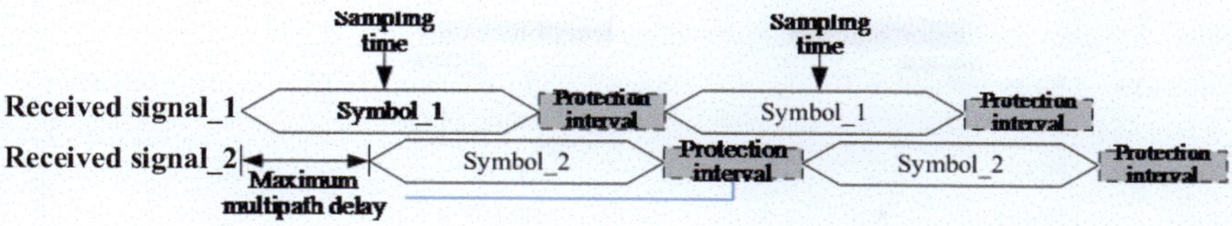

Fig. 7.4 Symbol signal with protection interval

The protection interval mentioned above implies that a signal is not sent during the protection interval. For the sake of understanding, the inter-code crosstalk caused by the multipath effect mentioned in the text is analyzed from the perspective of the baseband signal, that is, Symbol_1 and Symbol_2, corresponding to the baseband-signal waveform.

7.1.3 64-QAM Signal Modulation and Demodulation Simulation

(1) **64-QAM modulation simulation**

The modulation is divided into three main parts: serial/parallel conversion, level conversion, and constellation mapping.

a. **Serial-to-parallel conversion and level conversion**

Serial/parallel conversion is the transformation of a one-way serial sequence into two parallel subsequences, I and Q. The rate of the subsequences after conversion is half the rate of the original sequence, and each symbol is determined by a different number of binary bits h to determine a different order M. According to the value of M, 2-M level conversions are performed on the subsequences. The conversion rules are $a_k = (2i - 1 - M)$ and $i = 1, 2, \ldots M$, where $M = 2^{h/2}$ and i represents the corresponding binary data: 000—1, 001—2, 010—3, 011—4, 100—5, 101—6, 110—7, and 111—8.

Table 7.1 presents the relationship between the serial/parallel conversion and the level conversion for 64-QAM, where $h = 6$ and $M = 8$. That is, six-bit information represents a code element. First, three consecutive bits are taken from the six-bit code-element sequence to the I and Q sequences. Then, the corresponding i values are found according to these three bits of data, so a_k can take the level amplitude as follows: −7, −5, −3, −1, 1, 3, 5, and 7. Table 7.1 lists the conversion relationship when the I sequence is 000, the Q sequence changes between 000 and 111, and the conversion methods for the other 56 states are similar to those in Table 7.1.

Figure 7.5 shows the simulation results of the serial/parallel conversion, where the original one-way binary data become two-way binary data for the same phase branch and quadrature branch. Figure 7.6 shows the simulation results of the symbolic

Table 7.1 Level conversion and serial/parallel conversion

Sequences	Subsequence of I value of I/i	a_k of I sequence	Subsequence of Q value of Q/i	a_k of Q sequence
000000	000/1	−7	000/1	−7
000001	000/1	−7	001/2	−5
000010	000/1	−7	010/3	−3
000011	000/1	−7	011/4	−1
000100	000/1	−7	100/5	1
000101	000/1	−7	101/6	3
000110	000/1	−7	110/7	5
000111	000/1	−7	111/8	7

conversion; at this time, the same phase branch and quadrature branch respectively have eight levels of data.

b. **Constellation mapping**

Any combination of the magnitudes and phases of I and Q is mapped to a point within the polar coordinates. The state of each constellation point is represented by six bits of data, and there are 64 states after combining I and Q; that is, a 64-QAM constellation map is generated (Fig. 7.7).

c. **Up-conversion**

After completing the constellation mapping, the 64-QAM signal modulation is completed by multiplying the in-phase-mapped value by carrier $\sin(\omega t)$ and the quadrature-mapped value by carrier $\cos(\omega t)$, and then adding the two data streams. The data waveforms of the in-phase and quadrature branches and 64-QAM signal waveforms are shown in Fig. 7.8.

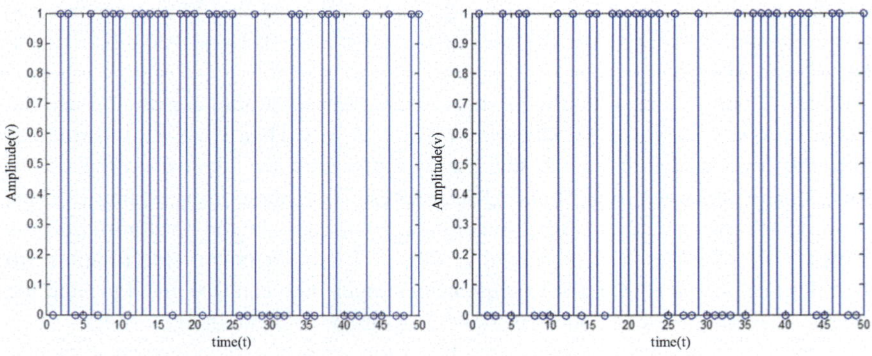

(a) In-phase components (b) Quadrature components

Fig. 7.5 In-phase and quadrature components after a serial/parallel conversion

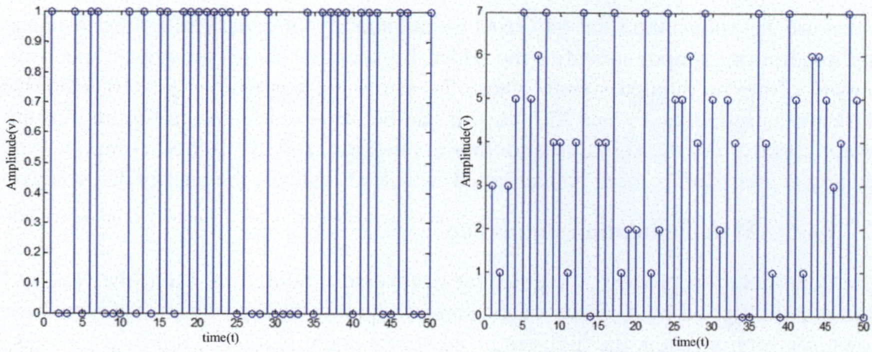

(a) In-phase components (b) Quadrature components

Fig. 7.6 In-phase and quadrature components after a level conversion

Fig. 7.7 64-QAM constellation map

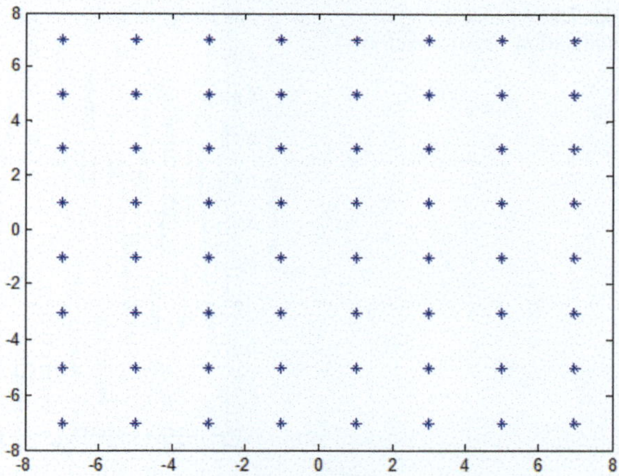

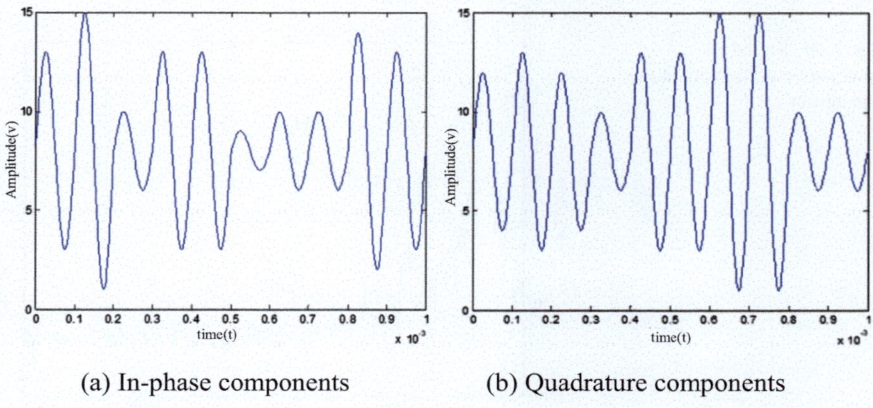

(a) In-phase components (b) Quadrature components

Fig. 7.8 Up-conversion

Figure 7.9 shows the final 64-QAM modulated signal generated by the in-phase and quadrature components after the adder. Figure 7.10 shows the power-spectrum density of the modulated signal, where the sampling frequency $F_s = 10{,}000$ (Hz) and carrier frequency $F_c = 1250$ (Hz). It can be observed from the Figure that the power spectrum of the signal is smoother at this time, and the highest component of the power spectrum is more concentrated, which also shows the energy distribution.

(2) **64-QAM demodulation simulation**

The demodulation process is equivalent to inverse modulation. First, the received signal is multiplied by a carrier of the same frequency as the transmitted signal for down-conversion; then, the high-frequency component is filtered out by a low-pass filter, leaving only the high-frequency signal. Next, the signal is inversely mapped, and then i is calculated, based on the amplitude of the electrical signal: $i = (ak +$

Fig. 7.9 64-QAM modulation signal waveform

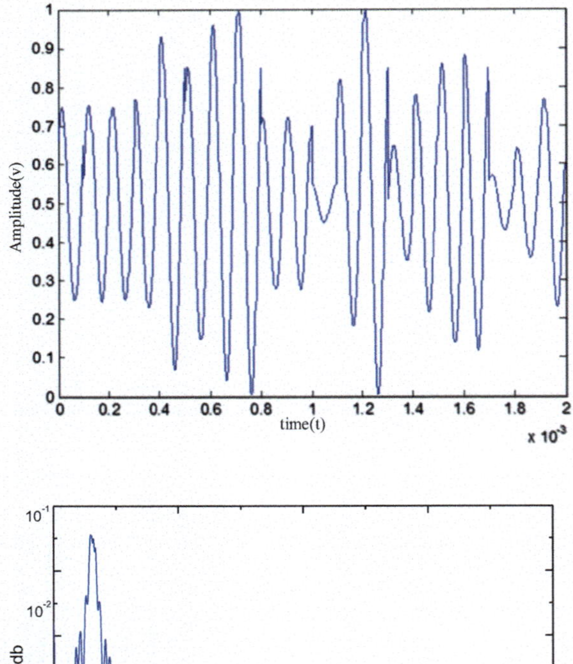

Fig. 7.10 Signal power-spectral density at the transmitter

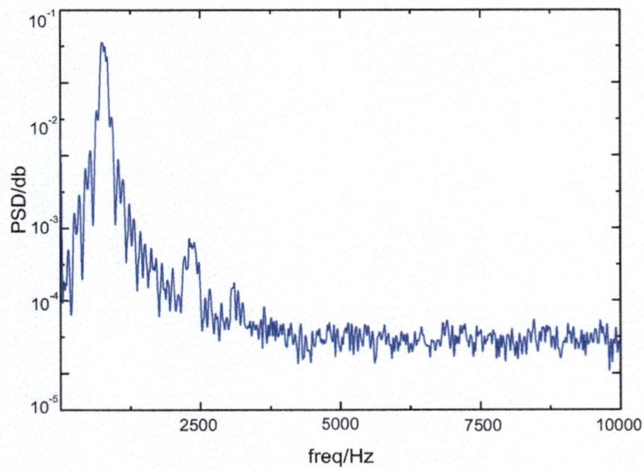

Table 7.2 Level conversion and parallel/serial conversion

a_k of I sequence	Encoding of subsequence I/i	a_k of Q sequence	Encoding of subsequence Q/i	Output sequence (I and Q combined)
−7	000/1	−7	000/1	000000
−7	000/1	−5	001/2	000001
−7	000/1	−3	010/3	000010
−7	000/1	−1	011/4	000011
−7	000/1	1	100/5	000100
−7	000/1	3	101/6	000101
−7	000/1	5	110/7	000110
−7	000/1	7	111/8	000111

1 + M)/2, where $M = 8$ and a_k is the level amplitude of the signal; the calculation yields $i = 1, 2, 3, 4, 5, 6, 7, 8$. The value of i is then converted into the corresponding binary data.

The next step is to obtain three consecutive bits of data from the I and Q sequences and combine them into a set of six binary digits to represent a symbol. Table 7.2 shows the binary sequence recovered when the magnitude of the I sequence was −7, and the magnitude of the Q sequence ranged from −7 to 7. The other seven amplitude levels were decoded in a manner similar to that in Table 7.2, and a total of 64 sets of data were decoded.

Figure 7.11 shows a comparison of the original transmitted code element and the demodulated code element at the receiver side for a channel noise of 0.

7.2 Channel-Equalization Technology

The laser-signal transmission process is significantly affected by atmospheric turbulence and atmospheric scattering, and the blind channel-equalization technique can suppress or compensate for the inter-code interference caused by atmospheric-channel distortion and atmospheric-turbulent light-intensity fluctuation. Therefore, it is important to study blind-equalization techniques in wireless optical 64-QAM modulation systems.

7.2.1 Equalizer Classification and Structure

In optical wireless communication, multipath propagation and limited bandwidth can cause inter-code interference, resulting in reduced communication-system performance and data-transmission speeds, increased BER, and ultimately, failure to restore

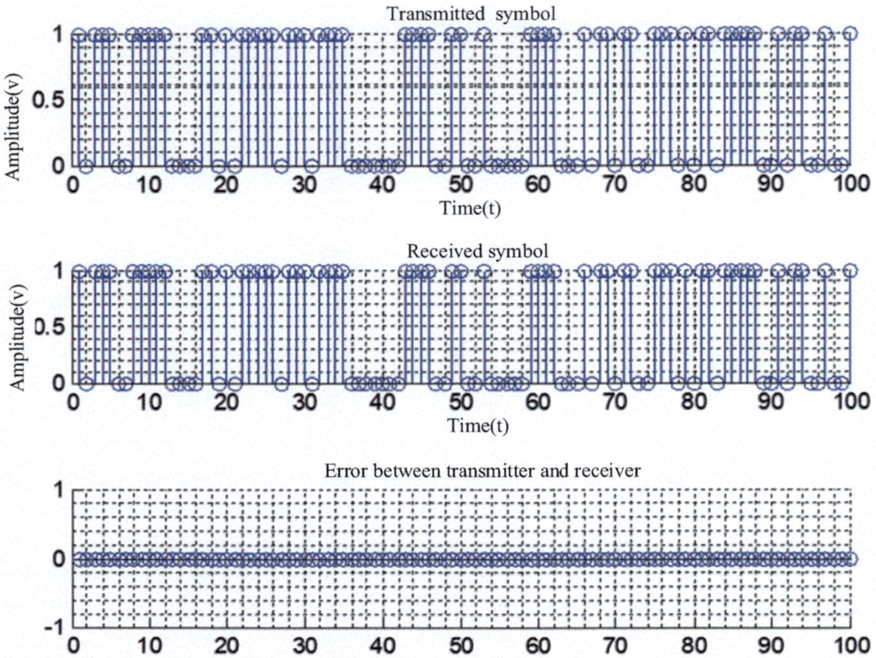

Fig. 7.11 Comparison of transmitted and received symbols

the original signal. The use of channel-equalization technology can reduce the impact of multipath propagation on the signal, while compensating for signal distortion.

Equalizers are classified according to two different classification criteria. Based on the data-sampling rate, an equalizer can be a baud-interval or fractional-interval equalizer. Based on the signal-processing method, they can be classified as linear or nonlinear. The equalizer classifications are shown in Fig. 7.12.

When compensating a channel, a common method is to equalize the channel based on a priori information on the channel or by sending a training sequence. When the communication system does not have training sequences or is unsuitable for using training sequences, a blind-equalization technique can also be used to ensure its reliability and validity.

Two main types of equalizers are commonly used in blind-equalization techniques: judgment feedback and linear. The linear equalizer is mainly applied to linear systems or nonlinear systems that can be equated to linear systems, whereas the judgment-feedback equalizer is mainly used for equalizing nonlinear systems. Figure 7.13 presents a diagram of the structure of a linear equalizer.

In the figure, the length of the linear equalizer is $N_f = 2L + 1$ and the output of the equalizer is given by the following equation:

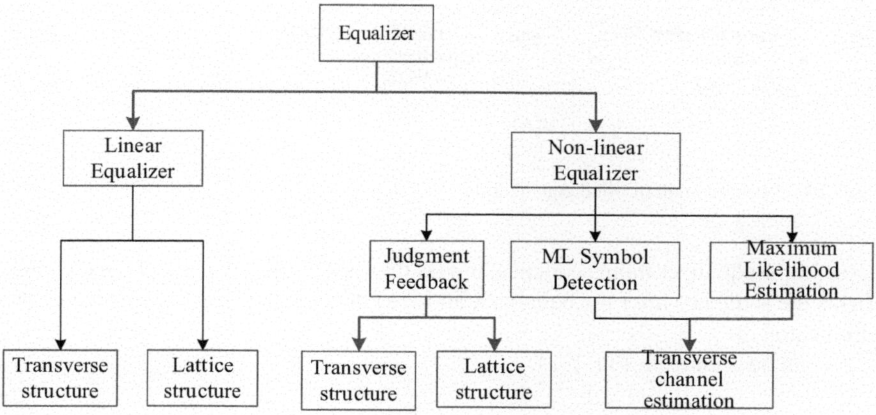

Fig. 7.12 Equalizer classifications

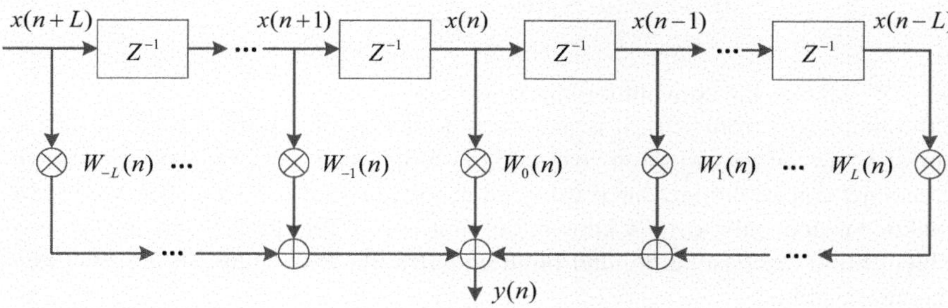

Fig. 7.13 Block diagram of a linear-equalizer structure

$$y(n) = X^T(n)W(n) = \sum_{i=-L}^{L} w_i(n)x(n-i). \tag{7.24}$$

7.2.2 *Blind Equalization*

(1) **Basic principles of blind equalization**

Blind equalization does not require the use of training sequences, and the channel characteristics can be equalized by sending only a priori sequence information. The principle of blind equalization is shown in Fig. 7.14.

In Fig. 7.14, $h(n)$ is the unit impulse response of the transmission channel, $w(n)$ is the blind-equalizer impulse response, $x(n)$ is the transmit sequence of the system, $y(n)$ is the sequence received at the receiver after being affected by channel noise, that

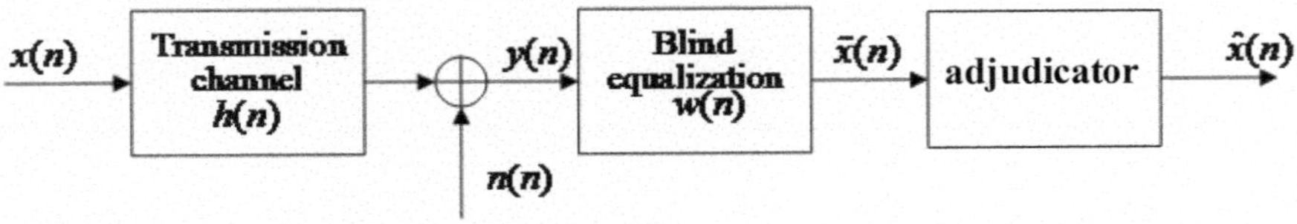

Fig. 7.14 Block diagram of blind equalization

is, the blind-equalizer input sequence, $n(n)$ is the superimposed noise of the channel, $\tilde{x}(n)$ is the sequence after the equalizer, and $\hat{x}(n)$ denotes the output sequence of the adjudicator.

As can be seen from Fig. 7.14,

$$y(n) = h(n) * x(n) + n(n) = \sum_i h(i) \cdot x(n-i) + n(n) \tag{7.25}$$

$$\tilde{x}(n) = w(n) * y(n) = \sum_i w(i) \cdot y(n-i), \tag{7.26}$$

where '*' denotes the convolution operation.

To obtain $x(n)$ from $y(n)$, it is necessary to deconvolve $y(n)$ or find the inverse channel $h^{-1}(n)$ of transmission channel $h(n)$. When both $y(n)$ and $x(n)$ are known, the channel can be compensated using an adaptive equalizer. However, when $x(n)$ is unknown and only $y(n)$ is known, solving it is difficult. We call this type of problem blind disconsolation, and blind equalization is its application in the field of communication.

If the effect of noise is not considered, the following equation holds:

$$y(n) = x(n) * h(n). \tag{7.27}$$

From Fig. 7.15, it can be seen that

$$\tilde{x}(n) = w(n) * y(n) = w(n) * h(n) * x(n). \tag{7.28}$$

The purpose of blind equalization is to use $\tilde{x}(n)$ as the best estimate of $x(n)$, thus, requiring

$$\tilde{x}(n) = x(n-D)e^{j\phi}, \tag{7.29}$$

where D denotes the integer delay and ϕ denotes a constant phase shift.

To satisfy Eq. (7.29), when the iterative addition of channel noise is not considered, it is required that

$$w(h) * h(n) = \delta(n-D)e^{j\phi}, \tag{7.30}$$

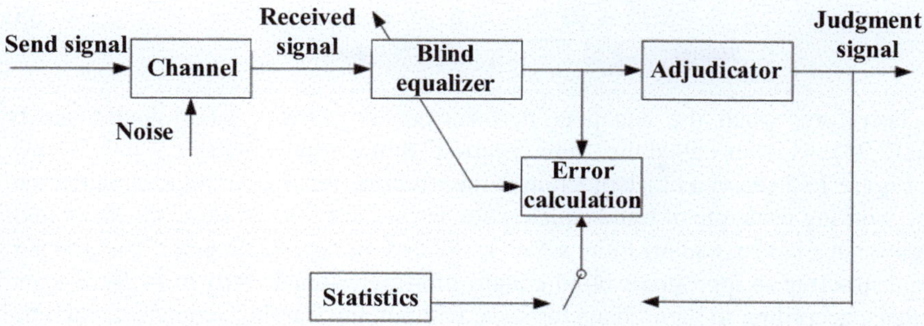

Fig. 7.15 Working-principle diagram of blind equalization

where $\delta(n)$ is the Kronecker δ function. By performing a Fourier transform on Eq. 7.30, we have

$$W(\omega)H(\omega) = \sum_{n=-\infty}^{+\infty} \delta(n-D)e^{j\phi}e^{-j\omega n} = e^{j(\phi - D\omega)} \tag{7.31}$$

or

$$W(\omega) = \frac{1}{H(\omega)}e^{j(\phi - D\omega)}. \tag{7.32}$$

Therefore, the purpose of the blind equalizer is to satisfy Eq. (7.32) between the channel and equalizer transfer functions. Usually, although the constant delay D and the constant phase ϕ are unknown quantities, D does not affect the recovery of the original input-signal sequence $\{x(n)\}$, whereas ϕ can be removed by the judgment device.

To implement the transfer function $W(\omega)$, one should first design the equalizer-tap coefficients w_i so that Eq. (7.30) holds. If $S(\omega) = H(\omega)W(\omega)$ and the tap coefficient of the joint system of the equalizer and channel is denoted as $s_i(n)$, then,

$$s_i(n) = h_i(n) * w_i(n) = \sum_j h_{i-j}(n)w_j(n). \tag{7.33}$$

However,

$$\tilde{x}(n) = \sum_{i=0}^{N} s_i(n)x(n-i) = x(n-D)e^{j\phi}. \tag{7.34}$$

Then, the finite dimensional vector $S(n) = [s_0(n), s_1(n), \cdots, s_N(n)]^{\mathrm{T}}$ is a vector containing only one nonzero element (module 1), that is,

$$S(n) = [\underbrace{0, 0, \cdots, 0}_{D-1}, e^{j\phi}, 0, \cdots 0]^{\mathrm{T}}. \tag{7.35}$$

Therefore, when the equalizer performance is optimal, $s_i(n)$ should satisfy Eq. (7.35), which is called the "zero criterion" of the blind equalizer.

Figure 7.15 shows a block diagram of the working principle of blind equalization. The working principle of blind equalization is similar to the working mechanism of traditional adaptive equalization, which is divided into two main parts. The first part is constructing an appropriate blind-equalization algorithm directly from the received signals, according to their characteristics, without any training sequence; when the algorithm converges, the inter-signal interference between the received signals at the receiver is minimized.

The second part is the judgment-guided mode, which is the same as traditional adaptive equalization. The judgment signal is mainly used as the desired signal, and the error signal is derived from it. Then, the judgment-guided algorithm is used to adaptively adjust the equalizer coefficients. Finally, the equalization algorithm is implemented.

(2) **Performance characterization of blind equalization**

The most important indicator of the application value of blind-equalization algorithms is the degree of their performance, which is generally evaluated using three criteria [7]:

- The convergence speed, which determines whether an algorithm can be applied to real-time systems.
- Whether the algorithm can obtain an optimal solution.
- The size of the steady-state residual error after algorithm convergence [8].

7.2.3 Multimode Blind Equalization (MMA)

(1) **Basic MMA algorithm**

The instantaneous cost function of the multimode algorithm (MMA) is formulated as follows [9]:

$$J(n) = E\{(y_R^2(n) - \gamma_R)^2 + (y_I^2(n) - \gamma_I)^2\}, \tag{7.36}$$

where $y_R(n)$ and $y_I(n)$ are the real and imaginary parts of the equalizer output $y(n)$, respectively. γ_R and γ_I are the real and imaginary parts of the difference constants of the source symbol $a(n)$, respectively, and $E(\cdot)$ denotes the mean value, which is expressed as

$$\gamma_R = \frac{E[a_R^4(n)]}{E[a_R^2(n)]}, \gamma_I = \frac{E[a_I^4(n)]}{E[a_I^2(n)]}. \tag{7.37}$$

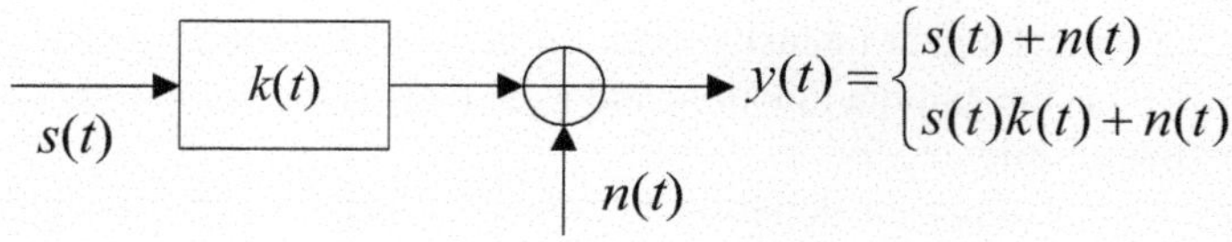

Fig. 7.16 Noise model

For QAM signals, we generally obtain $\gamma_R = \gamma_I \triangleq \gamma$. Then, Eq. (7.36) can be expressed as

$$J(n) = E\{(y_R^2(n) - \gamma)^2 + (y_I^2(n) - \gamma)^2\}. \tag{7.38}$$

Therefore, the multimode algorithm equalizer-tap update equation is

$$W(n+1) = W(n) + \mu \cdot e_{MMA}(n) \cdot X^*(n), \tag{7.39}$$

where $W(n)$ is the tapped power vector of the equalizer at moment n, $X(n)$ is the input vector of the equalizer at moment n, $(\cdot)^*$ is the complex conjugate, μ is the step size of the algorithm, and the expression of the error function $e_{MMA}(n)$ is

$$e_{MMA}(n) = y_R(n)\{\gamma - y_R^2(n)\} + j \cdot y_I(n)(\gamma - y_I^2(n)). \tag{7.40}$$

(2) **MMA equalization simulation**

To study the performance of 64-QAM systems, we first discuss the effect of noise in the channel.

The noise model is shown in Fig. 7.16. Assuming that the signal is $s(t)$ and the noise is $n(t)$, if the mixed iterative waveform is in the form of $s(t) + n(t)$, then $n(t)$ is said to be additive noise; if the iterative waveform is in the form of $s(t)\ k(t) + n(t)$, then $k(t)$ is said to be multiplicative noise.

Additive noise is noise that is iteratively added to the signal, and the noise $n(t)$ is always present, regardless of the presence or absence of the signal. Additive white Gaussian noise (AWGN) is the most basic noise; its amplitude distribution obeys the Gaussian distribution, while the power-spectral density is uniformly distributed.

Multiplicative noise exists with the signal and disappears when the signal disappears. In this study, a channel model obeying the gamma–gamma distribution was used.

We factorize the optical intensity of the channel model with a gamma–gamma distribution into the form of the product of two independent random processes, each containing a gamma probability-density function. The expression of the channel model with the gamma–gamma distribution is as follows:

$$f_I(I) = \frac{2(\alpha\beta)^{(\alpha+\beta)/2}}{\Gamma(\alpha)\Gamma(\beta)} \cdot I^{(\alpha+\beta)/2-1} \cdot K_{\alpha-\beta}\left(2\sqrt{\alpha\beta I}\right), \quad I > 0 \tag{7.41}$$

where I is the optical signal intensity, α and β are the probability-density function parameters, $\Gamma(\cdot)$ is the gamma function, and $K_{\alpha-\beta}(\cdot)$ B is the second-class modified Bessel function of order $\alpha - \beta$.

The convergence of the constellation diagrams before and after blind equalization at the receiver is discussed in the following three cases:

a. **Gaussian white noise with SNRs of 30 and 20 dB**

Figures 7.17 and 7.18 show the constellation changes of the signal before and after multimode blind equalization at the system receiver when the channel contains only additive noise. In the simulation, to optimize the bit error rate and steady-state mean-squared-error performance, the iterative step $\mu = 1 \times 10^{-6}$ is selected, and the equalizer order is 31. The figure shows that the constellation has an obvious convergence and a clear phase and amplitude after multimode blind equalization.

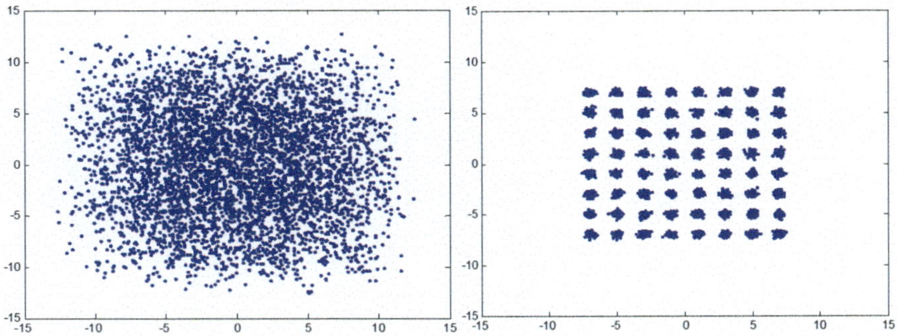

(a) Before equalization (b) After equalization

Fig. 7.17 Constellation diagram before and after blind equalization with SNR = 30 dB

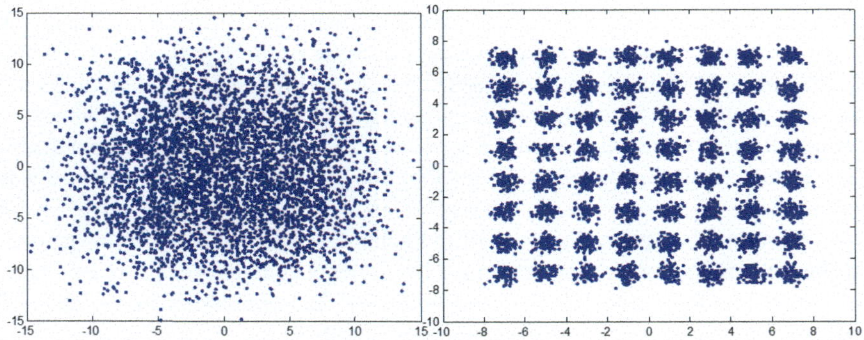

(a) Before equalization (b) After equalization

Fig. 7.18 Constellation diagram before and after blind equalization with SNR = 20 dB

b. **Multiplicative noise, where the variance of the light-intensity fluctuation is 0.1 and 0.05**

Figures 7.19 and 7.20 show the comparison of the signal-constellation diagrams before and after MMA equalization at $\sigma_R^2 = 0.1$ and $\sigma_R^2 = 0.05$. Observing the signal-constellation diagram after equalization, as the turbulence intensity increases, the individual constellation points become more difficult to identify, and when $\sigma_R^2 = 0.05$, the aggregation of the constellation points is already obvious.

c. **Additive Gaussian white noise and multiplicative noise**

Figures 7.21, 7.22, 7.23 and 7.24 represent the comparison of before and after MMA blind equalization when the SNR is 20 dB and 30 dB at $\sigma_R^2 = 0.1$ and $\sigma_R^2 = 0.05$, respectively. It is obvious from the constellation diagrams before and after equalization that the constellation diagrams are improved to different degrees by channel

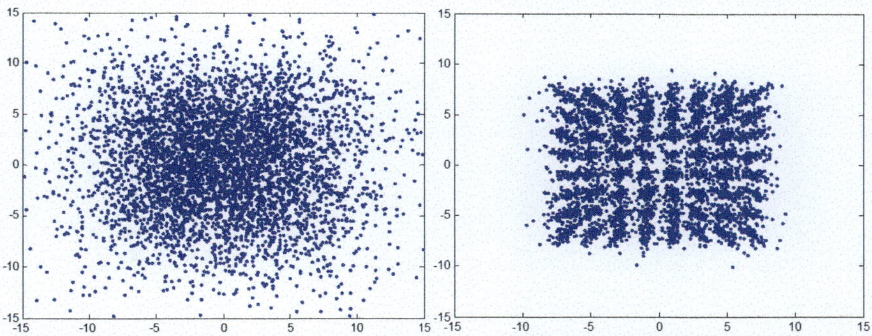

(a) Before equalization (b) After equalization

Fig. 7.19 Constellation diagram before and after blind equalization at $\sigma_R^2 = 0.1$

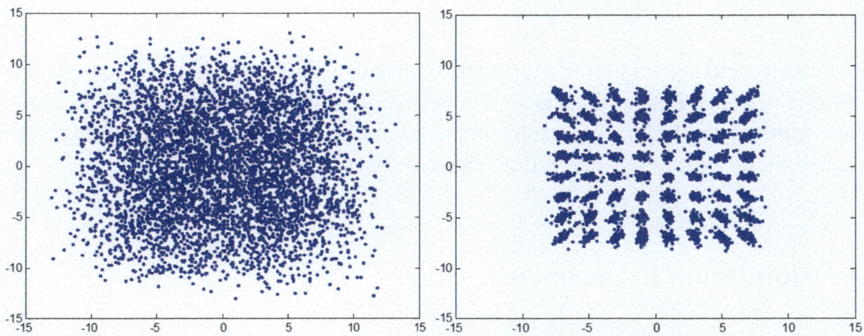

(a) Before equalization (b) After equalization

Fig. 7.20 Constellation diagram before and after blind equalization at $\sigma_R^2 = 0.05$

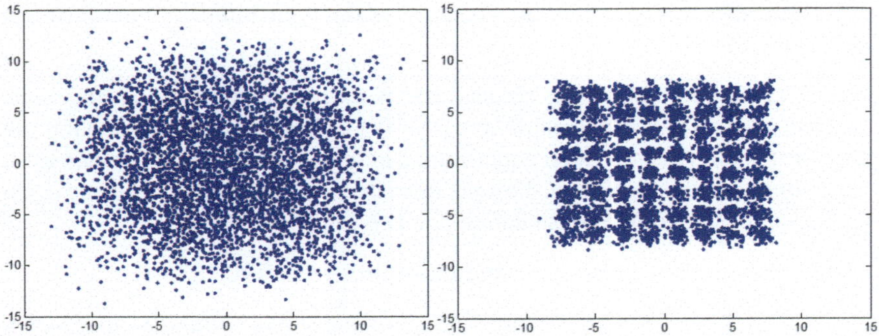

(a) Before equalization (b) After equalization

Fig. 7.21 Constellation diagram before and after blind equalization at $\sigma_R^2 = 0.05$ and SNR = 20 dB

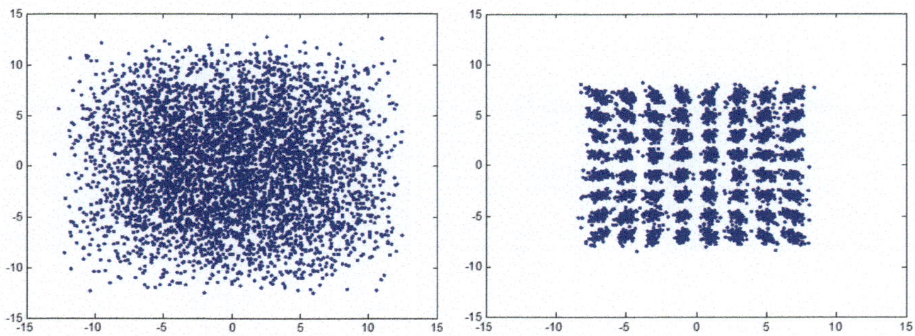

(a) Before equalization (b) After equalization

Fig. 7.22 Constellation diagram before and after blind equalization at $\sigma_R^2 = 0.05$ and SNR = 30 dB

equalization, and the smaller the turbulence intensity, the more obvious the improvement of the constellation diagrams. Therefore, it shows that the multimode blind-equalization algorithm has a certain suppression effect on atmospheric effects, such as atmospheric turbulence and atmospheric-molecule scattering.

7.3 Simulation Experiment

Wireless laser communication is a good means of communication, and many scholars have conducted extensive research on it since the introduction of lasers. Applying the orthogonal amplitude-modulation technique to wireless laser communication can

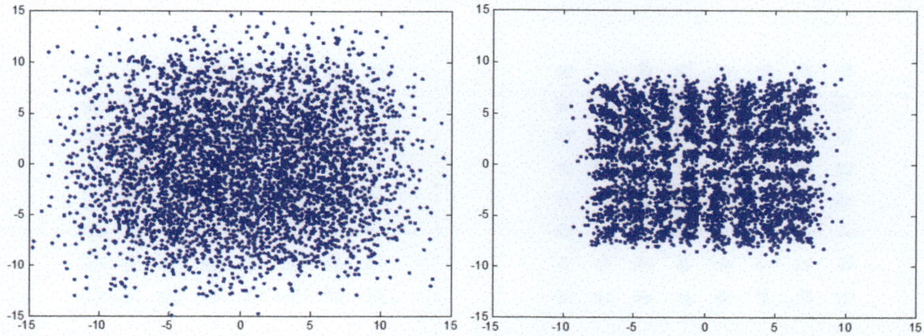

Fig. 7.23 Constellation diagram before and after blind equalization at $\sigma_R^2 = 0.1$ and SNR = 30 dB

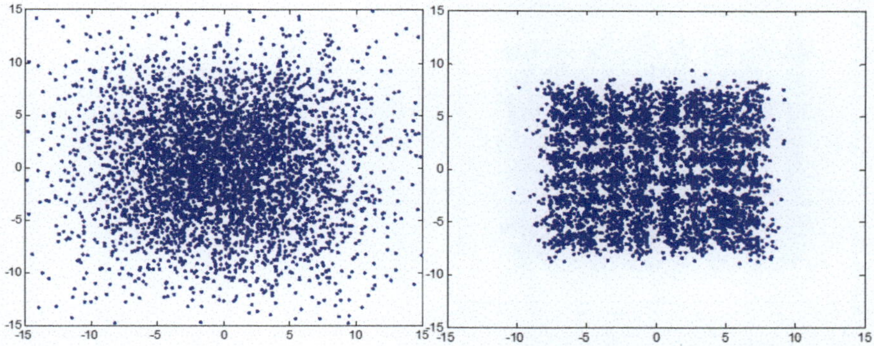

Fig. 7.24 Constellation diagram before and after blind equalization at $\sigma_R^2 = 0.1$ and SNR = 20 dB

improve the information-transmission rate and the reliability of optical communication. In this section, the simulation model and experimental system are reasonably built, according to the principle of orthogonal amplitude modulation, and the simulation and experimental results are analyzed.

7.3.1 Simulation Analysis

(1) **Analysis of the constellation diagram**

a. **Effect of Gaussian white noise on the system-constellation diagram**

Figure 7.25 shows the constellation diagram of the demodulation end of the system for SNR = 30, SNR = 20, and SNR = 15. The horizontal coordinate represents the amplitude of the I-way signal; the vertical coordinate represents the amplitude of the Q-way signal; both the horizontal and vertical coordinates are dimensionless

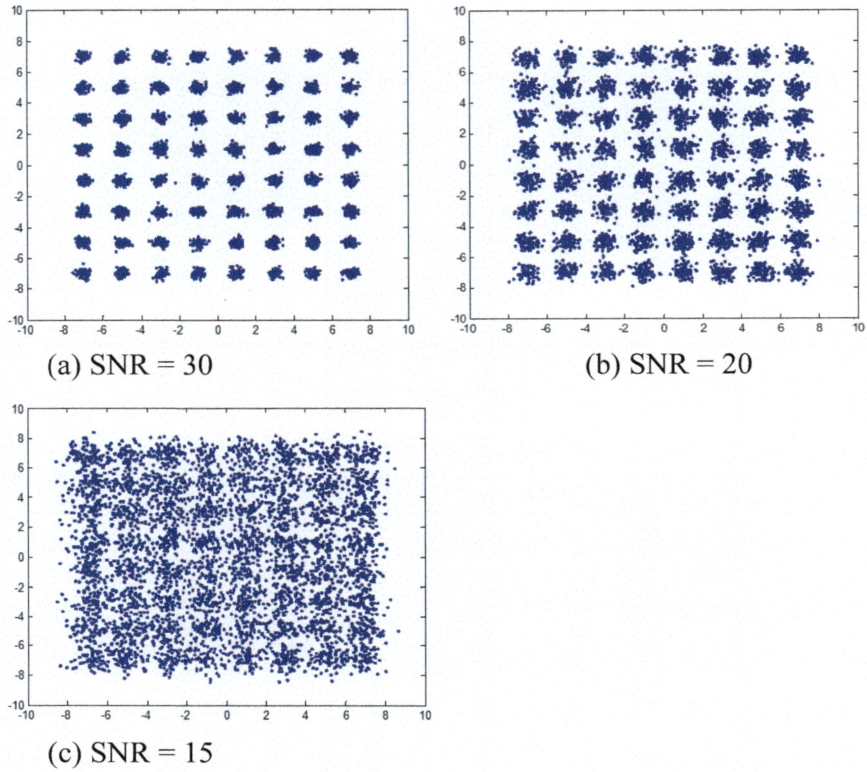

Fig. 7.25 Constellation diagram of 64-QAM demodulated signals with different SNRs

quantities. The constellation diagram had 64 points, and the amplitude of each point was taken from one of the following values: (−7, −5, −3, −1, 1, 3, 5, 7).

Figure 7.25a–c show that after the system is affected by Gaussian white noise in the channel, the degree of aggregation of each constellation point in the constellation diagram is affected to some extent; the larger the SNR is, the less the system is affected and the higher the aggregation degree of the constellation points.

b. **Effect of multiplicative noise on the system-constellation diagram**

The 64-QAM modulated signal passes through an atmospheric channel obeying the gamma–gamma distribution, and the signal is affected by multiplicative noise. In the gamma–gamma-distributed atmospheric-turbulence channel model under weak turbulence characteristics, the effect of 64-QAM modulation on the constellation diagram at the receiver side is analyzed, and Fig. 7.26a–f indicate the constellation diagrams at the demodulation side under different light-intensity undulation variances.

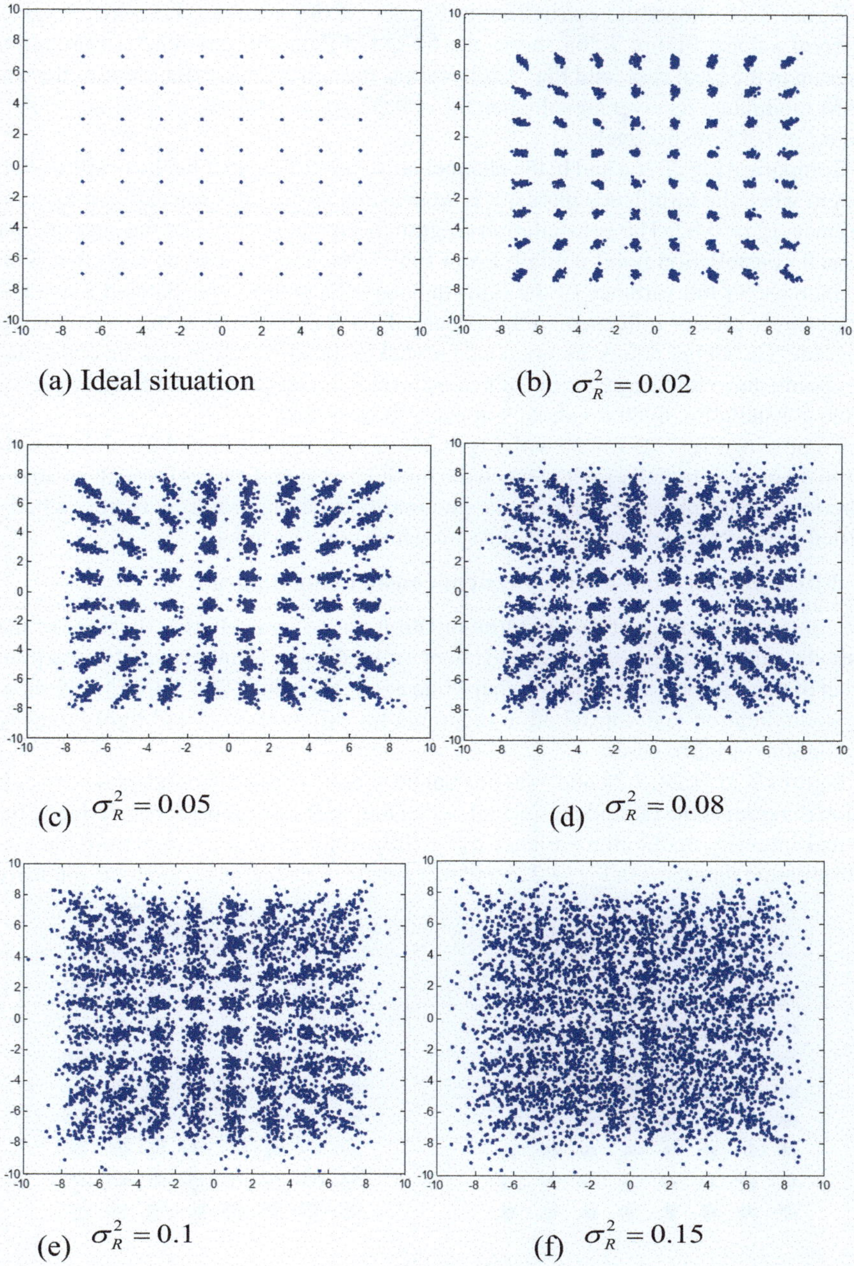

Fig. 7.26 Constellation diagram of a 64-QAM demodulated signal with different light-intensity undulation variances

Figure 7.26 shows the constellation diagram at the receiver side when A takes different values. Figure 7.26a shows the 64-QAM demodulation-end constellation diagram in the ideal case, and Fig. 7.26b–f show the constellation diagrams of the 64-QAM modulated receiver signal when $\sigma_R^2 = 0.02$, $\sigma_R^2 = 0.05$, $\sigma_R^2 = 0.08$, $\sigma_R^2 = 0.1$, and $\sigma_R^2 = 0.15$, respectively.

Comparing Fig. 7.26a and b, the channel noise has little effect on the transmission system when the turbulence intensity is weak, such as the light-intensity undulation variance $\sigma_R^2 = 0.02$. The constellation diagram is clearer overall and the aggregation of each constellation point is high. From Fig. 7.26e and f, it can be seen that with the increase of the variance of the light-intensity fluctuation, the channel noise has increasingly greater influence on the system. This is manifested in the constellation diagram: The signal points are not concentrated in the distribution and the constellation points have low aggregation. When $\sigma_R^2 = 0.15$, a large portion of the 64 points in the constellation diagram already appears to be mixed.

Comparing Fig. 7.25a and b with Fig. 7.26a–f, it can be concluded that the constellation diagram is radially shifted when the modulated signal passes through an atmospheric channel obeying the gamma–gamma distribution, while it is not radially shifted when it passes through a Gaussian white-noise channel.

c. **Effect of mixed noise on the system-constellation diagram**

We consider the effect of 64-QAM modulation on the constellation diagram of the demodulated signal at the receiver end under a mixed-noise channel model with a joint effect of weak turbulence and Gaussian white noise. Figures 7.27, 7.28, and 7.29 show the demodulation-end constellation diagrams for different SNR and light-intensity undulation-variance cases.

Figures 7.27, 7.28, 7.29 and 7.30 present the effects of mixed noise on the constellation diagram at the demodulation end, when the SNR is 30 and 20, respectively, for optical-intensity undulation variance $\sigma_R^2 = 0.02$ of the channel. Comparing the two constellation diagrams in Fig. 7.27a and b, it can be seen that when the light-intensity

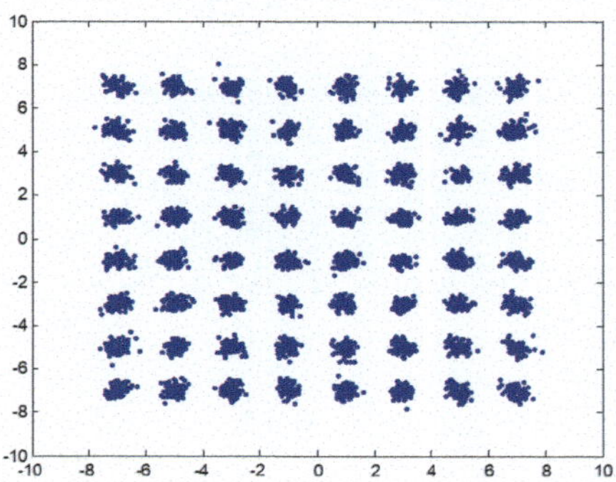

(a) SNR = 30

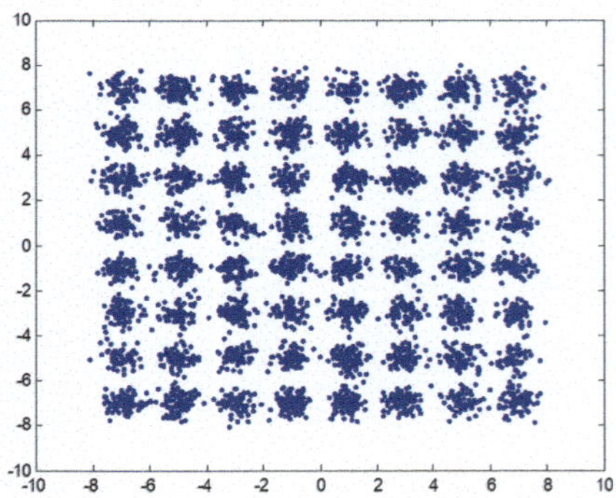

(b) SNR = 20

Fig. 7.27 Signal constellation at the receiver under 64-QAM modulation when $\sigma_R^2 = 0.02$

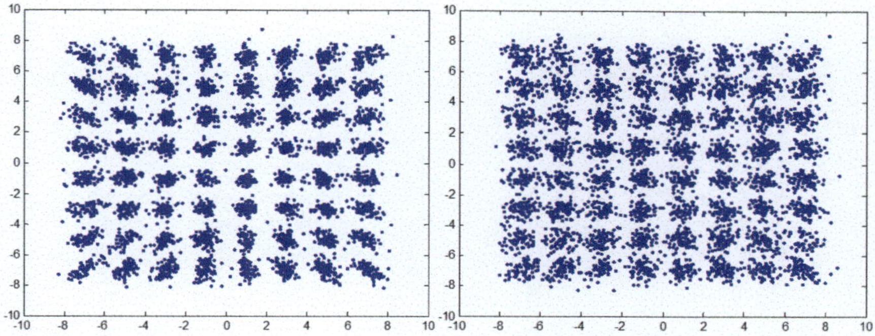

(a) SNR = 30 (b) SNR = 20

Fig. 7.28 Signal constellation at the receiver under 64-QAM modulation when $\sigma_R^2 = 0.05$

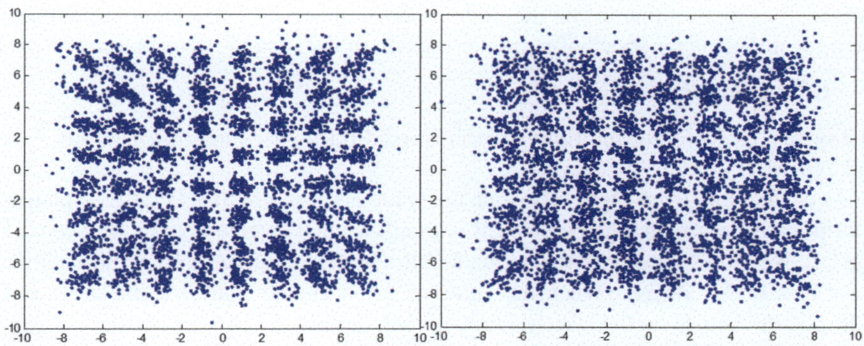

(a) SNR = 30 (b) SNR = 20

Fig. 7.29 Signal constellation at the receiver under 64-QAM modulation when $\sigma_R^2 = 0.08$

undulation variance is equal, the smaller the SNR, the higher the point aggregation in the constellation diagram, and the better the system performance. Comparing Figs. 7.27, 7.28, 7.29 and 7.30, when SNR = 30 and $\sigma_R^2 = 0.02$, the aggregation of constellation points is good and can be clearly identified, and each constellation point is not radially shifted.

When the SNR is kept constant and the light-intensity undulation variance is $\sigma_R^2 = 0.08$ and $\sigma_R^2 = 0.1$, the aggregation degree of the constellation points becomes increasingly lower as the light-intensity undulation variance increases, and the system performance becomes increasingly worse. Especially when $\sigma_R^2 = 0.1$, the constellation points are seriously mixed, and it is difficult to identify them.

The radial shift of each constellation point becomes increasingly obvious as the value of σ_R^2 increases, and the degree of change of the light-intensity undulation

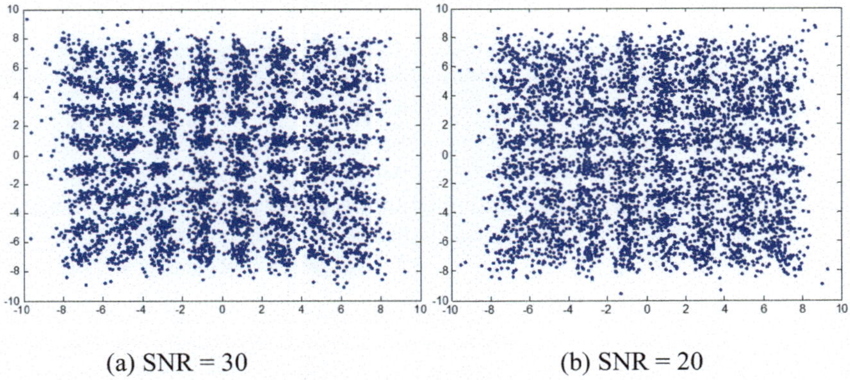

Fig. 7.30 Signal constellation at the receiver under 64-QAM modulation when $\sigma_R^2 = 0.1$

variance has more influence on the system performance. The degree of variation in the light-intensity variance has a greater impact on the system performance.

(2) **BER analysis**

The system BERs for the multiplicative and mixed-noise cases are shown in Figs. 7.31 and 7.32.

As can be seen from Fig. 7.31, when the value of σ_R^2 is less than 0.1, the impact of multiplicative noise on the system BER is small; when the value of σ_R^2 is between 0.1 and 0.4, the slope of the curve is large, and the system is affected by multiplicative noise to a higher degree; when the value of σ_R^2 is greater than 0.4, the impact of multiplicative noise on the system gradually decreases.

From Fig. 7.32, it can be seen that when the multiplicative noise to which the system is subjected is smaller than 0.01 and 0.05, the system BER is mainly affected by the additive noise. As σ_R^2 increases, the impact of multiplicative noise on the system is gradually prominent, and the system BER reaches more than 0.1 when

Fig. 7.31 BER curve for multiplicative noise

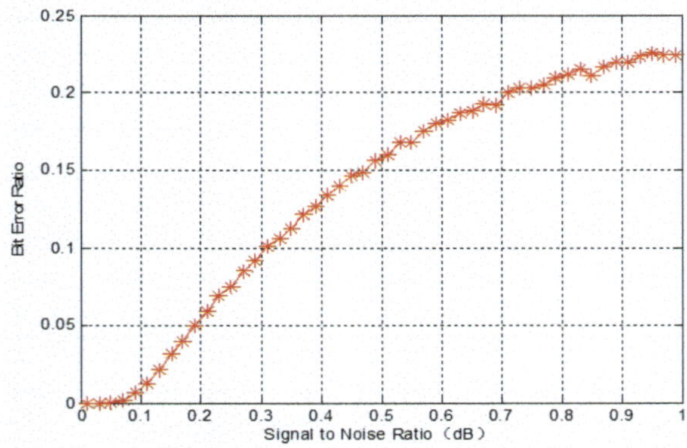

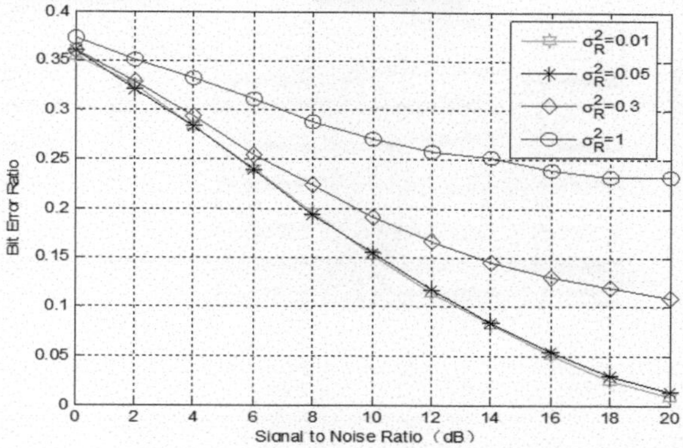

Fig. 7.32 BER curve for mixed noise

$\sigma_R^2 = 0.3$ and $SNR = 20$. It can also be seen that the system BER is mainly affected by the multiplicative noise when the value of σ_R^2 is greater than 0.05.

(3) **Eye-diagram analysis**

When the waveform of a serial baseband signal is displayed in the cumulative overlay of an oscilloscope's afterglow method, the shape of the displayed graph looks very much like an eye; hence, the name "eye diagram." From the eye diagram, it is possible to see the impact of noise and inter-code crosstalk on the transmission system and to estimate the degree of system performance. It can also be used to improve the transmission performance of the system by adjusting the filter at the receiver, according to the eye diagram. The eye diagram is shown in Fig. 7.33.

The inner region of the eye diagram is called the eye-diagram spread. In Fig. 7.33, the height of the eye-diagram spread at the specified sampling moment is defined as the noise tolerance of the system. If the "eye" is closed, owing to severe ISI or noise, then the noise tolerance is too low at the sampling moment and an error occurs.

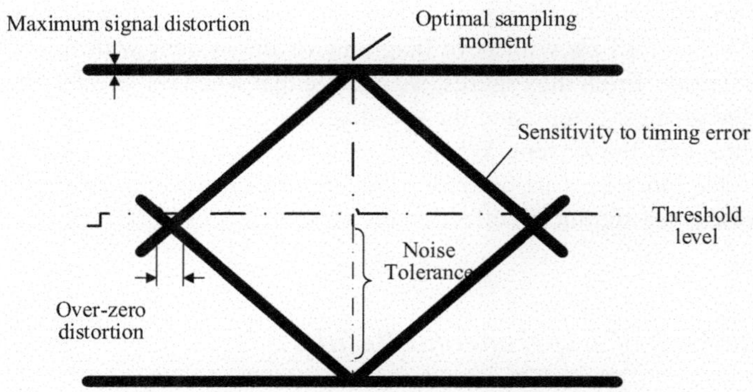

Fig. 7.33 Eye diagram

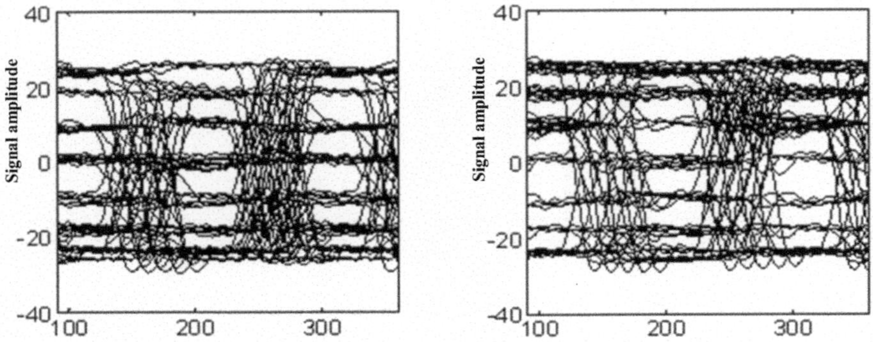

Fig. 7.34 Data eye diagrams at $\sigma_R^2 = 0.02$

The width of the eye-diagram spread determines the sensitivity of the sampling-time offset. A narrow eye-diagram spread means that the noise tolerance is small, and a small error in the sampling clock may result in a small sample with a very small noise tolerance; therefore, errors are more likely to occur. Thus, the degree of "eye" opening in the eye diagram can reflect the strength of the inter-code interference. The more open the "eye" is and the more correct the eye diagram is, the smaller the inter-code interference is; the less open the "eye" is or the less correct the eye diagram is, the greater the inter-code interference is Ref. [10].

When there is noise in the channel, the noise is superimposed on the signal, which leads to the blurring of the observed eye-diagram traces. If inter-code interference and noise are present simultaneously, the "eye" is opened to a smaller extent. Compared with the case without inter-code interference, the original clear line trace becomes a blurred band line and is incorrect. The greater the channel noise, the wider and more blurred the trace; the more serious the inter-code interference, the less correct the eye diagram.

By observing eye diagrams for different noise cases, we can analyze the impact of channel noise on the performance of an optical wireless communication system. Figures 7.34 and 7.35 show the eye diagrams of the system with different light-intensity variances. From the Figures, we can see that when $\sigma_R^2 = 0.02$, the "eye" is opened to the greatest extent, and the eye-diagram traces are also clear, which has the least impact on the system. When $\sigma_R^2 = 0.05$, the "eye" has basically closed, and the eye-diagram traces are blurred, which has the greatest impact on the system. Channel noise has a significant effect on the system.

7.3.2 Analysis of Experimental Results

A test signal with 60,000 points was used in the experiment. The collected data were demodulated at the receiver side using MATLAB, and the performance of

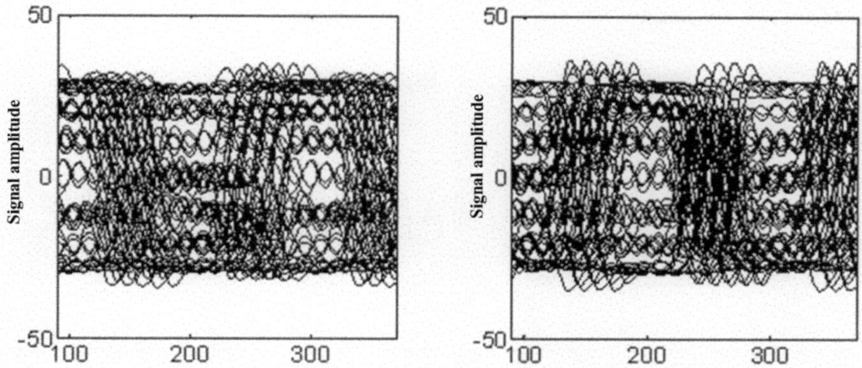

Fig. 7.35 Data eye diagrams at $\sigma_R^2 = 0.05$

the 64-QAM system was analyzed using power-spectrum estimation, BER, and eye diagrams.

(1) **Optical antenna power-spectrum estimation**

The Hamming window function was chosen to overlap the signal segments, and algorithms, such as a window function or fast Fourier transform (FFT), were added to calculate the self-power-spectrum estimation of a signal sequence [11]. The data collected in the experiment were divided into L segments. Then, the window function was multiplied by each segment in the time domain, the FFT was performed, and the squared spectral amplitude of each segment was summed and averaged to obtain the requested power-spectrum density.

$$P(f) = \frac{1}{MU}\sum_{i=1}^{L}\left|\sum_{n=0}^{M-1} x_i(n)d(n)e^{-j2\pi fn}\right|^2, \tag{7.42}$$

where U is the normalization factor, $d(n)$ is the data window, and M is the length of each segment.

Figure 7.36 shows the power spectrum of the signal at the receiver side. The sampling frequency of the sign al $F_s = 10{,}000$ (Hz), the carrier frequency $F_c = 1250$ (Hz), and the sampling period $T = 1/10000$. It can be seen from the figure that the power spectrum of the signal is relatively smooth, and the highest component of the power spectrum is more concentrated within the frequency range of 361–441 Hz with a peak of 0.04676 dB, which also shows the optical-energy distribution.

(2) **Analysis of the constellation diagram**

Figure 7.37 shows the constellation diagram of the received data, from which it can be seen that when the distance is small, the constellation points are not mixed, and the distribution is more concentrated and clear. As each constellation point has

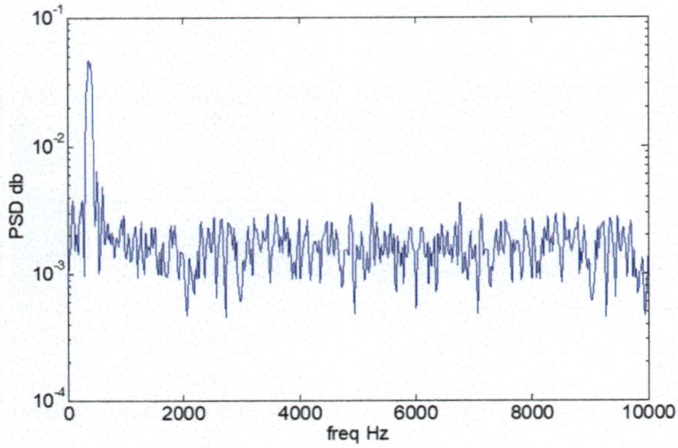

Fig. 7.36 Data power-spectrum estimation

a radial-offset phenomenon, the radial offset of the constellation point is smaller when closer to the center of the constellation diagram, and the radial offset of the constellation point is larger closer to the boundary. It can be seen that although the signal is affected by various noises in the transmission channel, multiplicative noise has a more obvious impact on the system.

The 64-QAM BER calculation formula is

$$
\begin{aligned}
BER_{64-QAM} &= \frac{P_e}{\log_2 64} = \frac{4(\sqrt{64}-1)}{(\log_2 64)\sqrt{64}} Q\left(\sqrt{\frac{3E_b(\log_2 64)}{(64-1)N_0}}\right) \\
&= \frac{7}{12} Q\left(\sqrt{\frac{2E_b}{7N_0}}\right),
\end{aligned}
\tag{7.43}
$$

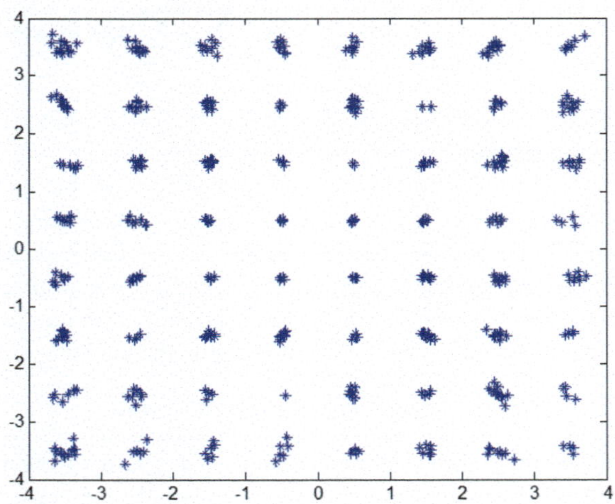

Fig. 7.37 64-QAM demodulation-signal constellation diagram

where P_e is the system-symbol error probability, E_b/N_0 is the SNR(Signal-to-Noise Ratio), E_b is the average energy/bit, and $Q(\cdot)$ is defined as

$$Q(u) = \int_u^\infty \frac{1}{\sqrt{2\pi}} e^{-\frac{x^2}{2}} dx, \quad u > 0. \tag{7.44}$$

The BER can be obtained using the above equation.

References

1. Qi YS, Lu P (1999) Design of a 16QAM-OFDM efficient modem. J Nanjing Instit Posts Telecommun, 1999, 19(4): 31–35.
2. Mohammed Farooque Mesiya (2014) Contemporary communication systems. Electronic Industry Press, Beijing, pp 519–522
3. Mesiya MF (2004) Contemporary communication systems. Tan Mingxin, Beijing, pp 519–520
4. Mesiya MF (2004) Contemporary communication systems. Tan Mingxin, Beijing, pp 521–522
5. Changxin F, Lina C (2010) Principles of communication. National Defense Industry Press, Beijing, pp 270–271
6. Yunbo X (2008) Analysis of pulse simulation modulation. J Qiqihar Normal College 6(106):123–125
7. Huan D, Yonggang W (2006) Blind balancing techniques in communication. Inf Secur 11(32):84–87
8. Xianda Z, Zheng B (1989) Analysis and processing of non-stationary signals. National Defense Industry Press, Beijing, pp 10–15
9. Xu SD, Dai XC, Xu PX (2007) J Electron Inf, Beijing 1352–1355
10. Fan C, Cao L (2010) Principles of communication. National Defense Industry Press, Beijing, pp 156–157
11. Chen D, Li JH (2005) Research on feature extraction and identification of acoustic signals based on principal component analysis. Acoustics Technol 24(1):39–41

Chapter 8
Dual-Amplitude Pulse-Position Modulation

This chapter introduces dual-amplitude pulse-position modulation (DAPPM) and discusses its symbolic structure, power-spectrum density, bandwidth requirement, and channel capacity in an ideal channel. It is compared with on–off keying (OOK), pulse-position modulation (PPM), digital pulse-interval modulation (DPIM), and dual-amplitude pulse-interval modulation (DAPIM).

8.1 DAPPM Principle and Performance Analysis

8.1.1 Modulation Principle

The modulation principle of DAPPM can be described as follows: When sending M-bit data, the symbol length after modulation by DAPPM is 2^{M-1}. The DAPPM pulse has two amplitudes, which are defined as A and aA $(0 < a < 1)$. Let k represent the decimal number of the sent binary. When $k \leq 2^{M-1}$, the amplitude is A and the pulse position corresponds to k. When $k > 2^{M-1}$, the amplitude is aA, and the pulse position corresponds to $k - 2^{M-1}$. Inverse amplitude correspondence can also be used. The DAPPM pulse-position modulation is shown in Fig. 8.1.

8.1.2 Transmission Efficiency

In atmospheric laser communication, the success or failure of a modulator/demodulator (modem) communication method is greatly influenced by the guarantee of the quality and reliable performance of the transmission efficiency and rate of the selected modulation mode.

X. Ke, *Coding Theory in Optical Wireless Communication Systems*, Optical Wireless Communication Theory and Technology, https://doi.org/10.1007/978-981-19-9837-9_8

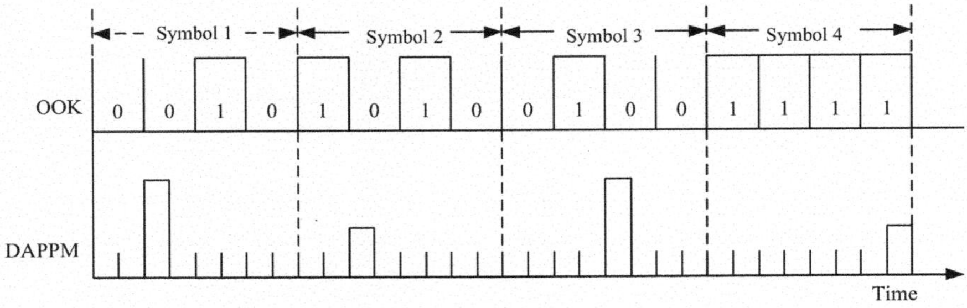

Fig. 8.1 DAPPM pulse-position modulation diagram

Transmission efficiency refers to the amount of information transmitted per unit of energy [1], and the information transmission rate refers to the amount of data transmitted over a channel in a unit of time. In an actual communication system, high efficiency and speed cannot be achieved simultaneously; therefore, one aspect should be emphasized, according to the situation. In atmospheric laser communication, the transmission power should be as small as possible or the transmission efficiency should be as high as possible, under the premise of guaranteeing a certain rate.

To reduce the interference between pulses during atmospheric laser communication, it is necessary to place a protection time slot between pulses, which is the minimum time interval between adjacent pulses. Figure 8.2 shows the DAPPM frame structure. A DAPPM frame is formed by mapping M bits of information into 2^{M-1} time slots and adding the corresponding protection time slot. The parameters that determine the DAPPM signal's structure are the transmission time slots L, time-slot width τ, and protection time slot D.

The data-transmission rate R of a DAPPM modulation signal is

$$R = (1 + \log_2 L)/T = (1 + \log_2 L)/(L + D)\tau \text{ bits/s} . \tag{8.1}$$

The DAPPM transmission capacity per time slot is

$$C_P = R\tau = (1 + \log_2 L)/(L + D)\text{bit/chip}. \tag{8.2}$$

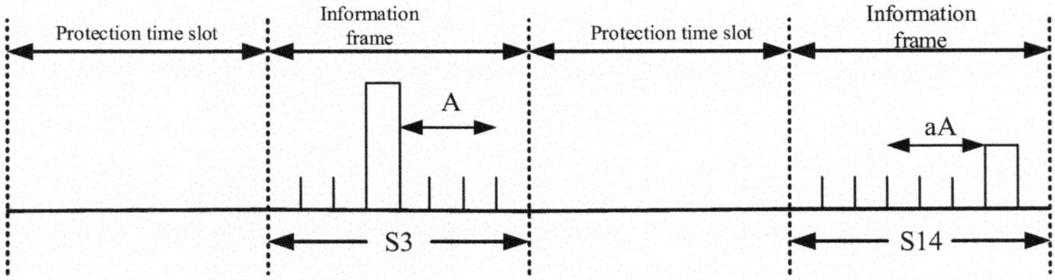

Fig. 8.2 DAPPM frame-structure map

8.1.3 DAPPM Ideal Channel-Capacity Analysis

The channel is the medium of signal transmission and can be divided into wireless and wired channels. Typically, generalized communication systems can be divided into coding and modulation channels. The function of a channel is to transmit information in the form of a signal. Studying a channel involves investigating the maximum amount of information that can be transmitted in a channel, that is, the channel capacity.

(1) **Shannon's formula**

Shannon's formula is a combination of communication and probability theory. When a symbol x is inputted into a channel, the conditional probability of the output symbol y is expressed as $W(y/x)$. In the channel, the source output with the $P_X(x)$ probability distribution is the input, and the channel output Y is the probability variable. The distribution of P_Y is given by $P_Y(y) = \sum_{x=\zeta} W(y|x)P_X(x)$, where ζ denotes the channel input symbol. In this case, C is called the channel capacity [2].

$$C = \max_{P_X} I(X; Y), \tag{8.3}$$

where $I(X; Y)$ is the amount of mutual information, which means that $\max_{P_X} I(X; Y)$ is the maximum of all possible probability distributions with respect to input X.

Shannon concluded from probability theory that the average amount of information is the difference between the average uncertainty $H(X)$ of the sender (source) and the uncertainty degree and measure of the source (that is, the amount of information lost due to channel interference) $H(X|Y)$ of the receiver after receiving the message Y[3].

$$I(X; Y) = H(X) - H(X|Y) \tag{8.4}$$

After studying the channel, Shannon found that the noise in the channel was mainly Gaussian white noise, that is, white noise whose power-spectral density function follows a Gaussian distribution. A Gaussian channel contains white Gaussian noise. The Shannon formula can be derived from a Gaussian channel:

$$C = B\ln(1 + S/N) = B\ln(1 + S/(N_0 * B)), \tag{8.5}$$

Where N_0 is the signal-to-noise ratio per unit frequency, B is the bandwidth, and S/N is the signal-to-noise ratio. Shannon's formula indicates that the amount of information C is related to the bandwidth and signal-to-noise ratio S/N. When the bandwidth B tends to infinity, the amount of information in the Gaussian channel tends to its limit.

$$C = \lim_{B\to\infty} B\ln(1 + S/(N_0 * B)) \approx 1.44S/N_0(\text{bits/s}) \tag{8.6}$$

(2) DAPPM ideal channel-capacity analysis

In a DAPPM optical wireless communication system, the receiver of a signaling session counts the number of photons in each of L time slots. Then, the time slot with the largest number of photons is judged to contain the signal pulse. With a noiseless channel, the laser is closed for L − 1 time slots. The probability of detecting no photons when the laser has an open time slot, assuming that the photon counting follows the Poisson distribution.

If the time-slot length is τ and the average photon-counting rate is s, the probability that the receiver detects no photons is [4]

$$q = e^{-s\tau}. \tag{8.7}$$

The channel capacity of DAPPM is given by the following formula:

$$C = (1 - q)(1 + \log_2 L) = (1 - e^{-s\tau})(1 + \log_2 L) = (1 - e^{-sT_s/L})(1 + \log_2 L). \tag{8.8}$$

DAPPM Each frame is $T = L\tau(s)$, so the channel capacity per second is

$$C_t = \frac{C}{T} = \frac{C}{L\tau} = \frac{(1 - e^{-s\tau})(1 + \log_2 L)}{T_D + L\tau}. \tag{8.9}$$

The average number of photons contained in each information frame is $s\tau$, so the channel capacity (capacity energy efficiency) of each received photon is C_p:

$$C_P = C/s\tau = MC_t/s \tag{8.10}$$

$$C_P = (T_D + L\tau)C_t/\left[\ln\left(1 - \frac{(T_D + L\tau)C_t}{1 + \log_2 L}\right)\right]. \tag{8.11}$$

In a channel with a constant pulse power and constant signal time T_S, C changes with L, as shown in Fig. 8.3. There exists an optimal number of time slots L or time-slot width τ to maximize the symbol capacity C and the time-transmission capacity of the channel. The two curves in the figure are calculated using the equation where $T_S = 100\ \mu\text{s}$ with the average photon count per second s (optical pulse power) as the parameter. When $s = 10^7$, the optimal number of time slots $L = 330$ corresponds to the maximum capacity of nine bits. When $s = 10^6$, the optimal number of time slots $L = 50$ corresponds to a maximum capacity of 5.8 bits.

With an average photon-counting rate s (constant optical-pulse power) and an unchanged time-slot width τ, Ct changes with L, as shown in Fig. 8.4. An optimal number of time slots L_T maximizes the transmission capacity Ct per unit time. The two curves in the figure were calculated using Eq. (8.9) for $s = 10^6$ and $\tau = 10\ \mu\text{s}$. When $T_D = 0.1$ ms, the optimal number of time slots $L_T = 8$ produces 22.5 kbits/s for the maximum channel-transmission capacity per second. When $T_D = 1$ ms, the

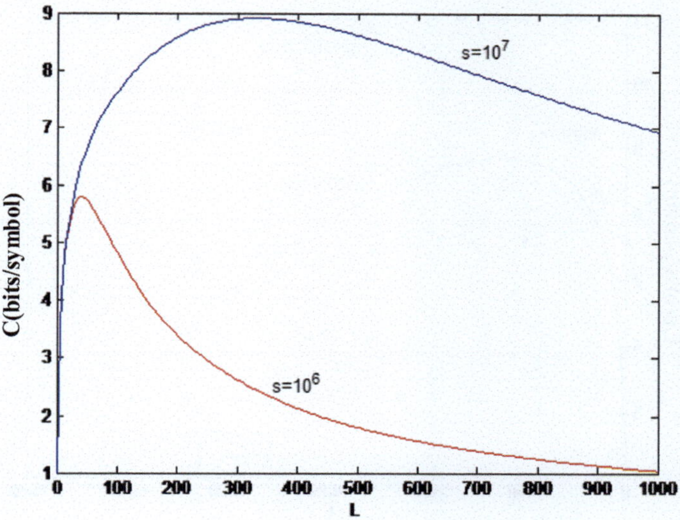

Fig. 8.3 Change of channel capacity C with the number of time slots L

maximum channel transmission capacity per second is 4 kbit/s, corresponding to the optimal number of time slots $L_T = 35$.

When the widths of C_t and the time slots remain unchanged, C_p changes with L and S, as shown in Fig. 8.5. There is an optimal number of time slots L_P or average

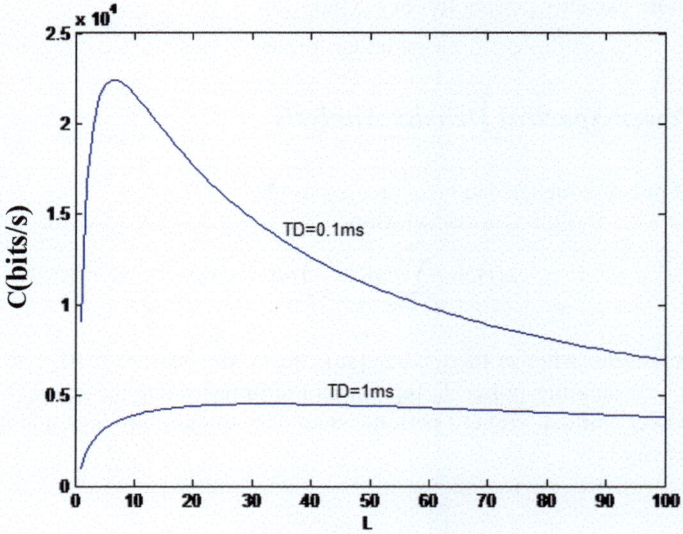

Fig. 8.4 Change of channel capacity C_t with number of time slots L per unit time

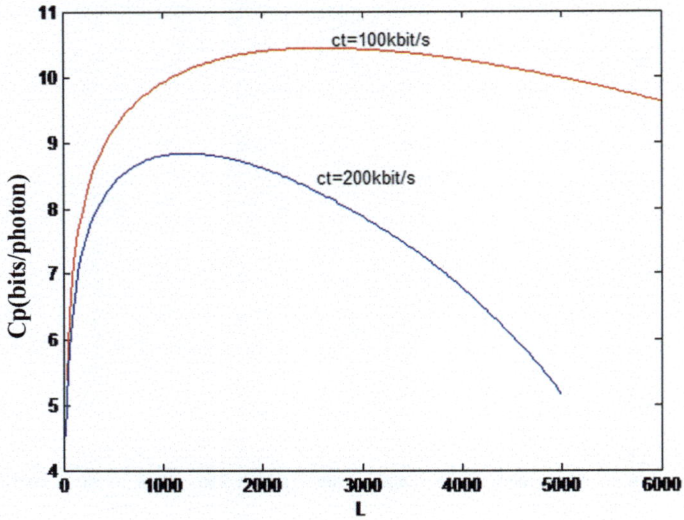

Fig. 8.5 Variation of the channel-capacity energy efficiency C_p with L time slots

photon-counting rate S_P to maximize the channel-capacity energy efficiency C_p. The two curves in the figure are the results calculated using Eq. (8.11) where $\tau = 10$ ns and $T_D = 5\ \mu$s with C_t as the parameter. When $Ct = 100$ kbit/s, the optimal number of time slots $L_P = 2500$ corresponds to a maximum capacity per photon of 10.5 bit s. When $C_t = 200$ kbit/s, the optimal number of time slots $L_P = 1150$ corresponds to a maximum capacity per photon of 8.8 bits.

8.1.4 Power-Spectral Density Analysis

A DAPPM pulse sequence can be expressed as [5]

$$x(t) = \sum_{n=-\infty}^{\infty} a_n \left(\frac{p_c}{A}\right) p(t - nT_c), \tag{8.12}$$

where a_n represents whether there is a pulse in the n_{th} timeslot, $a_n \in \{0, 1, \cdots A\}$, $p(t)$ represents a rectangular pulse, T_c represents the timeslot length, and p_c represents the peak power emitted. $x(t)$ is a periodic stationary random process, and its power-spectral density is

$$s(f) = \frac{1}{T_c} |p(f)|^2 s_a(f), \tag{8.13}$$

where $p(f)$ is the Fourier transform of the rectangular pulse $p(t)$:

$$p(f) = T_c \sin c(fT_c). \tag{8.14}$$

$s_a(f)$ is the power-spectral density of the time-slot sequence, namely, the discrete Fourier transform of its autocorrelation function R_k, which is expressed as follows [5]:

$$R_k = \frac{(A+1)(2A+1)}{3(L+1)} \quad k = 0 \tag{8.15}$$

$$R_k = \begin{cases} \dfrac{(A+1)^2(L+1)^{k-2}}{2L^k} & 1 \le k \le L \\ \dfrac{1}{AL}\sum_{i=1}^{L} R_{k-i} & k > L \end{cases}. \tag{8.16}$$

Studies show [5] that when k increases continuously, R_k approaches $E[a]^2$,$E[a] = (A+1)/(A(L+1))$, so the continuous and discrete power-spectral density of DAPPM can be approximately expressed as [5]

$$S_c(f) \approx \sum_{k=-5L}^{5L} [R_k - E[a]^2] \exp(-j2\pi kfT_c) \tag{8.17}$$

$$S_d(f) = \frac{E[a]^2}{T_c} \sum_{k=-\infty}^{\infty} \delta\left(f - \frac{k}{T_c}\right) \tag{8.18}$$

$$s_{a(f)} \cong \frac{2E(a)^2}{T_c} \sum_{k=-\infty}^{\infty} \delta\left(f - \frac{k}{T_c}\right) + \sum_{k=-\infty}^{\infty} (R_k - E(a)^2) \exp(-j2\pi kfT_c), \tag{8.19}$$

where $E(a)$ represents the mean value and the first term in $s_a(f)$ is the discrete spectral line of $f = k/T_c$. When $f = k/T_c (k \neq 0)$, $p(f)$ equals 0; hence, the power-spectral density of DAPPM only has discrete spectral lines at the DC component.

8.1.5 Performance Comparison of Several Modulation Modes

(1) **Comparison of several modulation modes**

Here, we compare several modulation modes (Table 8.1).

(1) On–off keying (OOK) modulation. OOK modulation is the simplest modulation mode, which turns optical signals on and off to transmit information. It is the bit-by-bit modulation of binary information. When the digital bit to be sent is

0, no pulse is sent in the corresponding time slot; when the digital bit to be sent is 1, a light pulse is sent in the corresponding time slot.

(2) Pulse-position modulation (PPM). PPM is essentially a phase modulation that sends information based on placing the pulse at different positions in the time slot. If a set of M-bit binary information is sent, then the number of time slots after modulation is 2^M, which corresponds to the decimal number of binary information; only one time slot in a frame of signals sends a pulse; the other time slots do not send a pulse.

(3) Digital pulse-interval modulation (DPIM). DPIM is also a type of pulse-position modulation. In contrast to PPM, the length of its modulation signal is not fixed. The pulse is in the first time-slot position of each frame signal, and the signal is sent using the time-slot interval between the pulse and the next frame-signal pulse. If a group of M-bit binary information is sent, the number of time slots after modulation is $k + 2$, where k is the decimal number corresponding to the binary information. The pulse is in the first time slot, followed by a protection time slot to reduce the influence of inter-symbol crosstalk, and then k empty time slots are added [6].

(4) Dual-amplitude pulse-interval modulation (DAPIM). In this modulation mode, the initial pulses of two amplitudes are defined as A and αA $(0 < \alpha < 1)$, and the symbol length of the modulation signal is not fixed. When sending M-bit information, the modulation signal can be mapped to 2^M symbols. Let S_k represent a symbol k $(2^M > k \geq 0)$, consisting of an initial pulse slot (slot), a protection slot (to reduce inter-symbol crosstalk), and the subsequent m space slots, where m satisfies the following: When $k < 2^{M-1}$, m equals k and the pulse amplitude is A; when $k \geq 2^{M-1}$, m equals $k - 2^{M-1}$, and the pulse amplitude is αA. Reverse correspondence is also possible, as DAPIM does not require symbol synchronization because each symbol starts with a pulse [3].

(5) Dual-amplitude pulse-position modulation (DAPPM). This modulation mode also uses signal pulses with two amplitudes. DAPPM is unlike DAPIM in that the pulses are in different time-slot positions, and the symbol length of the modulation signal is fixed. When sending M-bit binary information, the number of time slots of the modulation signal is 2^{M-1}; let k be the decimal number corresponding to the binary. When k is less than or equal to 2^{M-1}, the pulse amplitude of the modulation signal is A; when k is greater than 2^{M-1}, the pulse amplitude of the modulation signal is $aA(0 < a < 1)$. The opposite correspondence can also be used. Because the pulse of each frame is at a different position in the time slot, symbol synchronization is required.

(2) **Required-Bandwidth comparison**

In a laser communication system, if the transmission rate of the transmitter is R_b (bits/s), the bandwidth of the OOK modulation mode is the reciprocal of its pulse width: $B_{OOK} = 1/T_b = R_b$. For PPM, it is the reciprocal of the timeslot width [4]:

Table 8.1 Coding structures of several modulation modes

Source bits (OOK)	PPM	DPIM	DAPIM	DAPPM
000	10000000	10	100	1000
001	01000000	100	1000	0100
010	00100000	1000	10000	0010
011	00010000	10000	100000	0001
100	00001000	100000	a00	a000
101	00000100	1000000	a000	0a00
110	00000010	10000000	a0000	00a0
111	00000001	100000000	a00000	000a

$$T = \frac{m}{R_b \cdot L} = \frac{m}{R_b \cdot 2^m} \tag{8.20}$$

$$B_{PPM} = \frac{1}{T} = \frac{1}{m/(2^m \cdot R_b)} = \frac{R_b \cdot 2^m}{m}, \tag{8.21}$$

where T is the timeslot width in PPM, L is the symbol length in PPM, m is the number of bits sent, and R_b is the information rate. Similarly, the DPIM bandwidth requirement is also the reciprocal of its time-slot width.

$$T_{DPIM} = \frac{m}{R_b \cdot L_{DPIM}} = \frac{m}{R_b} \bigg/ \frac{2^m + 3}{2} = \frac{2m}{R_b(2^m + 3)} \tag{8.22}$$

$$B_{DPIM} = \frac{1}{T_{DPIM}} = \frac{(2^m + 3)R_b}{2m}. \tag{8.23}$$

Similarly,

$$T_{DAPIM} = \frac{m}{R_b \cdot L_{DAPIM}} = \frac{m}{R_b} \bigg/ \frac{2^{m-1} + 3}{2} = \frac{2m}{R_b\left(2^{m-1} + 3\right)} \tag{8.24}$$

$$B_{DAPIM} = \frac{1}{T_{DAPIM}} = \frac{(2^{m-1} + 3)R_b}{2m} \tag{8.25}$$

$$T_{DAPPM} = \frac{m}{R_b \cdot L_{DAPPM}} = \frac{m}{R_b \cdot 2^{m-1}} \tag{8.26}$$

$$B_{DAPPM} = \frac{1}{T_{DAPPM}} = \frac{2^{m-1} R_b}{m}. \tag{8.27}$$

The bandwidth requirements of the various modulation modes are shown in Fig. 8.6, and are normalized according to OOK. As can be seen from the graph, the bandwidth requirements of the various modulation modes gradually increase with an increase in bit resolution M. The bandwidth requirements of DAPPM are similar to

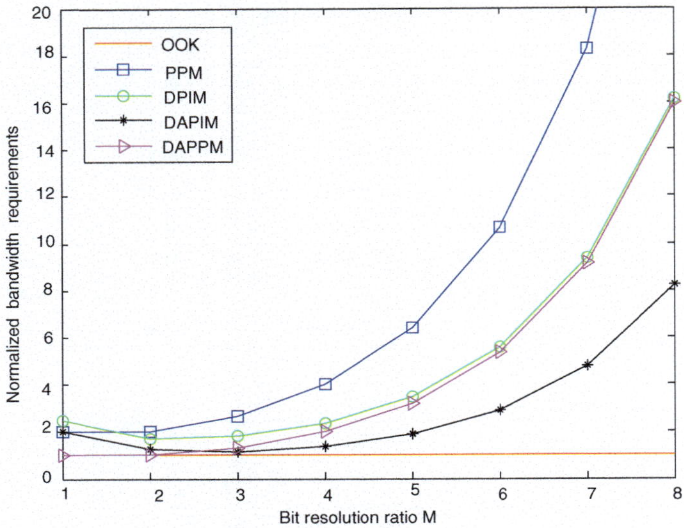

Fig. 8.6 Comparison of bandwidth requirements

those of DPIM, but are reduced by half, compared to PPM, thereby achieving better bandwidth efficiency than PPM.

(3) **Comparison of the unit-transmission rate**

The unit-transmission rate is defined as the number of bits transmitted per Hertz per second, and can be expressed as follows [3]:

$$\gamma = \frac{R}{B}(\text{bits s}^{-1}\text{Hz}^{-1}), \tag{8.28}$$

where R is the transmission rate and B is the signal bandwidth. Optical communication usually works on the laser pulse state; if the pulse duration is τ, the corresponding signal bandwidth can be defined as $B = 1/\tau$, with a duty ratio r_p (all of the following are symbols of duty ratio r_p), the slot width is T, and the signaling unit rate is [3]

$$\gamma_{OOK} = (m/T)/(1/\tau). \tag{8.29}$$

For a single-pulse PPM, if m bits are sent at a time, the number of timeslots is 2^m and the width of the 2^m timeslots is $T_{PPM} = 2^m \cdot \tau/r_p$. The corresponding unit-transmission rate is

$$\gamma_{PPM} = \frac{R_{PPM}}{B_{PPM}} = \frac{m/T_{PPM}}{(1/\tau)} = \frac{m \cdot \gamma_p}{2^m}. \tag{8.30}$$

Because the symbol lengths of the DPIM and DAPIM modulation modes are not fixed, we use the statistical average method to calculate their time-slot widths. Their time-slot widths and unit-transmission rates are as follows:

$$T_{DPIM} = \frac{2^m + 3}{2} \cdot \frac{\tau}{r_p} \tag{8.31}$$

$$\gamma_{DPIM} = \frac{R_{DPIM}}{B_{DPIM}} = \frac{m/T_{DPIM}}{(1/\tau)} = \frac{2m \cdot r_p}{2^m + 3} \tag{8.32}$$

$$T_{DAPIM} = \frac{2^{m-1} + 3}{2} \cdot \frac{\tau}{r_p} \tag{8.33}$$

$$\gamma_{DAPIM} = \frac{R_{DAPIM}}{B_{DAPIM}} = \frac{m/T_{DAPIM}}{(1/\tau)} = \frac{2m \cdot r_p}{2^{m-1} + 3}. \tag{8.34}$$

The time-slot width for DAPPM modulation is $T_{DAPPM} = 2^{m-1} \cdot \tau / r_p$, and for each unit, the transmission rate is (Fig. 8.7)

$$\gamma_{DAPPM} = \frac{R_{DAPPM}}{B_{DAPPM}} = \frac{m/T_{DAPPM}}{(1/\tau)} = \frac{m \cdot r_p}{2^{m-1}}. \tag{8.35}$$

(4) **Performance comparison of transmission capacity**

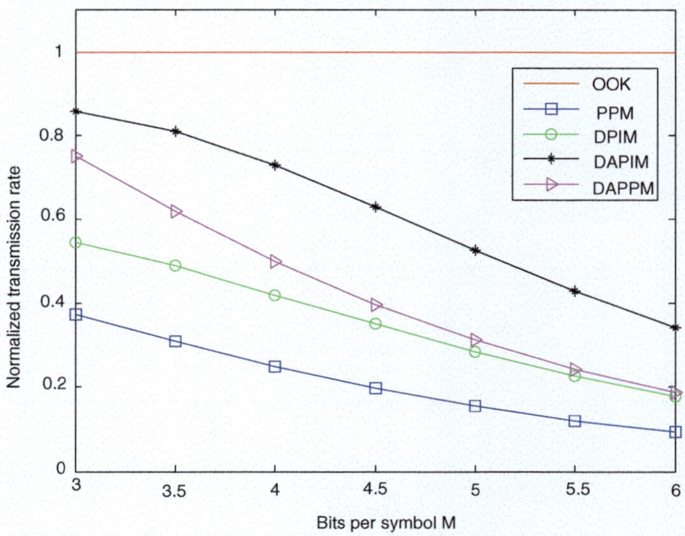

Fig. 8.7 Comparison of unit-transmission rates

The transmission capacity is an important performance index of wireless laser communication systems. It represents the size of the information transmission capacity of a communication system per unit time [1]. This section compares and analyzes the transmission capacities of the OOK, PPM, DPIM, DAPIM, and DAPPM modulation modes on the premise of an equal time-slot width.

Assuming that the modulation order is M, the transmission capacity is calculated using the bit rate, and the transmission capacity is equal to the ratio of the number of sent bits to the time required to send a frame signal. Suppose the time-slot width is τ and the time required to send a frame is T. Then, the transmission capacity of OOK is

$$C_{OOK} = M/T = M/(M \cdot \tau) = 1/\tau. \tag{8.36}$$

The transmission capacity of PPM is as follows:

$$T_{PPM} = 2^M \cdot \tau \tag{8.37}$$

$$C_{PPM} = \frac{M}{T_{PPM}} = \frac{M}{2^M \cdot \tau}, \tag{8.38}$$

where T_{PPM} is the time required to send a frame signal, τ is the time-slot width, and M is the number of bits sent at one time.

The transmission capacities of DPIM, DAPIM, and DAPPM are calculated similarly:

$$T_{DPIM} = \frac{2^M + 3}{2} \cdot \tau \tag{8.39}$$

$$C_{DPIM} = \frac{M}{T_{DPIM}} = \frac{2M}{(2^M + 3)\tau} \tag{8.40}$$

$$T_{DAPIM} = \frac{2^{M-1} + 3}{2} \cdot \tau \tag{8.41}$$

$$C_{DAPIM} = \frac{M}{T_{DAPIM}} = \frac{2M}{(2^{M-1} + 3)\tau} \tag{8.42}$$

$$T_{DAPPM} = 2^{M-1} \cdot \tau \tag{8.43}$$

$$C_{DAPPM} = \frac{M}{T_{DAPPM}} = \frac{M}{2^{M-1} \cdot \tau}. \tag{8.44}$$

Figure 8.8 shows the transmission capacities of several modulation modes. It can be seen that the transmission capacity of OOK is the highest, that of DAPPM is lower than that of DAPIM but significantly higher than that of DPIM, and that of PPM is

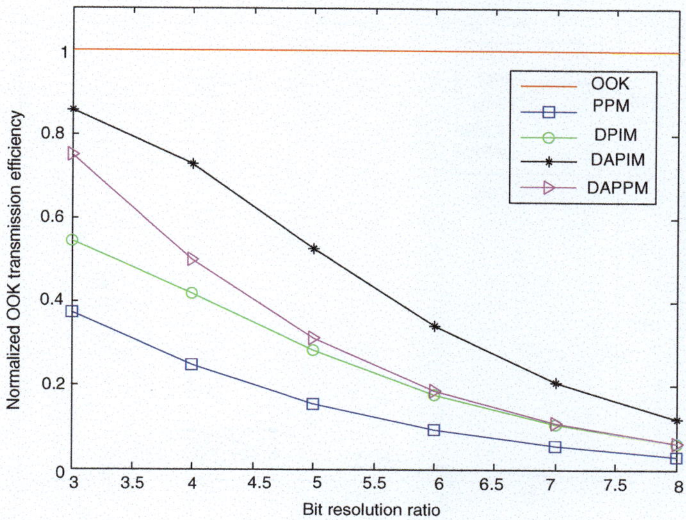

Fig. 8.8 Transmission-capacity comparison

the lowest. With an increase in bit resolution, the transmission capacities of the other modulation modes decrease gradually and tend to be the same, except that of OOK, which remains unchanged.

(5) **Comparison of the peak power required at the transmitter**

In a laser-communication system, the peak power P_t varies with different modulation modes. When comparing the peak power of several modulation modes, we assume that the average power (P) of all modulation modes is equal and the pulse amplitude is a. The peak power of OOK is $P_{tOOK} = 2P$ and that of PPM is calculated as follows:

$$P_{tPPM} = L \cdot P = 2^M \cdot P, \tag{8.45}$$

where L is the symbol length and M is the number of bits sent at one time.

Similarly, the peak power of DPIM, DAPIM, and DAPPM are as follows:

$$P_{tDPIM} = L_{DPIM} \cdot P = \frac{2^M + 3}{2} P \tag{8.46}$$

$$P_{tDAPIM} = \frac{2^{M-1} + 3}{1 + a} P \tag{8.47}$$

$$P_{tDAPPM} = \frac{2^M}{1 + a} P. \tag{8.48}$$

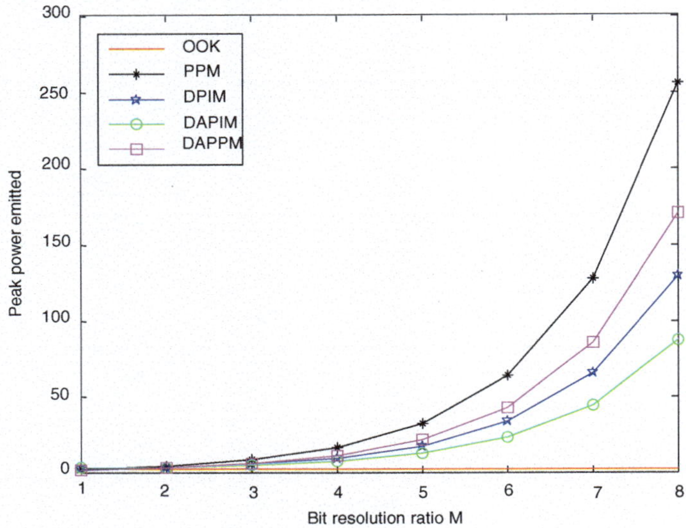

Fig. 8.9 Comparison of the peak power required at the transmitter

Figure 8.9 shows the peak power required by the transmitter for the five modulation modes. As can be seen from the figure, OOK has the lowest peak power, followed by DAPIM and DPIM. The peak power required by the DAPPM transmitter is higher than that of OOK, DAPIM, and DPIM, but lower than that of PPM. With an increase in the bit resolution M, the peak power of each modulation mode gradually increases.

8.2 Noise Analysis of a DAPPM Communication System

8.2.1 Composition and Influence of Atmospheric Molecules

The atmosphere mainly consists of atmospheric molecules, water vapor, and various complex impurity particles. The composition of atmospheric molecules mainly includes N_2, CO_2, O_2, H_2O, O_3, and inert gases [7]. The composition of impurities in the atmosphere is very complex, with various forms and wide distributions.

In general, impurity particles include both liquid and solid particles. Solid particles refer to dust, sand, smoke, and environmental pollution. Liquid particles mainly include raindrops, snow, fog drops, hail, and clouds. As the temperature of the atmosphere is constantly changing and compounded by wind, rain, snow, and other bad weather, the gas molecules and some particles are continually moving up and down, and the density, humidity, etc., have a great impact, and lead to turbulent movement in the atmosphere.

When a laser is transmitted through an atmospheric channel, the transmission quality of the laser beam is significantly affected. The influence of the atmosphere on a laser beam is reflected in its absorption, scattering, and refraction by atmospheric molecules, aerosols, and turbulence. Absorption and scattering will seriously reduce the transmission power of a laser signal. Atmospheric turbulence will cause the refractive index of the atmosphere to change constantly, causing the laser spot to drift and flicker, as well as the light intensity and phase to fluctuate, resulting in a turbulence effect. All these factors have a significant impact on the transmission of laser signals. For long-distance communication and when high accuracy is needed, the impact is more obvious.

8.2.2 *Attenuation Effect of the Atmosphere*

Because the atmospheric channel is affected by various factors, it is unstable. The temperature, density, pressure, and various gases and particles, such as carbon dioxide, smoke, fog, water vapor, and dust, are in a state of constant motion and change. In addition, if a laser signal passes through the channel during bad weather, such as rain, snow, or fog, it will be seriously affected.

These factors lead to atmospheric absorption and laser-beam scattering, resulting in the attenuation of the laser signal; this is called the atmospheric-attenuation effect. Atmospheric absorption will not affect the quality of the laser-beam transmission but will weaken the transmission power of the laser signal. Atmospheric scattering does not affect the energy transmitted by the laser but changes the light-intensity distribution and affects the profile of the light spot [8].

(1) **Absorption of atmospheric molecules**

Atmospheric-molecule absorption occurs when the laser signal produces a light-wave electric field while traveling through the atmospheric channel. The atmospheric molecules will be polarized by the light-wave electric field and will vibrate at a certain frequency, resulting in a loss of laser-signal energy.

Resonance absorption occurs when the vibration frequency is equal to the natural frequency of the atmospheric molecules, when their absorption capacity is the strongest. Therefore, the absorption capacity of atmospheric molecules is directly related to the frequency of the light waves, and the absorption capacity of the transmitting laser signals will differ according to the different atmospheric molecules.

In free-space optical (FSO) systems near the ground, the wavelength of the laser is usually in the near-infrared region, where absorption molecules include water and carbon dioxide. When the weather is clear, the absorption of the laser signal is more severe. Depending on the absorption spectrum, atmospheric molecules absorb certain wavelengths of light more severely and cause a communication failure. For some bands, the absorption capacity is weak. The usual communication windows are around the 0.8–1.6 and 8–13 μm bands. In a communication window with low attenuation, the light-source wavelength adopted by laser communication is usually

0.8, 106, 155 μm, etc., corresponding to the current photoelectric detector and laser technology.

(2) **Scattering of atmospheric molecules**

The atmosphere contains a variety of large and small particles, which are mainly gas molecules, dust, and water droplets. Pure scattering does not consume the energy of the laser signal, but changes the direction of the energy distribution. Scattering in the atmospheric channel includes both particle and molecular scattering. Near the ground, the scattering degree is much smaller, and the main factor leading to the loss of light energy is the scattering of floating particles to the light beam.

a. **Rayleigh scattering**

Rayleigh scattering occurs when the size of the scattered particles is smaller than the wavelength of the light waves. For Rayleigh scattering, the light-distribution scattering is almost uniform and symmetric. The volume scattering coefficients of Rayleigh scattering are expressed as follows [9]:

$$a_m(\lambda) = \frac{8\pi^3}{3}\frac{(n^2-1)^2}{N_g\lambda^4}\frac{6+3\delta}{6-7\delta}\left(\mathrm{km}^{-1}\right), \tag{8.49}$$

where n is the atmospheric refractive index related to the height and wavelength, N_g is the number of gas molecules per unit volume, and δ is the depolarization factor of scattered radiation, which is usually 0.035.

For Rayleigh scattering, if the beam intensity can be expressed by the function $f(\lambda)$, according to the distribution of the wavelength, $f(\lambda)\lambda^{-4}$ is the functional-distribution expression of the intensity of the scattered light. It can be seen that in Rayleigh scattering, the wavelength of the scattered light is strongly related to the intensity of the scattered light; therefore, shortwave frequencies are more suitable for Rayleigh scattering. Red light has better penetration than blue light when passing through a mist because red light is scattered less as it travels through the atmosphere. Therefore, many signal markers use red light.

b. **Mie scattering**

Mie scattering refers to the scattering of light signals by particles when the wavelength of the laser signal is smaller than that of the scattering particles. It is also called the aerosol attenuation of the signal. Mie scattering is mainly used for spherical particles such as haze, fog droplets, and small raindrops. The strength of Mie scattering depends on the particle size, location, distance, and weather conditions and is inversely proportional to the first power of the scattered light.

The intensities of Mie scattering and Rayleigh scattering are directly related to the wavelength. The shorter the wavelength of the signal, the stronger is the scattering. In general, Mie's law should be applied to particles with radius $r \geq 0.3\,\mu m$. For particles of radius $r \leq 0.3\,\mu m$ with wavelengths near 1 μm, the error of Rayleigh's law is less than 1%.

(3) **Influence of atmospheric attenuation on a laser signal**

Atmospheric attenuation of a laser signal is mainly caused by (1) molecular absorption, (2) molecular scattering, (3) aerosol scattering, and (4) aerosol absorption. The attenuation of a laser transmission in the atmosphere can usually be expressed as [9]

$$I(R) = I_0 \exp\left[-\int_0^R a(r, \lambda)dr\right], \tag{8.50}$$

where $I(R)$ is the light intensity of the laser with wavelength λ after transmitting over distance R in the atmosphere, $a(r, \lambda)$ is the atmospheric-attenuation coefficient at a distance, and I_0 is the beam intensity at the laser-emission point. The attenuation coefficient can be expressed as

$$a(\lambda) = A(\lambda) + S(\lambda), \tag{8.51}$$

where $A(\lambda)$ stands for absorption coefficient and $S(\lambda)$ is the scattering coefficient. Under the premise that the atmosphere is uniform, the gas-attenuation coefficient unit is 1/km. According to the derivation of the above equation, the characteristic attenuation of the atmosphere can be obtained:

$$r = -10\lg\frac{I}{I_0} = 10a(\lambda)/\ln 10 = 4.343a(\lambda)(\text{dB/km}). \tag{8.52}$$

On a clear day, the total attenuation of a laser caused by the atmosphere is $a = A_m + A_a + S_m + S_a$, where a is for aerosols, and m is for molecules.

When the laser signal is in the near-infrared band and the wavelength is in the atmospheric window, the absorption and scattering of atmospheric molecules can be ignored. Therefore, in good weather, attenuation caused by aerosols is the main cause of laser-signal attenuation. As the distribution of aerosol particles is random, it is difficult to obtain. Under normal circumstances, it can be approximated using the following equation [10]:

$$\mu = \frac{3.912}{V_b}\left(\frac{0.55}{\lambda}\right)^q, \tag{8.53}$$

where q is a correction factor related to visibility and V_b is atmospheric visibility. In different situations, the visibility can be expressed as follows [10]:

$$q = \begin{cases} 0.585V_b^{1/3} & V_b \le 6\,\text{km} \\ 1.3 & \text{Average visibility} \\ 1.6 & \textit{Excellent visibility} \end{cases}. \tag{8.54}$$

Average visibility is generally 10 $\sim$ 12 km; good visibility is generally 23 km.

Some scholars have modified the above values [11] to reduce the error and increase the number of applications. For the radiation loss of a wavelength $\lambda < 1\,\mu$m, the prediction error was relatively small; when the wavelength was long, the error increased.

When this q value was applied on the premise of poor visibility on foggy days, the error obtained was relatively large. The obtained attenuation coefficient was highly reliable for ordinary visibility. Nabouls et al. modified the q value for low visibility and obtained the following q values [11]:

$$q = \begin{cases} 0.16V_b + 0.34 & 1\,\text{km} < V_b < 6\,\text{km} \\ V_b - 0.5 & 0.5\,\text{km} < V_b < 1\,\text{km} \\ 0 & V_b < 0.5\,\text{km} \end{cases}. \tag{8.55}$$

Table 8.2 shows that, compared with rainy and snowy weather, fog has a more obvious attenuation effect on laser transmission. The main reason for this result is that the size of fog particles differs greatly from that of raindrops. According to research, the gap between raindrops is large, and the radius (μm magnitude) of a fog droplet is only about one thousandth of the radius (mm magnitude) of a raindrop, so the visibility with raindrops is relatively high, and light waves can more easily pass through. On the premise of fine weather, a communication system with a communication range of less than 10 km may not be able to communicate in fog.

Table 8.2 shows the corresponding correction factors for different meteorological distances.

For visible and infrared light, the molecules' scattering effect on the laser signal is relatively small, and sometimes it need not be considered [10]. However, the molecular absorption of laser signals can appear in any case. Regardless of the band, the spectral lines of a laser signal are absorbed by molecules in large quantities, and these spectral lines form spectral-band groups from the linear absorption or continuous absorption of light radiation; only a few absorption-intensity bands are relatively small.

When laser radiation travels through the atmosphere, the atmosphere absorbs it very well over a fixed range of wavelengths, making communication almost impossible. According to atmospheric characteristics, the near infrared is divided into several bands, and regions with relatively high transmittance are called "atmospheric

Table 8.2 Correction factors corresponding to different meteorological distances

Correction factor	Meteorological distance	Visibility level	Meteorological Condition
1.6	$V_b > 50$ km	9	Very clear
1.3	6 km < V_b < 50 km	6–8	Clear
$0.16\,V_b + 0.34$	1 km < V_b < 6 km	4–6	Fog
$V_b - 0.5$	500 m < V_b < 1 km	3	Light fog
0	$V_b \le 500$ m	< 3	Medium to thick fog

windows," in which the molecular absorption is weak, and the atmospheric absorption basically does not affect the attenuation coefficient. The working band of the semiconductor laser is 780–1550 nm. The current general laser bands work in these windows, and the commonly used bands are 850, 785, and 1550 nm. Therefore, the influence of the attenuation coefficient is mainly due to the scattering effect.

The atmospheric-transmission characteristics of horizontal light can be roughly expressed as follows. Depending on the different bands, the transmittance of infrared radiation is different. The transmittance of atmospheric bands 0.85–1.55 μm is relatively high. If the atmosphere is evenly distributed in the horizontal direction for horizontal light, according to the Beer–Lambert law, the transmittance expression is as follows:

$$\tau_a = \exp(-\mu L). \tag{8.56}$$

Here, μ is the attenuation coefficient and τ_a is the transmittance of the transmission optical path L, with dimension $[L^{-1}]$. The product of the index μL is called the optical thickness of the medium, and is denoted as l_τ, which is a commonly used dimensionless parameter. Through a large number of experiments, it was concluded that the wavelength is directly related to atmospheric transmittance [9]. This is because the atmosphere absorbs and scatters radiation, depending on the wavelength. According to the above analysis, the empirical formula for atmospheric transmittance and visibility

$$\tau_a = \exp\left[-\frac{3.912}{V_b}\left(\frac{\lambda}{0.55}\right)^q \times L\right]. \tag{8.57}$$

When the wavelength is fixed, the visibility is low, and the atmospheric transmittance increases with a decrease in the transmission distance. For fixed visibility, the longer the communication distance, the weaker the laser penetration through the atmosphere; the closer the communication distance, the stronger the penetration. When the communication range is constant, the penetration capacity increases with increasing visibility. The better the visibility is, the larger the communication range will be. The worse the visibility is, the faster the atmospheric-penetration capacity changes according to the communication distance. When the visibility is $V_b = 0.5$ km and $V_b = 1$ km, the signal crosses about 2 km in the atmosphere, and the transmittance is close to zero. Therefore, laser-signal transmissions are better in good weather conditions.

8.2.3 Atmospheric-Turbulence Effect and Its Influence on Laser Transmission

Because the atmosphere is composed of many components with different properties and coupled with the action of heat and wind, it is always in continuous motion. This results in many vortices. The temperature, density, velocity, and size of these vortices are all different, and they constantly change. Their motions are superimposed and crosslinked, resulting in random turbulent motion [12].

Atmospheric turbulence leads to a constant change in the atmospheric-refractive index, resulting in uncertain fluctuations in the phase and amplitude of the light wave, beam deformation, flickering, splitting, expansion, coherence reduction, polarization change, etc. The atmospheric turbulence influences the signal and the diameter of the optical signal and turbulence size are directly connected; if the turbulence size is larger than the diameter of the light signal, the turbulence will bend the beam. The beam as a whole is not qualitative, so the signal's center projection point (spot) will randomly jitter around a fixed point at the receiver. This is known as beam drift.

In the case of strong turbulence, the beam sometimes drifts out of the normal communication range, resulting in an inability to communicate. Under normal circumstances, the beam diameter should be greater than or equal to the turbulence size. At this time, the cross section of the beam contains many turbulence vortices. When the communication distance is long, the receiver will expand the sent laser signal.

In addition, various disturbance phenomena are observed at different positions in front of the signal light, which will cause the signal-light intensity to fluctuate constantly in time or space. In some locations, the light intensity varies between large and small, resulting in a flickering beam. These factors seriously interfere with the laser signal in the transmission process.

8.2.4 Influence of Background Light and Receiver Noise on the Modulation Signal

In an actual laser-communication system, when the signal is sent out the transmitting end, it is first affected by various noises in the atmospheric channel. Second, when the receiver receives the optical-pulse signal, the signal needs to pass through the photoelectric detector for photoelectric conversion. Because the signal is severely attenuated, it also needs to be preprocessed and undergo a series of processes before being restored.

In this process, the laser signal is not only affected by atmospheric-channel noise but also by the background light and internal noise of the receiver. Therefore, we must also consider the influence of background light and receiver noise on the laser-communication system. In atmospheric laser-communication systems, the photodetector is an important component of the laser receiver, and its noise directly affects the receiver's performance. If a PIN photodetector is selected, its noise sources can be divided into dark current noise, quantum noise, and internal-resistance thermal noise.

a. **Quantum noise**

When the photogenerated carrier moves in an uncertain or discontinuous manner, it leads to a continuous fluctuation of the carrier, thus causing quantum noise. The distribution function follows a Poisson distribution. A significant feature of the photodetector detection process is that the photogenerated carriers change qualitatively because of the photoelectric effect; its calculation formula is [13]

$$\left\langle i_Q^2 \right\rangle = \sigma_Q^2 = 2qI_pB, \tag{8.58}$$

where q denotes the electronic charge, B denotes the bandwidth, and I_p denotes the average photocurrent.

Quantum noise, also known as granular noise, is caused by the randomness of the photon or photoelectron generation and emission. Within any equal time interval τ, the probability n that the number of emitted photons is $p(n)$ satisfies the Poisson distribution:

$$p(n) = \frac{(\overline{n})^n}{n!} e^{-n}. \tag{8.59}$$

Photodetectors have three main sources of quantum noise: dark current, background light, and signal light. The average number of electrons generated by the dark current can be expressed as [14]

$$\overline{n_d} = \left[aAT\tau \exp\left(-\frac{\Phi}{kT} \right) \right] \Big/ e, \tag{8.60}$$

where a is the ratio coefficient determined by the photocathode material, A is the area of the photocathode, T is the temperature, Φ is the overflow work of the material, k is Boltzmann's constant, and e is the electronic charge.

Background-light noise is additional noise caused by random variations in the background-light intensity when the photodetector is working. This is generally unavoidable but can be reduced. Within the time interval τ, the average number of photoelectrons generated by background light can be expressed as

$$\overline{nb} = \frac{\tau\eta}{hv} P_b, \tag{8.61}$$

where P_b is the intensity of the background light, τ is the quantum efficiency of the photodetector, and h is Planck's constant.

Signal-light noise originates from the quantum property of the signal light, which is inherent in the light signal. Within the time interval τ, the average number of photoelectrons generated by the signal light is

$$\overline{n}_s = \frac{\tau\eta}{hv} P_s, \tag{8.62}$$

where P_s is the intensity of the signal light. The square of the total photoelectronic noise generated by these quantum noise sources on the photodetector photocathode is

$$(\overline{\Delta n})^2 = (\overline{n_b} + \overline{n_s} + \overline{n_d}). \tag{8.63}$$

This shows that the size of the quantum noise is closely related to the size of the signal, or that quantum noise is inseparable from the signal, which is a major feature of quantum noise. Quantum noise fluctuates with time, and the spectral characteristics of the noise can be analyzed using a Fourier transform. It can be seen from the specific changes that both quantum noise and thermal noise have uniform spectral characteristics, and both are white noise.

b. **Dark-current noise**

In the absence of light, the current flowing through the bias circuit inside the photodetector is called the dark current. Its noise is the sum of the surface dark current and the body dark current of the photodetector. Body dark current is caused by the photodiode's internal P–N-junction heating electrons and holes. Surface dark current is due to surface defects, gloss, bias voltage, and other factors. It is also known as the surface leakage current. The mean square value of the dark current is [13]

$$\langle i_D^2 \rangle = \langle i_{DB}^2 \rangle + \langle i_{DS}^2 \rangle = 2qB(I_d + I_1) = 2qBI_D, \tag{8.64}$$

where I_D denotes the dark current. The dark current is also natural particle noise.

c. **Internal-resistance thermal noise**

The noise caused by the irregular thermal movement of carriers in the conductor is called internal-resistance thermal noise, and its mean square value [7] is

$$\langle i_T^2 \rangle = \sigma_T^2 = \frac{4K_B T}{R_L} B, \tag{8.65}$$

where T denotes the absolute temperature, K_B denotes the Boltzmann constant, and R_L denotes the load resistance. The Brownian motion of charged particles in a conductive medium produces thermal noise; this occurs when the systems and circuits operate above absolute zero. This noise can be considered a mixture of many small independent current pulses.

The main task of the receiver is photoelectric detection, which converts the optical signal received by the lens into a current or voltage signal so that the subsequent circuit can process it and demodulate useful information from it. According to the transmission equation, the optical power of the signal received by the receiving system is [7]

$$P_1 = P_t \eta_t \eta_r e^{-\alpha L} \left(\frac{dr}{\theta L} \right)^2, \tag{8.66}$$

where η_t is the transmitting-antenna efficiency; P_t is the transmitting optical power; η_r is the receiving-antenna efficiency; α is the atmospheric-attenuation coefficient; $e^{-\alpha L}$ is the atmospheric-channel attenuation; dr is the diameter of the receiving antenna; and $(dr/\theta L)^2$ is the geometric attenuation of the optical power.

In stable tracking, the signal output power of the detector is [7]

$$P_c = G^2 I_s^2 R_L = G^2 \left(\frac{\eta e}{hv}\right)^2 P_r^2 R_L, \tag{8.67}$$

where R_L is the equivalent-load resistance of the detector, G is the current gain of the detector, η is the quantum efficiency, v is the incident-light frequency, and P_r is the optical power of the received signal.

The background-light noise, photodetector noise, and thermal noise directly affect the receiving sensitivity. The total power of these noises at the output end of the photodetector is [7]

$$P_n = (I_{ns}^2 + I_{nb}^2 + I_{nd}^2 + I_{nT}^2) R_L, \tag{8.68}$$

where $I_{ns}^2 = 2eG^2F\Delta f I_s = 2eG^2F\Delta f\left(\frac{\eta e}{hv}\right)P_r$, $I_{nb}^2 = 2eG^2F\Delta f I_b = 2eG^2F\Delta f\left(\frac{\eta e}{hv}\right)P_b$,

$$I_{nd}^2 = 2eG^2F\Delta f I_d, \text{ and } I_{nT}^2 = \frac{4kT\Delta f}{R_L}. \tag{8.69}$$

I_{ns}^2, I_{nb}^2, I_{nd}^2 and I_{nT}^2 are the mean square values of the signal photocurrent, background photocurrent, dark current, and noise current generated by the equivalent-load resistance of the photodetector, respectively. G is the noise multiplier factor, P_r and P_b are the signal and background-light power, respectively. T is the operating temperature of the photodetector and k is the Boltzmann constant. Therefore, the signal-to-noise ratio (SNR) of the direct-detection receiver is [7]

$$\frac{S}{N} = \frac{P_c}{P_n} = \frac{G^2\left(\frac{\eta e}{hv}\right)^2 P_r^2 R_L}{2eG^2F\Delta f\left[\left(\frac{\eta e}{hv}\right)(P_r + P_b) + I_d\right] + 4kT\Delta f}. \tag{8.70}$$

Assuming that the system uses an avalanche photodiode (APD) detector, we can use a Gaussian random process to simulate the signal-count fluctuation caused by thermal noise and the dark current of the receiver. The probability-distribution function of the background light count has been analyzed in the literature [7]. The avalanche effect of the APD can be used to generate extra free electrons from the excited electrons when the detector is photoelectric, and is used to complete the photomultiplier.

The conversion of optical signals into electrical signals and the regeneration of electrons by photomultipliers are usually random. The output of the photodetector can be regarded as a random current process. We found that regardless of the type of light field a detector responds to, random or non-random, when it is tested, randomness occurs.

First, we analyzed the initial counting probability of the APD under background-light conditions. Assuming that the background light is produced by a fixed-point light source, because of the signal-demodulation process to be carried out after the photoelectric detection, it is necessary to derive the background light after the photoelectric-detection probability-density function of the form.

The first calculation has a fixed-point light source caused by background light when the detector begins to count, and then calculating with the background light, the photoelectric detector's counting multiplies. The APD can enhance the current intensity; however, the statistical characteristics of the output count are affected by the random characteristics of the photomultiplier.

If k_1 is the electron count of the initial electron flow at a certain volume, then the probability of the output count is [15]:

$$P(k_2) = \sum_{k_1=0}^{\infty} P(k_2|k_1)P(k_1). \tag{8.71}$$

Here, $P(k_2|k_1)$ is the conditional probability function of the output count when the initial count is k_1, and $P(k_1)$ is the probability of the initial electronic count. For APD, the conditional-counting probability $P(k_2|k_1)$ after the initial count is multiplied by the avalanche. This was deduced by McIntyre and experimentally verified by Conradi. It has the following initial form [15]:

$$P(k_2k_1) = \frac{k_1\left(\frac{k_2}{1-keff}+1\right)}{k_2(k_2-k_1)\Gamma\left(\frac{keff\cdot k_2}{1-keff}+1+k_1\right)} \times \left[\frac{1+keff\,(G-1)}{G}\right]^{\frac{k_1+k_2}{1-keff}}\left[\frac{(1-keff\,)(G-1)^{k_2-k_1}}{G}\right], \tag{8.72}$$

where *keff* is the effective ionization rate, $\Gamma(x)$ is the Γ function, and G is the photomultiplier gain. Typical values of *keff* are 0.1 to 0.5 for germanium detectors and 0.01 to 0.1 for silicon detectors. G is usually in the 10 to 200 range. Because $P(k_1)$ is Poisson's count, Webb provided a useful approximation of $P(k_2)$ [15]:

$$P(k_2) = \frac{1}{\left(2\pi C_1^2\right)^{1/2}}\left[\frac{1}{1+\frac{(k_2-Gk_1)^{3/2}}{C_1C_2}}\right]\exp\left[\frac{-(k_2-Gk_1)^2}{2C_1^2\left(1+\frac{k_2-Gk_1}{C_1C_2}\right)}\right]. \tag{8.73}$$

The types are

$$C_1^2 = (Gk_1)^2F_e - 1 \tag{8.74}$$

$$C_2^2 = G(k_1 F)^{1/2}/(F_e - 1). \tag{8.75}$$

The additional noise factor F_e is

$$F_e = keffG + (2 - 1/G)(1 - keff). \tag{8.76}$$

If $G(k_1/F_e)^{1/2} >> 1$, Eq. (8.77) tends to a Gaussian probability envelope:

$$P(k_2) = \frac{1}{\left(2\pi G^2 k_1 F_d\right)^{1/2}} \exp\left[-\frac{(k_2 - Gk_1)^2}{2G^2 k_1 F_e}\right]. \tag{8.77}$$

When the k_2 value is low, Eq. (8.75) deviates from Eq. (8.74). When the k_2 value is greater than 50, the formula yields a more accurate approximation. Because the signal count typically exceeds this value, Eq. (8.75) can be regarded as a reasonable approximation of $P(k_2)$.

8.3 Bit Error-Rate Analysis of a DAPPM Modulation and Demodulation System

8.3.1 Influence of Atmospheric Turbulence on the Bit Error Rate of a Communication System

Atmospheric turbulence was analyzed in the previous section. Atmospheric turbulence constantly changes the refractive index of the atmosphere, resulting in irregular fluctuations in the refractive index of the laser beam, which is formed by the irregular motion of a vortex.

The DAPPM receiver produces an error code after being affected by atmospheric turbulence. The main reason is that the transmitted signal will produce some additional parasitic-amplitude modulation, owing to the influence of the light-intensity fluctuation, so that the signal cannot be correctly recovered. When the atmospheric-scintillation range is large, it saturates the photodetector, resulting in an error code. Alternatively, for a moment, the sensitivity of the detector may be much higher than its power, which will also cause an error code. In addition, atmospheric turbulence also leads to wavefront distortion of the laser beam, resulting in out-of-step bits or frames [16].

(1) **Probability distribution of the light intensity**

Depending on the random characteristics of the atmospheric channel, vortices of various scales will have different effects on the laser beam, which causes the light intensity of the receiver to change constantly. Usually, the probability-density function of the signal light intensity is used to describe the light intensity, and its size can be measured using a statistical average. To date, no suitable atmospheric-scintillation model has been established for the entire fluctuating region. The scintillation probability distribution of a single beam is typically used, and its distribution follows a lognormal distribution.

When the signal is transmitted over a long distance or through a strong turbulent channel, the variance is saturated, and flickering areas appear. In this case, the strong fluctuation in time follows an exponential distribution instead of a normal distribution. In general, communication systems are in weakly turbulent channels; therefore, this section mainly discusses the probability-density distribution of the light intensity under weakly turbulent conditions [17].

In [17], the probability-distribution function of the amplitude A in a weak turbulence region at a point on the receiving surface is

$$P(A) = \frac{1}{2\sqrt{2\pi}\sigma_x A} \exp\left\{-\frac{\left(\ln(A/A_0) - \langle x\rangle^2\right)}{2\sigma_x^2}\right\} \quad A \geq 0, \tag{8.78}$$

where $\langle x\rangle$ is the mean of the logarithmic amplitude X and σ_x^2 is the variance of the logarithmic amplitude. Suppose $A_0 = 1$; the probability density of light intensity I can be obtained from $I = A^2$:

$$P(I) = \frac{1}{2\sqrt{2\pi}\sigma_x I} \exp\left\{-\frac{\left(\ln(I/I_0) + 2\sigma_x^2\right)^2}{8\sigma_x^2}\right\}. \tag{8.79}$$

For a fixed mean value I_0, the curve shape of $P(I)$ differs for different values of σ_x^2. When $\sigma_x < 0.3$, the received normalized light intensity exhibits a lognormal distribution. With an increase in σ_x, the distribution range of I gradually increases, and the probability distribution follows an exponential distribution. The bit error rate of the communication system increases with increasing atmospheric scintillation.

When $\sigma_x = 0.1$, the probability density value of the transmitted signal light intensity $I = 1$ reaches a maximum. In this process, the probability of an error code appearing in the communication system is relatively small, as is the dispersion degree of deviation $I = 1$. When the signal light intensity varies over a large range with a certain probability distribution and the logarithmic amplitude's mean square error fluctuation is relatively large, the signal light-intensity value deviates from $I = 1$ with the increasing logarithmic-amplitude mean square error; thus, it increases the average error probability of the communication system. Therefore, additive white Gaussian noise (AWGN) can be used to simulate a weakly turbulent atmospheric laser-communication channel when $\sigma_x < 0.3$.

(2) **Irradiance-fluctuation impact on the error rate**

According to research [18], in a region with weak fluctuation intensity, the logarithmic amplitude and intensity fluctuations follow a normal distribution. If the logarithmic amplitude χ follows a normal distribution with a mean of $\langle\chi\rangle$ and variance of σ_x^2, then the logarithmic intensity follows a normal distribution with a mean of $2\langle\chi\rangle$

and variance of $\sigma^2 \ln_I$. The relationship between the logarithmic intensity-fluctuation variance and logarithmic amplitude-fluctuation variance is

$$\sigma^2 \ln_I = 4\sigma_x^2. \tag{8.80}$$

After many studies [18], Tartaskii proposed a theoretical model of light-intensity scintillation. For a region with weak fluctuations, we obtained the logarithmic-amplitude variance of the horizontal transmission based on this model:

$$\sigma_x^2 = 0.31 C_n^2 k^{7/6} l^{11/6} \quad \text{(plane wave)} \tag{8.81}$$

$$\sigma_x^2 = 0.124 C_n^2 k^{7/6} l^{11/6} \quad \text{(spherical wave)}. \tag{8.82}$$

From Eqs. (8.59), (8.60), and (8.61), the following can be obtained [18]:

$$\sigma^2 \ln_I = 4\sigma_x^2 = 1.24 C_n^2 k^{7/6} l^{11/6} \quad \text{(plane wave)} \tag{8.83}$$

$$\sigma^2 \ln_I = 4\sigma_x^2 = 0.496 C_n^2 k^{7/6} l^{11/64} \quad \text{(spherical wave)}, \tag{8.84}$$

where C_n^2 is the refractive-index structure constant of atmospheric turbulence representing the turbulence strength, $k = 2\pi/\lambda$ is the wave number, and l is the transmission distance.

In laser communication, after the laser is output from the transmitting end, it is collimated through an optical lens, and the waveform formed thereafter can be regarded as a plane wave. For a plane wave,

$$\langle \chi^2 \rangle = \sigma_x^2 = 0.31 C_n^2 k^{7/6} l^{11/6}. \tag{8.85}$$

It can be deduced from the above section that in a digital laser-communication system, when the optical receiver receives the optical signal, its bit error rate is

$$BER = \frac{1}{2}\left[erfc\left(\frac{Q}{\sqrt{2}}\right)\right], Q = \frac{I_1 - I_0}{i_1 + i_0}. \tag{8.86}$$

In terms of the light intensity, I_1 and I_0 are the average light intensities corresponding to bits 1 and 0, respectively, and i_1 and i_0 are the root mean square (RMS) of the light intensity caused by noise at bits 1 and 0, respectively. If the light intensity corresponding to bit 0 is set to 0, parameter Q can be expressed as

$$Q = \frac{I_1}{i_1 + i_0} = \left\langle \frac{1}{\chi^2} \right\rangle = \frac{1}{\sigma_x^2} = \frac{1}{0.31 C_n^2 k^{7/6} l^{11/6}}. \tag{8.87}$$

Substituting the above equation into Eq. (8.86), the relationship between the BER and intensity fluctuation can be obtained [18]:

$$BER = \frac{1}{2}\left[erfc\left(\frac{4}{\sqrt{2\sigma_{\ln I}^2}}\right)\right] = \frac{1}{2}\left[erfc\left(\frac{4}{\sqrt{2} \times 1.24C_n^2 k^{7/6} l^{11/6}}\right)\right]. \tag{8.88}$$

8.3.2 BER Analysis

(1) **System model**

To analyze the bit error performance of the DAPPM modulation mode, several assumptions are made in the modeling:

(1) The channel is a weak turbulence channel;
(2) The impact of scatter noise is considered;
(3) The transceiver can be accurately aligned without considering the beam jitter; and
(4) The interference of human light is ignored to reduce the complexity of the baseline jitter [19].

In a wireless laser-communication system, owing to the influence of various background lights, such as sunlight, electric light, etc. the total average power of the signal received by the photoelectric detector generates a DC inside the detector, causing scatter noise, which will have an important impact on the receiver's performance.

In general, the signal light intensity is much weaker than that of the background light; therefore, it can be considered that the scatter noise and the transmitted signal are independent of each other. In addition, there is leakage-current noise and dark-current noise caused by the photodetector and thermal noise of the internal resistance, which are independent of the transmitted signal. Therefore, the system noise can be regarded as Gaussian white noise, independent of the transmitted signal.

Therefore, the equivalent baseband model of the system can be expressed as [20]

$$y(t) = Rx(t) \otimes h(t) + n(t), \tag{8.89}$$

where $x(t)$ is the optical power, R is the detector response rate (A/W), $\otimes$ represents the convolution, and $n_0(t)$ represents the system noise. Because of human-eye safety requirements, the average transmitting power should be less than a certain maximum value $P_{\max}$; that is, it must satisfy [19]

$$P_{\max} \geq \lim_{T \to \infty} \frac{1}{2T} \int_{-T}^{T} x(t)dt. \tag{8.90}$$

As shown in Fig. 8.10, the information sequence $a_k \in \{0, 1\}$ passes through the encoder, b_k is the modulated sequence, and passes through the transmitting filter at the rate of $1/T_c$. Its impulse-response function is $Ab(t)$. Here, $A = 2RLP_{avg}(1 + a)$

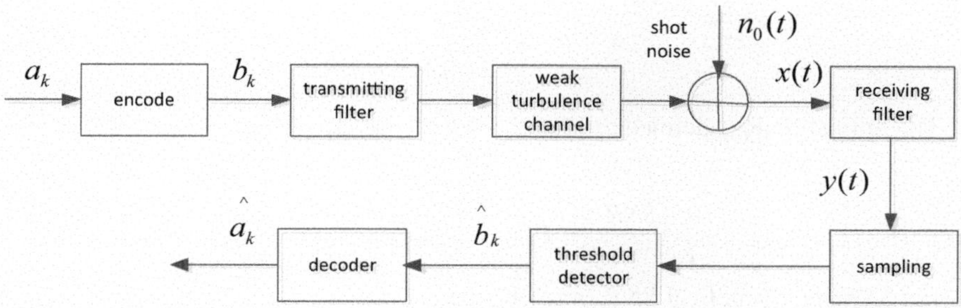

Fig. 8.10 System model

and P_{avg} is the channel where the average optical-power signal passes through the impulse-response function of $h(t)$. In addition, the white Gaussian noise has a the bilateral power spectral density of $N_0/2$.

Then, through the receiving filter, whose impulse response is $g(t)$, the signal is sampled to obtain the estimated value $\hat{b}_k$, and $\hat{a}_k$ is obtained through the decoder. It is assumed that the channel is ideal; that is, there is no multipath effect, and the impulse response function $h(t)$ is $\delta(t)$.

Because the outputs of the encoder are "0," "1," and "a", the input $x(t)$ of the received filter is $\sqrt{gP_t} + n_0(t)$ or $\sqrt{gaP_t} + n_0(t)$, when a pulse is present, and noise $n_0(t)$ when no pulse is present. P_t is the peak power of the transmitted pulse, $n_0(t)$ is the Gaussian white noise with a mean value of 0 and variance of σ^2. The convolution of the received filter and input is $x(t)$. Therefore, the output of the receiving filter at moment $T = T_c$ is in the form of energy, namely [21],

$$x(t) = \begin{cases} \sqrt{gP_t} + n_0(t) & when\text{ '1' }is\ sent \\ \sqrt{gaP_t} + n_0(t) & when\text{ "}a\text{" }is\ sent \\ n_0(t) & when\text{ '0' }is\ sent \end{cases} \tag{8.91}$$

$$y(t) = \begin{cases} E_p + n(T_c) & when\text{ '1' }is\ sent \\ aE_p + n(T_c) & when\text{ "}a\text{" }is\ sent \\ n(T_c) & when\text{ '0' }is\ sent \end{cases} \tag{8.92}$$

If the input waveform is a rectangular wave, then $y(t)$ is the convolution of input $x(t)$ and $\sqrt{gP_t}$ within a one-bit time. Therefore, $E_p = gP_tT_c$, $n(T_c)$ is Gaussian noise with a mean value of zero; however, its variance is $\sigma^2 = gP_tT_c^2\sigma_n^2$. If the probabilities of "1" and "0" occurring in the input bit information are equal, let P_0 be the probability of receiving "0" and P_1 be the probability of receiving "1". P_a is the probability of receiving "a"; then,

$$P_0 = (L-1)/L, \quad P_1 = P_a = 1/2L. \tag{8.93}$$

The average optical power P_n corresponding to the noise is

$$P_n = \frac{\sqrt{N_0}}{\sqrt{2T_c}R}. \tag{8.94}$$

The optical SNR is defined as [20]

$$SNR = \frac{P_{avg}}{P_n} = \frac{\sqrt{2T_c}P_{avg}R}{\sqrt{N_0}}. \tag{8.95}$$

Let SNR_0 be the optical SNR of OOK; then,

$$SNR_0 = \frac{\sqrt{2P_{avg}}R}{\sqrt{R_bN_0}}. \tag{8.96}$$

Because DAPPM uses two amplitudes to represent signal pulses, two threshold values $\beta_1 A$ (judgment threshold for "0" and "") and $\beta_2 A$ (judgment threshold for " " and "1") are needed, where $0 < \beta_1, \beta_2, < 1$. The probability of a judgment error for time slot "1," "0," and "a" are respectively,

$$P_{e0} = Q\left(\frac{\beta_1 A}{\sqrt{N_0/(2T_c)}}\right), \quad P_{e1} = Q\left(\frac{(1-\beta_2)A}{\sqrt{N_0/(2T_c)}}\right), \tag{8.97}$$

$$P_{ea} = Q\left(\frac{(a-\beta_1)A}{\sqrt{N_0/(2T_c)}}\right) + Q\left(\frac{(\beta_2 - a)A}{\sqrt{N_0/(2T_c)}}\right), \tag{8.98}$$

where $Q(x) = 1/\sqrt{2\pi}\int_x^\infty e^{-t^2/2}dt$. The DAPPM time-slot error probability is

$$\begin{aligned} P_{se} = {} & P_0 Q\left(\frac{\beta_1 A}{\sqrt{N_0/(2T_c)}}\right) + P_1 Q\left(\frac{(1-\beta_2)A}{\sqrt{N_0/(2T_c)}}\right) \\ & + P_a Q\left(\frac{(a-\beta_1)A}{\sqrt{N_0/(2T_c)}}\right) + P_a Q\left(\frac{(\beta_2 - a)A}{\sqrt{N_0/(2T_c)}}\right). \end{aligned} \tag{8.99}$$

We substitute Eq. (8.96) into Eq. (8.99) to obtain

$$\begin{aligned} P_{se} = {} & P_0 Q(C\beta_1) + P_1 Q[C(a-\beta_1)] \\ & + P_a Q[C(1-\beta_2)] + P_a Q[C(\beta_2 - a)], \end{aligned} \tag{8.100}$$

where C is

$$C = 2\sqrt{ML}SNR_0/(1+\alpha). \tag{8.101}$$

Because the lengths of DPIM and DAPIM symbols are not fixed, a time-slot error will affect subsequent symbols and produce error codes. To compare the error-code performance of several modulation modes, we used the error-packet rate for the analysis. If one bit in a packet is erroneous, the packet is determined to have an error code. If a packet has an N-bit signal, after DAPPM modulation, the number of time slots is NL/M and the packet error rate is

$$P_p = 1 - (1 - P_{se})^{LN/M} \approx LNP_{se}/M \,, \tag{8.102}$$

where N is the number of bits contained in a packet, M is the number of bits contained in the signal frame, and L is the number of time slots contained in a symbol. We substitute Eq. (8.100) into Eq. (8.102) to obtain

$$\begin{aligned} P_p = & \frac{N}{M}\{(L-1)Q(C\beta_1) \\ & +0.5\big[Q(C(1-\beta_2)) + Q(C(a-\beta_1)) + Q(C(\beta_2 - a))\big]\}. \end{aligned} \tag{8.103}$$

Assuming $a = 0.5$, $\beta_1 = 0.25$, and $\beta_2 = 0.75$, the above equation can be simplified as [20]

$$P_p = \frac{N(2L+1)}{2M} Q\left(\frac{C}{4}\right). \tag{8.104}$$

According to the above formula, the average optical power demand of DAPPM is

$$P_{req} = \frac{3\sqrt{N_0 R_b}}{R\sqrt{2ML}} Q^{-1}\left(\frac{2MP_p}{N(2L+1)}\right). \tag{8.105}$$

The average optical power demand corresponding to the bit error rate of the time slot can also be obtained as

$$P_{req} = \frac{3\sqrt{N_0 R_b}}{R\sqrt{2ML}} Q^{-1}\left(\frac{2LP_{se}}{(2L+1)}\right) \tag{8.106}$$

where R_b is the signal rate and R is the photodetector response rate.

(2) **Analysis of simulation results**

The simulation parameters are listed in Table 8.3. β_1 and β_2 are the decision thresholds of DAPPM and DAPIM, respectively, and β is the decision threshold of OOK, PPM, and DPIM.

As shown in Fig. 8.11, when M = 6, the curve of the error packet rate of the five modulation modes varies with the SNR. It can be seen from the figure that the error packet rate of each modulation mode decreases rapidly with an increase in the SNR. Although the bit-error performance of DAPPM is worse than that of PPM and DPIM, it is significantly better than that of OOK and DAPIM. DAPPM has two pulse amplitudes, and the bit error performance worsens because the smaller pulse amplitudes lead to an increase in the bit error rate.

Table 8.3 Simulation parameters

Parametric variable	Parameter values
Amplitude a	0.5
Scintillation index σ_x	0.25
β_1	0.25
β_2	0.75
β	0.5
g	0.02
Noise variance σ^2	$1e-8$
Number of symbols in a package	1024

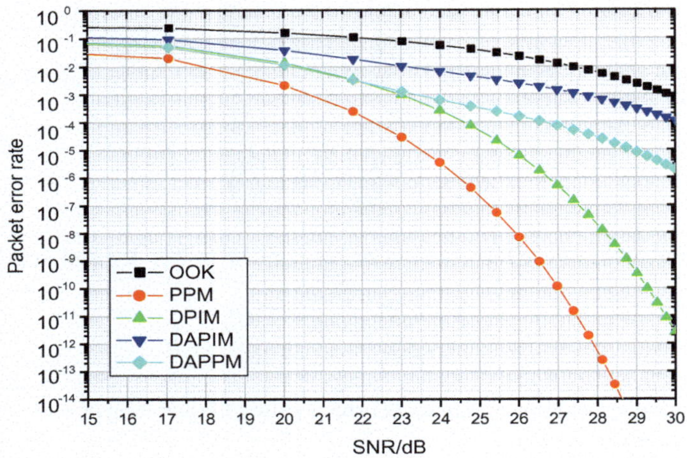

Fig. 8.11 Error-packet rates for several modulation modes

At the same bit resolution, the packet error rate of DAPPM is higher than those of DPIM and PPM but is obviously better than those of DAPIM and OOK. Because DAPPM removes the redundant time slots in PPM, the symbol length is reduced by half over PPM, resulting in higher bandwidth utilization than PPM.

References

1. Pang Z, Piao D, Zou C (2002) Performance comparison of several modulation modes in optical communication. J Guilin Univer Electron Technol 22(05):1–4
2. Ke X (2004) Introduction to wireless laser communication. Beijing University of Posts and Telecommunications Press, Bei Jing
3. Wang H, Zhu Y, Zhang T et al (2006) Performance analysis of modulation mode in optical wireless communication. Prog Laser Optoelectron 43(6):37–39

4. Lesh JR (1983) Capacity limit of the noiseless, energy-efficient optical PPM channel. IEEE Trans Commun 31(4):546–548
5. Sethakaset U, Gulliver TA (2004) Differenrial amplitude pulse-position modulation for indoor wireless optical channels. IEEE Trans Communications Society, Globecom, pp 1867–1870
6. Zhang K, Zhang H, Zhang M et al (2003) Performance analysis of infrared double-amplitude pulse-interval modulation communication system. J Infrared Millimeter Waves 22(6):411–414
7. Wang J (2009) Performance analysis of optical wireless communication system based on PPM and RS codes. Xidian University, Xi 'an
8. Wan L, Jiang L (2002) Influence of light intensity scintillation in turbulent atmosphere on optical communication link. Opt Commun Technol 2(2):17–20
9. Jian D (2008) Study on atmospheric channel characteristics of optical wireless communication system. Xidian University, Xi 'an
10. Meyer RR, Kirkland A (1998) The effects of electron and photon scattering on signal and noise transfer properties of scintillators in CCD cameras used for electron detection. Ultramicroscopy 75(1):23–33
11. Levoni C et al (1997) Atmospheric aerosol optical properties a database of radiative characteristics for different components and classe. Appl Opt 136(30):8031–8041
12. Rao R, Wang S, Liu X et al (1998) Intensity fluctuation and spot statistics of laser in turbulent atmosphere. J Quantum Electron 15(8):155–163
13. Zhou X, Chen W, Xiong Z et al (1994) Research on performance of optical PPM coding system for direct detection under background noise limit. Telecom Sci 10(2):28–31
14. Zongming Tao Y, Chen DC et al (2010) Study on quantum noise characteristics of photodetectors. J Atmos Environ Opt 5(1):63–66
15. Shen K, Zeng L (2006) Noise analysis of 40GPIN/TIA photodetector receiver. Opt Commun Technol 30(9):7–9
16. Nan M, Wang H, Sun X (2008) Analysis and comparison of error performance of atmospheric optical wireless communication systems under turbulent channel conditions. J Naval Aeronautical Eng College 23(5):518–519
17. Kun Y (2008) Study on bit error rate characteristics of wireless laser communication system. Henan Normal University, Xinxiang
18. Qinggong H (2004) Channel research of atmospheric laser communication system. Harbin Engineering University, Harbin
19. Zhang T, Wang H, Zhu Y, Sun X (2006) Analysis of influence of decision threshold on modulation error rate of optical wireless communication. J Project Arrows Guidance 26(4):323–325
20. Zhang K (2004) Study on dual-amplitude pulse-interval modulation in optical wireless communication and its application in optical ETC. Tsinghua University, Beijing
21. Zhang T, Wang H, Zhu Y et al (2007) Dual amplitude pulse position modulation in optical wireless communication. Laser Mag 28(6):71–13

SPRINGER NATURE

GPSR Compliance

The European Union's (EU) General Product Safety Regulation (GPSR) is a set of rules that requires consumer products to be safe and our obligations to ensure this.

If you have any concerns about our products, you can contact us on ProductSafety@springernature.com

In case Publisher is established outside the EU, the EU authorized representative is:

Springer Nature Customer Service Center GmbH
Europaplatz 3
69115 Heidelberg, Germany

The manufacturer's authorised representative in the EU is Springer Nature Customer Service Centre GmbH, Europaplatz 3, 69115 Heidelberg, Germany. If you have any concerns regarding our products, please contact ProductSafety@springernature.com

Printed and bound by CPI Group (UK) Ltd, Croydon, CR0 4YY
30/09/2026
02225298-0002